Textbook of Industrial Microbiology

Renu Agrawal

# Textbook of Industrial Microbiology

Springer

Renu Agrawal
Central Food Technological Research Institute
Mysuru, Karnataka, India

ISBN 978-981-97-9581-9 ISBN 978-981-97-9582-6 (eBook)
https://doi.org/10.1007/978-981-97-9582-6



This Springer imprint is published by the registered company Springer Nature Singapore Pte Ltd.
The registered company address is: 152 Beach Road, #21-01/04 Gateway East, Singapore 189721, Singapore

*This book is dedicated in the memory of my mother, Dr. Sarla Agrawal, who always inspired me with her knowledge, writings, and dedication to the family. She was highly talented and an all rounder, a lovely mother, a fine teacher, supportive of her children, and a loyal friend. She is greatly missed by me.*

# Preface

The book entitled *Industrial Microbiology* presents a microbiology book starting from isolation of microorganisms to the microorganism prompted process technologies in fermenters/bioreactors for large-scale production of many useful products. The technologies use the metabolic ability of microorganisms to produce useful and health improving metabolites. The yields are enhanced by optimization of various parameters of microorganism, fermentation media along with fermenter/bioreactor parameters, which have been discussed in detail in this book. In order to obtain higher yields of the metabolites the microorganism can be improved by mutation and other genetic methods, which have been explained in this book. The technologies of preservation of microorganisms have been discussed and a thorough knowledge of culture collection centers has been given. Furthermore, the beneficial probiotic microorganism have been discussed which improve the GUT health and combat many harmful diseases including cancer. Due to increasing antibiotic resistance, an alternative is needed. Microorganisms can produce bacteriocins which have anti microbial activity and come under GRAS category. They have been approved by FDA. These have been discussed in detail. The book discusses well on downstream processing till the purification of the products.

I am extremely thankful to the staff of Springer who helped me with my queries time and again, and to the Springer team for their substantial teamwork and for printing this book.

Mysuru, Karnataka, India

Renu Agrawal

# Acknowledgments

I would like to thank Wikipedia for allowing me to use figure on Bioreactor in this book. I am thankful to the staff at Springer for assisting me from time to time during the writing of the book. I would like to thank the Springer Publishing staff for putting their inputs in publishing the book.

# Contents

# About the Author

**Renu Agrawal** is Ex. Chief Scientist and Rural Program Coordinator for outreach activities at CSIR-CFTRI (Council of scientific and Industrial research-Central Food Technological Research Institute), Mysuru, India. She has contributed for more than 30 years in the field of Microbiology. She has over 20 patents to her name with 67+ publications in international and national journals in the field of microbiology. She has presented more than 160 papers in various national and international conferences, seminars, workshops, and symposium with many best paper awards and has delivered more than 1000 talks. She has guided a stream of students for MSc, MPhil, and PhD in Biotechnology and Microbiology. She had teaching assignments as faculty of MSc Food Technology at CFTRI and a professor at ACSIR in Microbiology. She has represented India as a team leader taking scientific delegation for the Asia meet at Bali, Indonesia, nominated by DST, Government of India. She completed her PhD at University of Rajasthan, Jaipur. Her specializations are in microbiology specifically in Biotransformation and Probiotics. She has been awarded by multiple scientific associations. Her scientific fellow awards include FAMI, FNABS, FSAB, and FISBT. She has been awarded Karamveer Chakra by REX-UN, Best woman scientist by NABS, Woman achiever by International Guild of Women Achievers, Reflections Excellence Award by All India Association of Rotary Inner wheel, Kissan Ratna Puraskar by JCI International, Abdul Kalam Lifetime Achievement Award, Nari Shakti Award by Random Kindness and the Ministry of Women and Child Development, and Bharat Shiksha Gaurav Puraskar. She has published many book chapters and many books with international publishers (CRC press, Taylor and Francis, and Marcell Dekker). The microbial processes of her team on Ready to make idli and dosa batter are the most downloaded processes from CFTRI. She has drafted the Food and Nutrition Policy for Karnataka state and National Policy on Vegan Foods under APEDA. She has been a member of various science and technology committees as an expert member at national and international levels. She has been serving as an expert member in committees of APEDA, DST, DBT, NCERT, and UPSC by the Government of India. She is an executive committee board member of "National Academy of Biological Sciences." She has successfully completed many national and international research projects with different government organizations as PI on her proposed ideas in the field of microbiology. She is a reviewer and on the editorial board of several high-impact international scientific journals from Elsevier, Springer Nature, Taylor and Francis, and Cambridge University Press. She is an expert member in the high steering committee of the Ministry on Forestry, Climate Change and Environment. She is nominated to the Consultative Group by the Principal Scientific Advisor, GoI. She has served as the president of the "Association of Microbiologists of India" and treasurer in the "Society of Biological Chemists," Mysore chapter. Her scientific interviews on various aspects of Microbiology have been telecasted on Doordarshan (DD science). She is also involved in the upliftment of rural education in Karnataka through "Pratham" schools. Currently she is a science communicator, reviewer, author, and visiting professor in many universities and colleges and a National Expert Advisor in many committees with the Ministry of Science and Technology and has been imparting training to KAS and KPS officers for Sustainable Development Goals, in Karnataka.

# Abbreviations

| | |
|---|---|
| AcSIR | Academy of Scientific and Innovative Research |
| ADY | active dry yeast |
| AFLP | amplified fragment length polymorphism |
| AIIMS | All India Institute of Medical Sciences |
| ANOVA | analysis of variance |
| API | analytical profile index |
| ARI | Agharkar Research Institute |
| ASM | American Society for Microbiology |
| ASTM | American Society for Testing and Materials |
| ASV | American Society for Virology |
| ATCC | American Type Culture Collection |
| ATM | atomic force microscopy |
| ATP | adenosine triphosphate. |
| BGC | biosynthetic gene cluster |
| BGC's | biosynthesis gene clusters |
| BOD | biological oxygen demand |
| BRC | Biological Resource Centre |
| C | cytosine |
| CAS | cell alive system |
| CATCH | Cas9-assisted targeting of chromosome segments |
| CCINFO | Culture Collection Information |
| CDC | Centers for Disease Control and Prevention |
| CEMS | continuous emission monitoring system |
| CIEF | capillary isoelectric focusing |
| CLSM | confocal microscopy |
| CRISPR | clustered regularly interspaced short palindromic repeats |
| DESS | ethylene-diamine-tetra acetic acid, saturated salt |
| DHA | docosahexaenoic acid |
| DiPac | direct pathway cloning |
| DMSO | dimethyl sulfoxide |
| DNA | deoxyribonucleic acid |
| DSW | distillery spent wash |
| EC | European Commission |
| ED | Enforcement Directorate |

| | |
|---|---|
| EFSA | European Food Safety Authority |
| EIFFF | electrical field-flow fractionation |
| ELISA | enzyme-linked immuno serum assay |
| ELSD | evaporative light scattering detector |
| EMBO | European Molecular Biology Organization |
| EMP | electromagnetic pulse |
| ESI | electrospray ionization |
| FACS | fluorescence activated cell sorter |
| FDA | Food and Drug Administration |
| FISH | fluorescence in situ hybridization |
| FN | Fingerprint Number |
| FTIR | Fourier transform infrared spectroscopy |
| G | guanine |
| GCM | greatest common multiple |
| GCMS | gas chromatography mass spectroscopy |
| GEM | genetically engineered |
| GMO | genetically modified organism |
| GOS | galacto-oligosaccharides |
| GRAS | Generally Recognized As Safe |
| IATA | International Air Transport Association |
| IBC | Institutional Biosafety Committee |
| IDA | International Depository Authority |
| IISER | Indian Institute of Science Education and Research |
| LCMS | liquid chromatography-mass spectrometry |
| LDI | laser desorption/ionization |
| MALDI | matrix-assisted laser desorption ionization |
| MALDI-TOF | matrix-assisted laser desorption/ionization -time-of-flight |
| MCC | microbial culture collection |
| miRNA | micro RNA |
| MoEF | Ministry of Environment and Forests |
| MORF | Multi-Omics Research Factory |
| MPN | most probable number |
| MS | mass spectrometry |
| MSA | The Mycological Society of America |
| MSG | monosodium glutamate |
| MSGE | multiplexed site specific genome engineering |
| NADH | nicotinamide adenine dinucleotide |
| NCCS | National Centre for Cell Science |
| NCIM | National Collection of Industrial Microorganisms |
| NCL | National chemical laboratory |
| NCTC | National Collection of Type Cultures |
| NIAID | National Institute of Allergy and Infectious Diseases |
| NMR | nuclear magnetic resonance |
| ncRNA | non-coding RNA |
| NSF | National Science Foundation |

| | |
|---|---|
| NTG | N-methyl-N-nitrosoguanidine |
| OTR | oxygen transfer rate |
| PBS | phosphate-buffered saline |
| PCA | principal component analysis |
| PFGE | pulsed-field gel electrophoresis |
| pH | hydrogen ion concentration |
| PHA | poly hydroxyalkanoate |
| PI | propidium iodide |
| RAPD | random amplified polymorphic DNA |
| RNAseq | RNA sequencing |
| RFLP | restriction fragment length polymorphisms |
| RNA | ribonucleic acid |
| RPM | Revolution per minute |
| RQ | respiratory quotient |
| RT-PCR | real time-PCR |
| SBR | Semibatch reactor |
| SELDI | surface-enhanced laser desorption/ionization |
| SEM | scanning electron microscopy |
| SIMB | The Society for Industrial Microbiology and Biotechnology |
| SLF | solid liquid fermentation |
| SLF | substrate liquid fermentation |
| SM | secondary metabolite |
| SmF | submerged fermentation |
| SOP | standard operating procedures |
| SOTR | standard oxygen transfer rate |
| SPE | solid-phase extraction |
| SPME | solid-phase microextraction |
| SSF | solid state fermentation |
| SSF | substrate surface fermentation |
| TAR | transformation assisted recombination |
| TEM | transmission electron microscopy |
| US FDA | United States Food and Drug Administration |
| USGS | US Geological Survey |
| UV | ultra violet |
| WGS | whole-genome sequencing |
| WHO | World Health Organization |
| WIPO | World Intellectual Property Organization |

# Introduction and History of Industrial Microbiology

# 1

**Abstract**

Microorganisms are very important for human kind. They are utilized in the formation of many beneficial secondary metabolites. This chapter deals with the history of industrial microbiology. The various fermentation processes and the ability of microorganisms to ferment and convert sugars into alcohol which, started in the fourteenth century AD. This chapter also deals with the development in improving the microbial strains for higher yields of the metabolites by mutation and genetic modification. Classification of microorganisms, their taxonomy, and their basic metabolic pathway have been explained.

**Keywords**

Microbes · Pathogens · Antibiotic · Strain improvement · Classification

## 1.1 The Beginning of Industrial Microbiology

Microorganisms are very important form of life on this earth. As compared to the green plants, it is the microbes (60%) that conduct higher level of photosynthesis. The human body contains 90% of microbial cells. Yeasts have the ability to convert sugar to alcohol. Yeasts are used for beer making and in the preparation of bread and cheese by fermentation. Wines are fermented with molds and bacteria. Milk is fermented to make yogurt and other products such as Kefir and Koumiss using *Kluyveromyces* species. Saccharification of rice with molds for the formation of Koji got started in 700 AD. In the fourteenth century AD, the distillation of fermented grains to alcohol and vinegar manufacture became common. In the nineteenth century, it was Pasteur who discovered two optical types of amyl alcohol, D and L. However, he could not separate the two. Koch (1876), from Germany, discovered that bacteria from anthrax infections cause this disease. Gaffky and Loeffler


R. Agrawal, *Textbook of Industrial Microbiology*,
https://doi.org/10.1007/978-981-97-9582-6_1

(1884) studied the role of infectious bacteria for typhoid and diphtheria, which could be prevented by the process called as pasteurization. Pasteur and Koch discovered that it is the body's own defense system that is responsible to fight the pathogenic microbes. This brought the invention of vaccines. Ehrlich, discovered the use of synthetic dyes in the prevention of relapsing fever, syphilis, and trypanosomiasis. Ereky (1919) termed the word "Biotechnology" for the production of many biochemicals (solvents, alcohols, organic acids, and enzymes) by using starch/sugar fermenting microbes. Culture collections for various yeast species were established in Berlin (Prague, Delft) followed by fermentation and brewing research institutes (Pasteur Institute, Paris; Carlsberg Institute, Copenhagen; Institut fur Gärungsgewerbe, (Institute for Fermentation Industries), Berlin). In 1915, Seligman introduced the plate heat exchanger to ferment potato starch along with grains and anaerobic bacteria for the production of acetone. Fermentation processes for glycerol, lactic acid, and yeast were developed by Connstein (1870) in Germany. It was the work of Van Niel (1929) who discovered the aroma compound of butter as diacetyl, important as a bioflavor for the growing margarine industry. Bremus et al. (2006) explained the role of microorganisms for the production of L-ascorbic acid. Currie, worked on the production of citric acid using the mold *Aspergillus niger*. Northern Regional Research Laboratories (NRRL) in Peoria, Illinois, was set up in 1939–1940. High yielding strain of *Clostridium acetobutylicum* was obtained by Weizmann (patent, 1994) that produced acetone, butanol, and ethanol (ABE). Buchholz and Collins (2010) have written an article on the concepts in biotechnology, the history, science, and business through microorganisms. Poulson and Buchholz (2003) have explained the history of enzymology with special emphasis on food production. Commercial production of lactic acid from glucose-rich starch hydrolysates and corn steep liquor was started in 1930s by the American Maize Products Co., DuPont, and Clinton Corn Syrup Refining Co. After this, many microbial companies came up. The notables have been in Europe; Byk-Guldenwerke and Boehringer Sohn (Germany), Société Normande de Produits Chimiques (France), Bowmans Chemicals (UK), Schiedamsche Melkzuur Fabriek (SMF, later named CCA) (The Netherlands), and Kemisk Vaerk Koge (KVK) (Denmark). Other companies were established in Italy, Hungary, Czechoslovakia, Poland, Romania, and Russia. Many industries came up in Japan as Takeda Chemical Industries, Tanabe, Sankyo, Dai Ichi Seiyaku, Dai Nippon Industries, which produced lactic acid.

Glycerol was produced by adding sodium bisulphite into yeast ethanol fermentation process using beet sugar. A British patent has been obtained by Fernbach and Strange (1911) on the fermentation process producing amyl, butyl/ethyl alcohol, butyric, propionic, and acetic acid. The production of acetone and butyl alcohol by a bacteriological process was initially demonstrated by Speakman (1919). Using surface culture, many industrial processes were started in Belgium using microorganisms (1919). Qureshi et al. (2006) were successful in producing butanol from corn fiber xylan using *Clostridium acetobutylicum*. Later companies such as Citrique Belge S.A. Company, John and Sturge in the UK, Montan industrie in Czechoslovakia, and Boehringer in Germany were started to produce different microbial products. Fermentations from fungi were studied for the production of itaconic acid, kojic acid, fumaric acid, and gallic acid in Japan, the United States, and in Europe. Many industrial enzymes were also produced from yeasts, molds, and bacteria such as amylase, lipase, protease, trypsin, pepsin, and invertase. It was Michaelis and Menten (1913) who studied the kinetics of enzyme action. Later in 1950s, detergent enzymes were invented.

## 1.2 The Antibiotic Era

**Penicillin**

Fleming (1929) found that the plates containing *Staphylococcus aureus* were contaminated with the mold, *Penicillium notatum*, which led to the invention of penicillin. Abraham and Chain (1940) discovered the enzyme from bacteria that had the ability to destroy penicillin. The history of penicillin production has been explained (AIChE 1970). Bud (2007) wrote an article on the triumph and tragedy of penicillin. A strain selection program was established at the USDA in Peoria, the Carnegie Institution at Cold Spring Harbor in New York, Stanford University, and the University of Wisconsin. Demain (2004) discovered that lysine was a potent inhibitor of penicillin biosynthesis and it could be reversed with the use of L-α-aminoadipic acid. Commercialization of antibiotics (cephamycin, cefoxitin, cefmetazole, temocillin, cefotetan, and other semisynthetic cepahlosporins) was made by Merck company. Later, many new antibiotics were discovered from the filamentous bacteria, actinomycetes, which included actinomycin, neomycin, and streptomycin. Ozcengiz and Demain (2013) studied the advances in the biosynthesis of penicillins, cephalosporins, and clavams. They also studied the regulation of these antibiotics.

## 1.3 Mutagenesis and Strain Improvement

Secondary metabolite production was improved by mutagenesis. Okanishi (1974) discovered the protoplast fusion in actinomycetes using polyethylene glycol. In 1985, many strain improvement programs were conducted, which included transposition mutagenesis, targeted deletions and duplications by genetic engineering, and genetic recombination by using techniques such as protoplast fusion and plasmid transformation. Combinational biosynthesis was used for the discovery of new hybrid drugs by rDNA technology. Hotchkiss (1979) identified that nucleic acids are the genetic determinants. Semisynthetic erythromycins such as clarithromycin, roxithromycin, azithromycin, and telithromycin were developed. Schegloff (1968) studied the sequencing in conversational practices. Then, inventions were made for hypocholesterolemic agents (compactin, mevinolin, pravastatin, enzyme inhibitors), antitumor agents (taxol, ergot alkaloids, bioherbicides), plant growth stimulants (gibberellins), and antiparasitic agents such as monensin, polyethers, avermectins, and bioinsecticides. Scientists have made it possible to produce microbial products such as beer, cheese, citric acid, amino acids, enzymes, antibiotics, ethanol, starch products, and other organic acids through biotechnology. In 1980s, many inventions in genetic engineering and recombinant DNA (rDNA) technology took place with the manipulation of genetic material and recombinant biotechnology. Bailey (1996) studied the metabolic engineering and advances in molecular and cell biology. The products were produced that included rDNA human insulin, interferon, and monoclonal antibody (mAb) diagnostic kits. First patent in this direction was filed by Cohen et al. (1973). Berg and Mertz (2010) gave their perspectives on the anecdotal and historical commentaries of genetics. During this time, personal reflections on the origin and emergence of recombinant DNA technology was given by many scientists. Chang et al. (1973) conducted in vitro studies on the construction of biologically functional bacterial plasmids. Later, Boyer and Swanson started the first gene cloning company. This helped in the modeling of the processes and quantification of many biotransformation reactions. Following this, the bioreactors and downstream operations were developed. Optimization of a microorganism in a process was made using transcriptome, proteome, and metabolome investigations. These were developed with DNA-arrays in transcriptomics as analytical methods. Other tools were developed such as 2D-gel electrophoresis for qualitative analysis, ELISA for obtaining quantitative data in proteomics, MFA (metabolic flux analysis via mass balance including isotope materials), FBA (flux balance analysis via modeling) for fluxomics, GC-MS for the study of metabolomics, and other classical analysis for substrates, products, pH, $O_2$, $CO_2$, temperature, time, and power input.

The proteome contains around 4000 proteins. This metabolome helps in the production of more than 2000 metabolites. Metabolome describes the metabolic activities in a cell. Amino acids, citric acid, lactic acid, propanediol, penicillin G, synthetic drug intermediates, and therapeutic proteins are very important for the industries. Dupont company has utilized *Corynebacterium glutamicum* for lysine and tryptophan production. Bioinformatics provide critical information. Using the high yielding strains, scientists cloned the first industrial enzymes, penicillin amidase, and α-galactosidase in *Escherichia coli*. Resistance against diseases were studied in genetically modified (GM) plants and for reducing the usage of chemical pesticides. Guidelines for recombinant DNA (rDNA) regulations were made in 1975. In 2006, cultivation with GM crops was started in land (102 million ha), which majorly included soybean (64%), cotton (43%), maize (24%), and rape (20%).

## 1.4 Nucleic Acid Sequencing

Much before the discovery of gene cloning, small synthetic oligonucleotides and RNA viruses were utilized for sequencing (Schegloff 1968) or the phage Lambda (Wharton et al. 1984). It was in the 1960s that small RNAs, such as tRNAs and RNA viruses, were purified. DEAE-cellulose ion exchange chromatography was developed for the isolation of a tRNA and countercurrent chromatography was utilized for sequencing. Initiation of translation was determined by sequencing the fragments of the RNA phages along with ribosome binding and then degradation with the help of ribonuclease. The development of PCR and enzymes are important prerequisites in the sequencing technologies, which are utilized for sequencing reactions or novel chemistries when used as reversibly terminating deoxynucleotides along with different colored fluorescent compounds. This may vary with each type of base and microarray scanning. The microarray scanning is a laser detection method. Millions of sequencing reactions can be scanned with mega sequencing technology. Avery et al. (1944) demonstrated the transfer of DNA (and not RNA, proteins, or lipids) and correlated it with the inheritance of an altered property (phenotype) in the recipient. Later, a crystal data of DNA was produced by Franklin, which was used by Watson and Crick (1953) for the preparation of a tentative model of DNA structure. This helped to understand the processes of replication, mutation, and repair. In the 1970s, DNA transformation was discovered and it was found that the plasmids usually carry antibiotic resistance genes. Mermelstein et al. (1993) studied the metabolic engineering of *Clostridium acetobutylicum* ATCC 824 for the increased acetone production by increasing the acetone enzyme with the use of synthetic acetone operon. Szybalski (2013) made non-reverting mutants by transferring DNA from normal cells to the mutant cell lines. The cell fusion gave rise to hybridomas to produce monoclonal antibodies or cell lines carrying one or a limited number of human chromosomes. In order to locate any heritable diseases, the DNA clones were used as probes. This was initiated with the discovery of restriction fragment length polymorphisms (RFLPs) (Botstein et al. 1980). This was helpful in the diagnosis of causal gene for Huntington Chorea, an incurable disease with neuronal degeneration (Macdonald et al. 1993). Later, a method called as DNA shuffling was developed by Stemmer (1994). The importance of recombination and point mutation was discovered. The in vitro recombination system was used effectively to generate combinatorial library fragments. Harrington et al. (1997) discovered the human accessory chromosome vectors (HAC). In this, the clones had the ability to carry up to 10,000 kb of DNA. Initially, for isolation and manipulation of DNA, *E. coli* has been usually taken for studies. Then, it is transferred to a new host (*Agrobacterium*) that uses the T-DNA transfer system for transferring the DNA fragment to the plant host (Gelvin 2003). These vectors are helpful for cloning in prokaryotes (gene mining), especially for antibiotics, primary and secondary metabolites. The overlapping sequences and clones are known as contigs. Bridging contigs have been obtained by the sequencing of clone libraries (Soderlund et al. 2000). The shotgun approach helped to obtain the total genome sequence in a microorganism (*Haemophilus influenzae* Rd in 1995). The first sequence of a eukaryote genome was of the *Drosophila melanogaster* genome (Rubin et al. 2000). The early phase of human genome sequencing tracked the genes for many heritable diseases. Many new recombinant antibodies have been produced to cure diseases. Innovative transgenic animals were produced with recombinant DNA in their genome, the stem cells in the developing embryo were differentiated, and new viral vectors were developed, which have many applications. Some examples are the production of knock-out mice for fundamental research and the study of various diseases using animal models (Hall et al. 2009).

**Gene Splicing** Berget et al. (1977) discovered that at the mRNA level, many of the coding regions get interrupted by intervening sequences that needs to be removed by a splicing process. This led to the discovery of RNA-enzymes, and SELEX technology was developed. Hwang et al. (2017) studied the transfer of genes from bacteria (*Agrobacterium tumefaciens*) to plants. They found out that the DNA chip arrays could be used for gene copy number variation (GCNV). It has been possible to find genes that affect many syndromes and cancer. When methylation at the DNA level is changed, it can cause changes in the gene expression, which is a new area of research and diagnostics, especially, in the health care. The microbial diversity has helped the biotechnological industry for the production of many secondary metabolites and it is termed as gene mining. The diffraction of X-rays by crystals was discovered by Von Laue (1912). Bragg discovered the X-ray spectrometer to be used with X-rays of defined wavelengths. Electrophoresis was discovered by Tiselius (1937) based on Faraday's work on electrolysis in the eighteenth century and on the studies of moving boundaries in electrolytes. It was the work of Smithies (1955) to detect the cause of inherited diseases due to altered protein structure. Sanger (1943) studied the gene sequencing after which the genetic code was studied.

## 1.5 Taxonomy and Classification of Microorganisms

(a) **Basic Nature of Cells in Living Things**

All living beings are made up of cells, which are of two types: the prokaryotic cell and the eucaryotic cell. Prit (1975) had brought out the principles and cell cultivation methods for microbes.

(b) **Cell Wall** In prokaryotes, the cell wall is made up of glycopeptides and in eukaryotes it has chitin, cellulose, and other sugar polymers that give rigidity to the cells. Growth is related to the doubling time of the cells, which is around 12 years in humans, 2 years in cattle, 1.5 years in pigs, 6 months in chicken, and 15 min in the bacterium (*E. coli*).

(c) **The Proteobacteria** This group is named after Proteus, the Greek god. It contains many pathogens such as *Escherichia*, *Salmonella*, *Vibrio*, *Helicobacter*, and free-living bacteria. Some of these bacteria can fix nitrogen. The red pigment in purple bacteria helps in using the solar light for photosynthesis. All Proteobacteria are Gram-negative, with an outer membrane mainly composed of lipopolysaccharides. The motile bacteria use flagella and the nonmotile bacteria use gliding. Proteobacteria are divided into five groups: alpha, beta, gamma, delta, and epsilon. These can be facultative or obligate anaerobes and heterotrophs. *Acetobacter* and *Gluconobacter* have industrial importance. *Zymomonas* bacteria belong to the "alpha" proteobacteria and it is industrially very important as it produces high yields of alcohol. The acetic acid bacteria are *Acetobacter* (peritrichously flagellated) and *Gluconobacter* (polarly flagellated). They produce acetic acid and are used in the manufacture of vinegar. *Gluconobacter* does not have the ability to oxidize acetic acid as the citric acid cycle is incomplete in these bacteria. However, *Acetobacter* has all the citric acid enzymes and can therefore oxidize acetic acid to $CO_2$ and are acid resistant at pH 5.0 or at lower pH. They help in the production of glucuronic acid from glucose, galactonic acid from galactose, and arabonic acid from arabinose. Ascorbic acid (Vitamin C) can be produced from sorbitol by acetic acid bacteria. All Gram-positive bacteria have been classified under Firmicutes, whereas all Gram-negative bacteria come under Proteobacteria. The mycoplasmas have no cell walls and therefore do not respond to Gram staining. Spore forming Firmicutes have the ability to form internal spores, whereas in the Actinobacteria, the spores are produced externally. This group is divided into two. The first one is the *Bacillus* spp., which comes under aerobic and grows in the presence of air. The other, *Clostridium* spp., is anaerobic. The anaerobic bacteria grow in the absence of oxygen. Many of the *Bacillus* spp. are used in the production of various enzymes. They also have the ability to kill insects and are specified as insect-repellent bacteria. They have a lot of importance in agriculture. *Bacillus papilliae* infects and kills the larvae of the beetles in the family *Scarabaeidae*, while *B. thuringiensis* is used to kill mosquitoes. Using genetic engineering techniques, the genes in the toxin of *B. thuringiensis* are being bioengineered into plants to have the quality of pest resistance.

(d) **The Lactic Acid Bacteria (LAB)** These bacteria do not form spores and are known as nonspore formers. They have low G + C ratio. The LABs are very important for industries. They are used in the production of foods and pharmaceutical products, and these can be rods or cocci. The various genera include *Enterococcus*, *Lactobacillus*, *Lactococcus*, *Leuconostoc*, *Pediococcus*, and *Streptococcus*. These do not contain porphyrins and cytochromes and do not undergo electron transport phosphorylation. Lactic acid bacteria are divided into two major groups: the homofermentative group and the heterofermentative group. In the homofermentative group of LABs, the bacteria produce lactic acid as the sole product of the fermentation of sugars. In the heterofermentative group, the bacteria produce many other acids along with lactic acid, ethanol, and $CO_2$. They lack enzyme aldolase, which decomposes hexose glucose in the EMP pathway into sugar moieties. In the EMP pathway, the homofermentative lactic acid bacteria converts glyceraldehyde 3-phosphate to lactic acid. The heterofermentative lactic acid bacteria receives five-carbon xylulose-5-phosphate from the pentose pathway. This gets decomposed into glyceraldehyde 3-phosphate (3-carbon), which later forms lactic acid and ethanol. The lactic acid bacteria have the ability to ferment low-cost raw materials, they require low amounts of nitrogenous substances, and can grow under low pH and high temperature conditions. The fermentation of sugar varies with the species.

(e) **The Actinomycetes** These bacteria have branched filamentous hyphae. Morphologically, these bacteria look like fungi as their cells are elongated. These cells branch into filaments or hyphae. However, the hyphae of actinomycete bacteria are very different from that of fungi as the hyphae they possess are very small in size as compared to the fungal hyphae. The cell walls contain peptidoglycan. The cell walls are usually around 1.0 to 1.5 inch in diameter, which is much smaller than that of the fungi. They help in the production of many pharmaceutical products. These bacteria have the ability to utilize and degrade a large variety of substrates found in the soil, including polymers such as chitin, cellulose, and hemicellulose present in the insects and plants that are very difficult to be degraded otherwise. The best-known genus is *Streptomyces*, from which many antibiotics as well as non-antimicrobial drugs are obtained.

(f) **Eucarya** These are fungi that are mainly used in the industrial production of metabolites. These are categorized into four groups: Phycomycetes, Ascomycetes, Fungi Imperfecti, and Basidiomycetes. Industrially important ones are Phycomycetes (Zygomycetes) and Ascomycetes, which include Yeasts. Yeasts are used in the production of ethanol and for alcoholic beverages. *Claviceps purperea* is used for the production of ergot alkaloids.

*Phycomycotina* are lower filamentous fungi and include genera such as *Rhizopus* and *Mucor*, which are used for the production of esterase enzyme and for the production of cheese flavors.

*Ascomycotina* are the organisms that possess a special sac-like structure (ascus). This sac contains sexual spores (ascospores). The spores are obtained by the fusion of two nuclei, which is then followed by meiosis.

*Basidiomycotina*: Under this category the fungi are quite evolved and contain spores, which are called as basidiospores. These are present externally in a basidium. Mushrooms, bracket fungi, rusts and smuts, and jelly fish come under this category.

*Deuteromycotina*: These are also known as fungi imperfecti. It is named as there is no perfect sexual stage present in them. Some examples are *Aspergillus*, *Penicillium*, and *Botrytis*.

*Yeasts*: These are single-cell microorganisms and multiply by the process called as budding/fission.

*Chemoautotrophic bacteria*: These bacteria grow by the oxidation of inorganic compounds. Some bacteria that fall in this category are hydrogen bacteria, thiobacilli, nitrifying bacteria, *Pseudomonas*, *Mycobacterium*, and *Nocardia*.

*Phototrophic bacteria*: These bacteria carry an anaerobic photosynthesis. Purple, green, and brown bacteria come under this category.

*Cyanobacteria*: These are prokaryotic bacteria that lack a nucleus. Photosynthesis takes place with chlorophyll *a* and phycobilin proteins. Blue-green algae come under this category.

*Other bacteria*: These are higher forms of life and come under the category of eukaryotes. The bacteria have two domains: the Bacteria and the Archaea. Each is distinct from one another.

## 1.6 Metabolism

All the biochemical reactions that take place in a cell or organism are called as metabolism. There are many reactions by which the substrate molecules get broken down. During these reactions, energy is generated. There are two types of reactions that take place in a cell. These are exergonic (produces energy) and endergonic (needs energy) reactions. In any living bacterial cell, the reactions take place with the help of enzymes present in the cell. During the metabolism of the microorganism, the cell grows depending on the doubling time of that particular microorganism. The cells multiply and form new metabolites depending on the reactions undergoing in the cell. Energy gets generated when substrate oxidation takes place. This energy is stored in the form of high-energy compounds such as adenosine diphosphate (ADP) and adenosine triphosphate (ATP) or compounds containing the thioester bonds. This energy is utilized along with the enzymes, cofactors, coenzymes, vitamins, and minerals in the synthesis of new metabolites (Jurtshuk Jr. 1996).

## 1.7 Genetic Modification

Bacteria can be identified by genetic techniques that are based on the profiling of genetic material of an organism (DNA). The phenotypic techniques are based on profiling either the metabolic attributes and/or its chemical composition. The genotypic techniques are better as they are independent of the physiological state of an organism and are not influenced by the composition of the growth medium or the phase of growth of the organism. Phenotypic techniques, provide functional information about the metabolic activities for the survival, growth, and development of the organism.

Genetic modification improves the characteristics of the produced enzyme and also inhibit the production of toxic metabolites like mycotoxins. Microorganisms (MO) play an important role in the production of beer, yoghurt, kefir, cheese, soy sauce, wine, vinegar, and many more. Microorganisms have the ability to produce a new protein. This could improve/enhance the production of an existing protein/ingredient or the existing protein can have a new application. Genetically engineered microorganisms (GEMs) are microorganisms (bacteria, fungi, and yeasts) that have been modified using in vitro molecular biology techniques toward a specific function, especially for the production of medical and food substances. These processes are cost-effective and do not cause any environmental pollution. Microbial insulin, trypsin, and chymosin are such examples. Higher yields of riboflavin, vitamins, amino acids, functional proteins, nutritional proteins, oligosaccharides, flavors, and sweeteners have been produced by GEM successfully (Holmes and Jobling 1996).

**FAQs**

1. What is known as the golden age of microbiology?
2. What are the primary mechanisms within a host that mediate microbe–microbe and host–microbe interactions?
3. How do microbial communities function to affect the phenotype of the host?
4. Can microbes prevent infections?
5. How can a "systems biology" approach improve our understanding of host–microbe interactions?
6. Can compositional or evolutionary changes in microbiomes help hosts adapt to environmental change within the lifetime of the host?
7. Which scientist divided the organisms into three domains?
8. Which microorganism was discovered first?
9. What is the difference between microbes and microorganisms?

## References


Abraham EP, Chain E (1940) An enzyme from bacteria able to destroy penicillin. Nature 146:837

AIChE (1970) The history of penicillin production. Chem Eng Progr Symp Ser No 100:66. American Institute of Chemical Engineers, New York

Avery OT, Colin MM, Maclyn Mc C (1944) Studies on the chemical nature of the substance inducing transformation of pneumococcal types: induction of transformation by a deoxyribonucleic acid fraction isolated from *Pneumococcus* Type III. J Exp Med 79(2):137–158

Bailey JE (1996) Metabolic engineering. Adv Mol Cell Biol 15:289–296

Berg P, Mertz JE (2010) Perspectives: anecdotal, historical and critical commentaries on genetics. Personal reflections on the origins and emergence of recombinant DNA technology. Genetics 184:9–17

Berget SM, Moore C, Sharp PA (1977) Spliced segments at the 5' terminus of adenovirus 2 late mRNA. Proc Natl Acad Sci USA 74(8):3171–3175. Bibcode:1977PNAS.74.3171B

Botstein D, White RL, Skolnick M, Davis RW (1980) Construction of a genetic linkage map in man using restriction fragment length polymorphisms. Am J Hum Genet 32(3):314–331

Bremus C, Herrmann U, Bringer-Meyer S, Sahm H (2006) The use of microorganisms in L-ascorbic acid production. J Biotechnol 124:196–205

Buchholz K, Collins J (2010) Concepts in biotechnology—history, science and business. Wiley-VCH, Weinheim

Buchholz K, Collins J (2013) The roots—a short history of industrial microbiology and biotechnology. Appl Microbiol Biotechnol 97:3747–3762

Bud R (2007) Penicillin—triumph and tragedy. Oxford University Press, Oxford

Chang A, Boyer H, Helling R (1973) Construction of biologically functional bacterial plasmids in vitro. Proc Natl Acad Sci USA 70:3240–3244. https://doi.org/10.1073/pnas.70.11.3240

Cohen SN, Chang ACY, Boyer HW, Helling RB (1973) Construction of biologically functional bacterial plasmids in vitro. Proc Natl Acad Sci USA 70(11):3240–3244

Connstein W (1870–?) Worked for many years at the Vereinigte Chemische Werke AG at Charlottenburg, Berlin 5

Demain AL (2004) Pickles, pectin, and penicillin. Annu Re Microbiol 58:1–42

Ereky IC (1919) Biotechnologie der Fleisch-, Fett- und Milcherzeugung im landwirtschaftlichen Grosbetriebe. Verlag Paul Parey, Berlin, p VII. 84

Fernbach A, Strange EH (1911) Fermentation process producing amyl, butyl, or ethyl alcohol, butyric, propionic, or acetic acid, etc. British Pat 15:203

Fleming A (1929) On the antibacterial action of cultures of a penicillium, with special reference to their use in the isolation of *B. influenzæ*. Br J Exp Pathol 10(3):226–236

Gaffky G, Loeffler (1884) Zur Ätiologie des Abdominaltyphus. Mitteilungen aus dem Kaiserlichen Gesundheitsamt 2:372–420

Gelvin SB (2003) *Agrobacterium*-mediated plant transformation: the biology behind the "gene-jockeying" tool. Microbiol Mol Biol Rev 67(1):16–37. https://doi.org/10.1128/MMBR.67.1.16-37.2003

Hall B, Limaye A, Kulkarni AB (2009) Overview: generation of gene knockout mice. Curr Protoc Cell Biol Chapter 19:Unit19.12:1–17. https://doi.org/10.1002/0471143030.cb1912s44

Harrington JJ, Van Bokkelen G, Mays RW, Gustashaw K, Willard HF (1997) Formation of de novo centromeres and construction of first-generation human artificial microchromosomes. Nat Genet 15(4):345–355

Holmes RK, Jobling MG (1996) Genetics. In: Baron S (ed) Medical microbiology, 4th edn. University of Texas Medical Branch at Galveston, Galveston (TX). Chapter 5. https://www.ncbi.nlm.nih.gov/books/NBK7908/

Hotchkiss RD (1979) The identification of nucleic acids as genetic determinants. In: Srinivasan PR, Fruton JS, Edsall JT (eds) The origins of modern biochemistry. New York Academy of Sciences, New York, pp 321–342

Hwang HH, Yu M, Lai EM (2017) *Agrobacterium* mediated plant transformation: biology and applications. Arabidopsis Book 15:e0186. Imprint CRC Press: 40. eBook ISBN 9780429105647. ISSN 0092-8674, https://doi.org/10.1016/0092-8674(84)90491-4

Jurtshuk P Jr (1996) Bacterial metabolism. In: Baron S (ed) Medical Microbiology, 4th edn. University of Texas Medical Branch at Galveston, Galveston (TX). Chapter 4

Koch R (1876) Robert Koch-Institut Die Ätiologie der Milzbrand-Krankheit, begründet auf die Entwicklungsgeschichte des *Bacillus Anthracis*. Cohns Beiträge zur Biologie der Pflanzen (in German) 2(2):277 (1–22)

Macdonald ME, Christine MA, Mabel PD, Richard HM et al (1993) A Novel gene containing a trinucleotide repeat that is expanded and unstable on Huntington's disease chromosomes. Cell 72:971–963. Cell Press

Mermelstein LD, Papoutsakis ET, Petersen DJ, Bennett GN (1993) Metabolic engineering of *Clostridium acetobutylicum* ATCC 824 for increased solvent production by enhancement of acetone formation enzyme activities using a synthetic acetone operon. Biotechnol Bioeng 42(9):1053–1060

Michaelis L, Menten ML (1913) Die Kinetik der Invertinwirkung Biochemische Zeitschrift. 49:333–369

Okanishi M, Suzuki K, Umezawa H (1974) Formation and reversion of *Streptomycete* protoplasts: cultural condition and morphological study. J Gen Microbiol 80:389–400

Ozcengiz G, Demain AL (2013) Recent advances in the biosynthesis of penicillins, cephalosporins and clavams and its regulation. Biotechnol Adv 31:287–311

Pirt J (1975) Principles of microbe and cell cultivation. Blackwell, Oxford

Poulson PB, Buchholz K (2003) History of enzymology with emphasis on food production. In: Whitaker JR, Voragen AGJ, Wong DWS (eds) Handbook of food enzymology. M Dekker, New York, pp 11–20

Qureshi N, Li XL, Hughes S, Saha BC et al (2006) Butanol production from corn fiber xylan using *Clostridium acetobutylicum*. Biotechnol Prog 22:673–680

Rubin GM, Hong L, Brokstein P, Evans-Holm M, Frise E, Stapleton M, Harvey DA (2000) *Drosophila melanogaster* genome. A Drosophila complementary DNA resource. Science 287(5461):2222–2224

Sanger (1943) Studies for a doctorate at Cambridge University. Cambridge university

Schegloff EA (1968) Sequencing in conversational openings. Am Anthropol 70(6):1075–1095. http://www.jstor.org/stable/669510

Seligman (1915) Plate heat exchangers and spiral plate heat exchangers. In: Heat exchanger design handbook, 2nd edn. First published: 2013
Smithies O (1955) Zone electrophoresis in starch gels: group variations in the serum proteins of normal individuals. Biochem J 61:629–641
Soderlund C, Humphray S, Dunham A, French L (2000) Contigs built with fingerprints, markers, and FPC V4.7. Genome Res 10(11):1772–1787. https://doi.org/10.1101/gr.gr-1375r
Speakman HB (1919) The production of acetone and butyl alcohol by a bacteriological process. J Soc Chem Ind 38:155T–161T
Stemmer W (1994) Rapid evolution of a protein in vitro by DNA shuffling. Nature 370:389–391
Szybalski W (2013) The 50th anniversary of gene therapy: beginnings and present realities. Gene 525(2):151–154. ISSN0378-1119
Tiselius (1937) Electrophoresis of serum globulin I. Biochem J 31:313–317
Van Niel CB, Kluywr AJ, Derx LG (1929) Uber das Butteraroma. Biochem Z 210:234–251
Von Laue M (1912) Physics forum then & now. Max von Laue and the discovery of X-ray diffraction in 1912. In: Michael Eckert (ed). Ann Phys (Berlin) 524(5):A83–A85 (2012). https://doi.org/10.1002/andp.201200724
Watson JD, Crick FH (1953) Molecular structure of nucleic acids; a structure for deoxyribose nucleic acid. Nature 171(4356):737–738. https://doi.org/10.1038/171737a0
Weizmann C, George B, Kauffman IM (1994) Chemist, biotechnologist, and statesman: the fateful interweaving of political conviction and scientific talent. J Chem Edu 71(3):209
Wharton RP, Brown EL, Ptashne M (1984) Substituting an α-helix switches the sequence-specific DNA interactions of a repressor. Cell 38(2):361–369

# Emerging Technologies for Improvement of Microbial Strains Using Omic Sciences and Measuring Methods for Maximum Product Formation

2

**Abstract**

Microorganisms are very beneficial to human kind. They produce primary and secondary metabolites that are useful to combat many harmful diseases caused by different microorganisms. These metabolites are used in foods, pharmaceuticals, textiles, cosmetics, and many other industries. The enzymes produced by microorganisms are used in waste water treatment, production of alcohol, and other fermented products. The antibiotics are harmful to the human body when used repeatedly as it causes antibiotic resistance. Around 700,000 deaths around the world are caused due to antibiotic resistance every year and this has become a major health concern. Therefore, it has become vital not only to diagnose but also to implement the corrective measures. For this purpose, rapid and accurate methods are needed.

The most common bacterial diseases are diabetes, sepsis, and gonorrhea. Many biochemical and serological techniques are time-consuming and not equipped to differentiate similar species. Mass spectrometry (MS), laser desorption/ionization (LDI), matrix-assisted laser desorption ionization (MALDI), or surface-enhanced laser desorption/ionization (SELDI) and electrospray ionization (ESI) are being used. However, with latest advances and discoveries, MALDI-TOF MS is combined with "omics" approach. This is helpful in detecting the genes in mRNAs (transcriptomics), proteins (proteomics), lipids (lipidomics), and metabolites (metabolomics) in biological samples. This chapter deals on the emerging technologies for the improvement of microbial strains using omic sciences and other methods for the maximum product formation and combating pathogenic microorganisms. This can be done by classical or by molecular genetics methods. The techniques to measure the betterment have also been discussed in this chapter.



R. Agrawal, *Textbook of Industrial Microbiology*,
https://doi.org/10.1007/978-981-97-9582-6_2

**Keywords**
Genetic methods · Metabolite · Antibiotic · Characterization · Product · Improvement · Omic science · Proteomics · Lipidomics · Epigenomics · Microbial · Strain

## 2.1 Introduction

The products that are produced during cell metabolism, respiration, and photosynthesis are known as metabolites. The metabolites that help in the growth, development, and reproduction come under the category of primary metabolites. The end products of primary metabolites form secondary metabolites. These are not required by the organism for basic metabolism (Armstrong et al. 2001).

Most of the physiological processes are carried out with the help of metabolites. These include alcohols such as ethanol, lactic acid, and certain amino acids.

In industrial microbiology, alcohol is produced at large scale and is a very common primary metabolite. Using alcohol, products such as beer and wine are produced by fermentation processes. The microorganism *Corynebacteria glutamicum* produces primary metabolites such as amino acids (L-glutamate and L-lysine). Citric acid is produced by *Aspergillus niger*, which is a very important primary metabolite. These metabolites are used commonly in food, pharmaceutical, and cosmetic industries. In industrial microbiology, these are important for the production of many antibiotics, food products, enzymes, amino acids, vaccines, and fine chemicals. The various uses of these metabolites include biotransformation, formation of biofuels, utilization of organic and industrial wastes, removal of heavy metals, microbial biomass production (microbial protein or single cell protein) for food and feed, production of biocontrol agents, and fermentation of food products (Bérdy 2005). The pathways of metabolism can be changed to enhance the metabolite yield by using metabolic engineering, which involves genetic techniques including recombinant DNA. While growing the culture, the physical parameters such as pH, time, temperature, and the growth requirements of microorganisms play a vital role. These are specific and are involved in the breakdown of large molecules into the simpler ones.

The immobilization of enzymes or using them as such has been found to maintain the microbial processes. For example, when wine is heated during processing, it produces a carcinogen known as ethyl carbamate (urethane). However, if yeast strain is used along with the urease enzyme, it helps in reducing the accumulation (Ough and Trioli 1988).

Many metabolites from microorganisms have antioxidant, antitumor, vasorelaxing, diuretic, and laxative properties. Some produce colorants, pesticides, and are also known to be growth promoters for animals and plants (Kusale et al. 2021).

Due to the awareness among the global population, there has been a paradigm shift toward the use of natural products. It has been found that FDA has approved more than 53% of such products that are produced from the microorganisms (Miranda et al. 2003). However, to enhance the metabolite yields, various physical and chemical parameters should be optimized as they are the source of nutrition, growth conditions, feedback control, and enzyme induction/inactivation. These are regulated by transfer RNA, σ factors, and gene products formed during post-exponential development. They are coded by clustered genes on chromosomal DNA and not by plasmid DNA. A lot of research is underway to understand the pathways.

Recently, cell factories have shown a new and innovative way to enhance these metabolites. It is important to understand the pathway of the metabolites. The strains can then be improved with a multi-omic-based (MOB) approach, which can be measured from scalable fermentations (Multi-Omics Research Factory, MORF; Springthorpe et al. 2020; https://morf-db.org/projects/DETOX/styrene). In this method the fermentation and the multi-omic (transcriptomic, proteomic, metabolomic, and lipidomic) data can be accessed. This allows the response of the cell factory for chemical biosynthesis. The interventions can thus be identified and productivity can be improved. Lian et al. (2016) have used this approach in the synthesis of styrene in *E. coli* as cell factory. Joseph et al. (2022) explained the multi-omic-based production and strain improvement (MOBpsi) for biomanufacturing of toxic chemicals.

The secondary metabolites are specialized metabolites that have the ability to improve health, nutrition, and economics in the society (Monciardini et al. 2014). Industrial microorganisms offer many new opportunities for strain discovery, characterization, and their improvement using omic sciences, which include the application of genomic, transcriptomic, or metabolomic sciences. This gives a great potential toward the development of novel compounds in pharmaceutical, nutraceutical, dyeing, and agricultural industries. Nawaz et al. (2020) wrote an excellent review on the marine bacteria that produce bio-pigments at the industrial level. A basic bioreactor/fermentor is shown in Fig. 2.1 in which metabolites can be produced in large scales. The details on fermentor/bioreactor have been discussed in Chap. 7 of this book.

Demain (1999) worked on the pharmaceutically active metabolites from microorganisms.

## 2.2 Increase in Metabolite Yield by Various Methods

### 2.2.1 OSMAC (One Strain Many Compounds) Approach

Insect-associated bacteria are known to produce many secondary metabolites. Many new bioactive molecules have been produced by changing the culture conditions of a strain that helps in the activation of biosynthetic gene clusters. This is called as OSMAC approach (one strain many compounds). The production of cyclic pentapeptides pentaminomycins C, D, and E from *Streptomyces* sp. GG23 strain have been increased many folds by this approach.

The analysis of the non-ribosomal peptide biosynthetic gene cluster suggested that the unicity of the compounds, based on the structural variations, originates from the low specificity of the adenylation domain in the non-ribosomal peptide

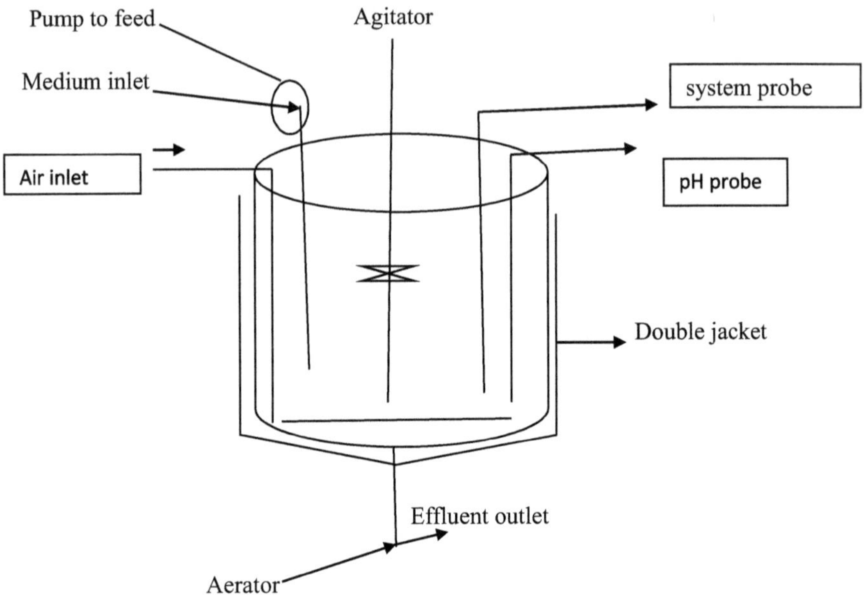


**Fig. 2.1** A basic bioreactor/fermentor

synthetase (NRPS) module 1, indicating that macrocyclization can be catalyzed non-canonically by penicillin binding protein (PBP)-type TE. Additionally, pentaminomycins C and D exhibited significant autophagy-inducing activities and were cytoprotective against menadione-induced oxidative stress in vitro.

### 2.2.2 Enhancing the Product Yields by Coculturing

Many a times the metabolite yield can be increased by using microbial cocultures, with metabolome tools. According to a study, three actinomycete strains have been isolated from the Red Sea sponge, namely *Coscinoderma mathewsi* (23 isolates). These are species of *Micromonospora*, *Nocardia*, and *Gordonia* (Shamikh et al. 2020) and have shown to produce important metabolites such as chlorocardicin, neocopiamycin A, and chicamycin B. These metabolites were absent in the respective monocultures. This study has shown a way that silent biosynthetic gene clusters can be activated to produce secondary metabolites by their internal interaction when the cultivation conditions are modified. However, it is important to know the stage of growth phase of cultures, at what stage of development the culture belongs to, and at what proportions should the cultures be mixed. This is to ensure that all the cultures grow together and one does not interfere or takes over the growth of the other culture. The ultimate result is in maximizing the yield of the secondary metabolite. Therefore, either the cell number (cfu/mL) must be optimized for each microbe or the biomass levels can be fixed before doing the coculturing.

In case of spores, it is vital to fix the developmental stage of the culture before coculturing. The relative growth of all the cultures has an effect while coculturing.

The marine microorganisms are important in metabolite production. They are known to produce many therapeutic molecules. Trabectedin is an antineoplastic drug isolated from *Ecteinascida turbinata* that has been approved by the EU. Other important drugs are salinosporamide A, obtained from *Salinospora tropica*, which is a proteasome inhibitor and is used to treat multiple myeloma (Potts and Lam 2010). With the general awareness of health, most of the population around the globe are using PUFA (poly unsaturated fatty acid) in place of saturated fats. *Aurantiochytrium* sp., a marine microalga, has been known to produce large quantities of docosahexaenoic acid (DHA), which is very helpful in cellular processes of human beings.

Recent inventions of using textiles having antimicrobial activity can help in reducing infections. These are being used extensively and the approach is on the forefront. Endophytic fungi, isolated from marine organisms, have antimicrobial or antioxidant activities and the metabolite yields have been improved (Hamed and Tawwab 2021).

There are specialized compounds, known as siderophores. These compounds get attached to the ferric iron and form complexes. Iron is a cofactor in many enzyme reactions and in oxygen transport, and it is used during respiration, for fixing nitrogen, during photosynthesis, and in the synthesis of DNA, amino acids, and fatty acids. However, the bioavailability of iron is usually poor. The best pH for iron oxidation is 7.0. As per the literature, free ferric is toxic and therefore a specific amount, below $10^{-24}$ M, in humans is recommended (Raymond et al. 2003). Siderophores can cure iron overload diseases. They are useful in agriculture as green iron chelators. They are also used in medicines. The uptake of antibiotics can be improved with the use of a drug conjugate of siderophore antibiotic. From a marine environment, it has been possible to extract siderophore named as serratiochelin A (which kills *S. aureus*) in high yields from an iron-depleted coculture of *Serratia* sp. and *Shewanella* sp. (Schneider et al. 2020). This could be due to the competition for iron among the two strains in the culture medium. It is also toxic to eukaryotic cell lines, namely A2058 and MRC5.

Bacteria, fungi, or yeasts have different pathways to form siderophores. This depends on the concentration of iron in the surrounding environment. In bacteria, with low content of purines, iron acts as a cofactor and gets bound to a specific DNA operator. This reduces the transcription of genes. When the purines (G + C) are high in bacteria, the repressor proteins are responsible for iron pathway.

Microorganisms have thus shown various new avenues to produce new biomolecules.

Bacteria or fungi, in both the cases, the method of coculturing has taken over monocultures to produce new molecules. The major products include medicines and biopesticides. Among fungi, they can be terpenes, alkaloids, or polyketides. The compounds impart a large number of beneficial properties as antimicrobial and anticancer activities. The technique of gene mining has been useful to know the gene clusters responsible to synthesize these metabolites.

Among bacteria, *Streptomyces* is known to produce many bioactive compounds. It has the ability to produce three angucyclines along with rabelomycin and 5-hydroxy-rabelomycin (Voitsekhovskaia et al. 2020). Biologically, they have anti-cancer and antiproliferative properties. In many cases, the gene clusters have been identified.

Bioactive properties of microorganisms have been increased in the fungus *Curvularia papendorfii*, which is known to have antiviral activity against the human coronavirus HCoV 229E, and the feline coronavirus FCV F9 has antibacterial activity against *Staphylococcus* spp. It is antiproliferative and can prevent breast cancer cell line MCF7 in human beings. It also contains Kheiric acid, which can prevent the growth of methicillin-resistant *Staphylococcus aureus* (MRSA).

Biofilms are produced by many microorganisms as bacteria and fungi. These biofilms are formed by the pathogenic microorganisms. Commonly, they are present in the epithelial or endothelial lining of the lungs, intestine, and in the vaginal mucus layer. The plaque on teeth and infections on medical instruments and implants are basically due to the infections by pathogens present in the biofilms. The biofilms consist of microorganisms that are live and still reproducing. They continue to grow at a rapid speed and therefore they have the ability to form communities in a short time that get attached to the surfaces of organs, instruments, and so on. Even the treatment of antibiotics does not work as they are resistant to most of them. Most of these biofilms cannot be removed as they produce an extracellular polymeric substance (EPS) to protect themselves.

When the pathogenic microorganism gets attached, it multiplies by secreting a slime. This has been found on metals, plastics, natural materials, medical implants, kitchen counters, contact lenses, walls of a hot tub, and even the swimming pool among many others. Majorly, the contamination in the food industries happens due to the formation of biofilms. This biofilm-associated infection (BAI) is the main cause of cross-contamination in the food products. This has been resulting in big losses to the food industries. They are present in the soil, agricultural lands, fruits and vegetables, seafood, dairy products, milk products, and most of the meat products. With recent advances and a lot of scientific inventions, many new techniques have been developed to combat the biofilm matrix. This is known as quorum sensing (QS).

At the same time, the biofilms have been found to be beneficial to protect the plants, in the treatment of wastewater, and also in reducing corrosion. The techniques such as cold plasma, ultrasound, light-assisted technologies, pulse field, and high-pressure processing have been found successful to inhibit biofilms. The crop production could be increased by using biofilms as biofertilizers. Papaianni et al. (2020) found that the lytic phages of Xccφ1 (*Xanthomonas campestris* pv. *campestris*-specific phage) can remove/inhibit biofilm formation.

## 2.3 Identification of Microorganisms by Biochemical Tests and Other Methods for Quantification of Metabolites

As per the food regulations, whenever a microorganism is used for industrial application or for any disease treatment, it should be initially characterized, identified, and classified on the basis of taxonomy.

Systematics is the study of diversity of organisms and their relationships. In taxonomy, the organisms are classified in a hierarchical structure of homogeneous groups. Commonly, this classification is made on the basis of kingdom, division, class, family, genus, species, and strain. Research work in the field of classification, systematics, and identification of microorganisms are interconnected and have an impact on each other. Accurate identification affects the taxonomic classification of microorganisms. If the identification is precise, then the classification and systematics go well together. The polyphasic method deals with the morphological and biochemical data, which is complemented with the molecular technique data. When the classical approach is combined with 16S rRNA genes/molecular fingerprinting techniques and/or other molecular markers, it helps in the identification and classification of microbes and quantification of metabolite. Traditionally, the time required to identify a microorganism includes long procedure that involves various steps such as characterization of morphology, physiology, chemistry, and biochemistry, which takes at least 2–5 days and can last up to 12 days in the case of molds. These methods are labor intensive and need a lot of material. They do not identify the microorganisms up to the species or the strain level. However, with molecular fingerprinting techniques, the identification is up to the strain level and has been found to be very useful, effective, and rapid. They are helpful in quantification of the metabolites.

Mutations can be obtained using chemicals such as nitrous acid, ethyl methane sulphonate, diethyl sulphate, *N*-methyl-*N*-nitro-*N*-nitrosoguanidine (NTG), base analogues, and acridine compounds and radiations such as ionizing radiations or ultraviolet light. The use of hybridization methods have also been found useful. The identification of different microorganisms and quantification of metabolites can be done on the basis of the tests as given in Table 2.1.

**Table 2.1** Different biochemical tests for the identification of microorganisms

| Method | Types of tests |
|---|---|
| Staining | Gram, Albert, Acid Fast, and special |
| Motility | Inoculate bacteria in motility medium |
| Biochemical | Catalase, coagulase, oxidase, sugar fermentation, indole, citrate, urease |
| Phage type | To identify strains of the same species grown in agar and dried |
| Serological | Antigen–antibody test (ELISA) |
| Disc | Kirby Baur disc diffusion method to check antibiotic susceptibility |
| Molecular | G + C content, DNA hybridization and sequencing, real-time qPCR, nucleic acid sequencing, ribotyping |
| Spectroscopical | LCMS, MALDI-TOF, FTIR, Microfluidic chips |

### 2.3.1 Drug Discovery Programs

After Covid-19 epidemic, a lot of research work is being carried out around the globe to combat the disease. Protein Mpro, which is a protease, has proved to show antiviral activity on human beings infested with severe respiratory syndrome (Sayed et al. 2020). Out of the many thousand compounds screened, six of them had high activity against the corona virus designated as anti-SARS-CoV-2 drug candidates.

### 2.3.2 Challenges in Production Steps

While producing secondary metabolites from microorganisms, there are many challenges that need attention. When any microbial process is in progress or even after the process is completed, the following steps should be looked into.

It is important that cleaning of the equipment, pipe lines, and premises are done very precisely. This avoids any form of contamination. In order to maintain good quality, it is vital that the cleaning of various containers is done separately. Also, all established methods should be followed. The entire process should be monitored. It is vital to test the final product in the presence of any endotoxins by AOAC-established methods. Checking for bioburden needs should be undertaken. While using detergents/disinfectants, care must be taken to check that they are devoid of any contamination. It is always better to follow the guidelines as mentioned in ISO 14644-1 (2). While performing aseptic processing, it is important to check the environment, personnels, all surfaces involved, container sterilization, and the filters. The use of isolator technology helps in reducing microbial contamination. It is important to design them with a material that does not allow any puncture or leak. If the system units are continuous, then in such cases, steps should be taken, starting from microbial inoculation, product formation, filling, and then sealing. All these steps help in reducing the chances of contamination. Having minimum manpower during the process also helps in controlling the contamination. Inspections and other controls done either with probes or from outside, usually helps in avoiding contamination. Too many washing areas and drains increase the chances of contamination, which must be avoided. Any type of backflow should not be present in the system.

The conveyor belt should be sterilized with steam/dry heat. It is important that the design is made in such a way that the maintenance and repairs could be done from outside. In any microbial production unit, timely checking of all equipment such as sterilizers, air filters, air vents, water treatment, product, storage and distribution of the product are necessary. The maintenance needs good planning and execution. For water treatment plants, it is important to check the source of water, which should be devoid of any contamination and water quality should be good. The use of water in any processing plant should be done as per the designed capacity and should never be used beyond it. It is advisable to circulate water at a temperature of above 70 °C.

Once the product is formed, it has to be taken into a container. Closing of the containers must be done by using standard methods as per the rules set by the Food Safety and Standards Authority of India (FSSAI).

Automation in sealing is the best. However, if the containers are closed by fusion method, they must be tested before marketing. Sterilized capping system needs to be undertaken. The use of corrective lenses for inspection and any correction to be done during the process is helpful. Making a record of each step and the immediate correction helps in the production of a good microbial product (Edward et al. 2019).

## 2.4 Role of Genetic Modification in Metabolite Production and Omic Sciences

### 2.4.1 Common Methods Used for Microorganisms Improvement to Produce Secondary Metabolites at Industrial Scale

There are some commonly used simple methods to modify microorganisms genetically for increasing the production of metabolites. These are given below.

Improvement for the productivity of a microorganism can be done as follows:

- *Mutants with enhanced yield of the product*: The plating is done on a medium that will grow colonies for high yields against the parent ones. This variant high-yielding colony is selected after primary screening. Then they are grown in a liquid media.
- *Strain improvement for higher yield by genetic engineering*: In this method the colony of interest is taken and the sequence of genes are modified in such a way that it improves the productivity. High yielding genes are also incorporated from other organisms.
- *Modification in metabolism*: In this engineering method the metabolic pathways are modified, which produce higher yields of secondary metabolite. The genes that encode for enzymes being utilized in the product formation are either inserted or the undesired genes are knocked off.

(i) **Classical Genetic Techniques to Improve Microorganism Strain**

Initially, a random screening of the colonies is made. The high-yielding colonies are selected. There are various mutagens used for enhancing the product yield. The physical mutagens are usually used in this method, which are mostly UV light or chemical mutagens such as *N*-methyl-*N′*-nitro-*N*-nitrosoguanidine or ethyl methane sulphonate (Baltz 1999).

This method is simple and does not involve any expensive equipment. Therefore, it is used commonly to obtain improved strains. This method has been utilized for the production of polyketides. The improved strains have been used to overcome glucose repression and to overcome repression due to ammonium and phosphate ions (Vinci and Byng 1999). The modification in metabolic pathways has helped in the overproduction of metabolites. Amino acid precursors have been supplemented to the media to get enhanced amounts of penicillin and cephalosporin obtained from the fungi *Penicillium chrysogenum* and *Acremonium chrysogenum*, respectively. In case of secondary metabolites like an antibiotic, culture strains are isolated and modified to produce higher yields of the metabolite (Schwab 1988).

There are other methods in fungi and actinomycetes to improve their strains for higher productivity. These are mainly sexual or parasexual crosses by conjugation. In another method the protoplasts of two cells can be fused to obtain a new modified cell with higher productivity. This method of protoplast fusion has given high productivity for many metabolites (Elander and Lowe 1992).

The methods used for improving the microbial strains should be reliable and tested. With the recent advances of molecular and genetic methods, the study of biosynthetic pathways has become very precise. The study of vectors and transformation methods have guaranteed the improvement of industrial strains. The study of regulatory mechanism has helped the scientists to regulate the promoter region and insert active sequences wherever required. In this regard, a study has been carried out (Pérez-Llarena et al. 1998) in which a regulatory gene (ccaR) has been amplified to obtain high yields of cephamycin and clavulanic acid in *Streptomyces clavuligerus*.

Studies have shown that a new enzymatic activity can also be incorporated in the strain of interest (heterologous gene) with which the yields can increase many folds and can be of industrial interest. In a study, for increased yields of cephalosporin C, a bacterial protein was introduced into *A. chrysogenum*, which is an oxygen transporter. The modified colony was isolated and used for the production of higher yields of antibiotics (DeModena et al. 1993).

### (ii) **Genetic Modification of Culture Strains**

*Corynebacterium glutamicum* produces essential amino acids such as L-methionine, L-lysine, and L-tryptophan, as well as the nonessential amino acid such as L-glutamic acid. With the help of genetic engineering, the organism (*C. glutamicum*) has been found to produce L-lysine and L-glutamic acid at much higher yields. L-glutamic acid is used for the production of monosodium glutamate (MSG), which is a food flavoring agent and is in demand. The production of organic solvents such as acetone, butanol, and isopropanol uses a series of *Clostridia* bacteria species by fermentation. Scientists have achieved much higher yields after using a genetically altered strain.

Human diseases are caused by more than 1400 pathogens. The main pathogens are *Campylobacter jejuni*, enterotoxigenic *Escherichia coli*, *Shigella* spp., *Vibrio cholerae*, *Aeromonas* spp., enterotoxigenic *Bacteroides fragilis*, *Clostridium difficile*, and *Cryptosporidium parvum*. Many of these pathogens can be killed by beneficial microorganisms and the products produced by them.

(a) **Gene Editing for Metabolites by Forming Biosynthesis Gene Clusters (BGCs)**

When the secondary metabolites are formed, they involve few genes. These are called as biosynthesis gene clusters (BGCs). These genes contain the information for the synthesis of SMs, how the genes are assembled, sequenced, and how they can be regulated for maximum product formation. Microorganisms genome contains a variety of cryptic or silent genes that help in the production of secondary metabolites. Most BGCs are silent and are not expressed under laboratory conditions. It is possible to carry out this method of genetic engineering in heterologous as well as homologous host to enhance the production of SMs.

Microbial enzymes or proteins are being commonly used in food. These are mainly α-amylase, lipase, protease, ice structure protein, and hemoglobin. Many microbial enzymes also help another microorganism or the same microorganism to produce food ingredients such as riboflavin, steviol glycoside, and oligosaccharides. Additional modifications can be done by the deletion of endogenous genes or by the insertion of transporters. The promoter and terminator sequences help in controlling the expression of functional genes in a microorganism. The DNA expressed by the genetically engineered microorganism may arise endogenously from the same organism or it could be exogenous from a different organism. Self-cloning is a method when the DNA is sourced exogenously from a closely related microorganism capable of natural DNA exchange. Usually, the exogenous DNA is from a distant microorganism, which cannot be grown efficiently under industrial conditions. When applied to food, the term bioengineered is a regulatory term. Therefore, it is necessary to label them for the knowledge of the consumer. In earlier times, the DNA was isolated from one microorganism, then amplified and transferred into another microorganism. However, with recent advances, DNA sequences are being made synthetically, which can be inserted into the microorganisms. This produces a genetically engineered microbe at a faster rate, which has the ability to produce many variants of enzyme proteins. The synthetic DNA does not allow the unintentional sequences, and pathogens are very specific.

(b) **Gene Manipulation**

When the organisms cannot be cultured in heterologous hosts, the genes can be managed or manipulated to activate the biosynthesis of gene clusters (BGCs). However, in homologous host, the genes are manipulated for all natural factors, which are necessary to produce secondary metabolites. Using newer genome techniques, it has been possible to produce metabolomics for the clustered regulatory interspaced short palindromic repeats (CRISPR-Cas9) (Alamillo et al. 2023).

(c) **Gene Cloning**

With recent scientific developments, newer cloning techniques such as Cas9-assisted targeting of chromosome segments (CATCH), transformation-assisted recombination (TAR), and TAR-CRISPR are being used (Jiang et al. 2015). In this method, initially, a suitable heterologous host is located in which the target BGC has to be cloned. Then the BGC is transferred into the selected host and is expressed in the host system. Different parameters need to be optimized for the maximum yield production. Cas9-assisted targeting of chromosome segments (CATCH) is a cloning technique that utilizes the CRISPR-Cas9 gene editing system for direct BGCs cloning into the host. A study has been carried out on a new TAR cloning method that can analyze the structure and function of the genomes (Noskov et al. 2002). Recently, TAR and CRISPR-Cas9 have been coupled in a new approach called TAR-CRISPR (Hiyoung et al. 2020), which has increased the clone efficiency and has produced high yields of secondary metabolites.

(d) **Gene Refactoring**

This method helps to activate the biosynthetic gene clusters (BGCs), which in turn produce higher yields of novel SMs. Scientists have worked toward the refactoring of silent BGCs for higher yields of natural products. For this purpose, the BGC promoter is replaced, which helps in the production of secondary metabolites (Liu et al. 2021). Multiplexed CRISPR-Cas9 and transformation-associated recombination (TAR)-mediated promoter (mCRISTAR) are newer methods in gene refactoring (Kang et al. 2016). These methods have the advantages of the CRISPR-Cas9 system and TAR. In mCRISTAR method, first the CRISPR-Cas9 breaks the double-strand present in the promoter region of the BGC, and then the fragments produced are again reassembled by TAR using synthetic gene-cluster-specific promoter cassettes. Gene refactoring can be done at a faster rate by cloning and refactoring of BGCs in a direct pathway cloning (DiPaC).

Recent development of CRISPR technology has brought revolution. It facilitates precise cutting and pasting of DNA by specialized proteins. There are three types of gene editing proteins: zinc fingers, TALENs, and CRISPR-Cas. CRISPR-Cas helps in the treatment of genetic diseases, growing climate-resilient crops, and in developing designer materials, foods, and drugs. CRISPR stands for clustered regularly interspaced palindromic repeats and these are chunks of regularly recurring bits of DNA that arose in certain bacteria as an ancient defense system against viral invasions (Barrangou et al. 2007). CRISPR-Cas is an enzyme complex (Cas proteins) along with genetic guides (CRISPR sequences) that together finds and edits DNA. RNA sequences can recognize specific DNA code in a living cell with a genetic defect that can be replaced with nucleic acids into the exact place of the target sequence. This can repair the DNA by either modifying the sequence or introducing a new gene. When deletion, restoration, or modification is done by this technique, it is called as gene editing. The introduction of new DNA sequences into the genome by CRISPR-Cas is called genome editing. This technique is being used for enhanced production of secondary metabolites at industrial levels. The work of scientists (Martín-Valmaseda et al. 2023) introduced CRISPR/Cas-mediated genome into fruit tree species and produced better and improved fruit crops with respect to the development of the tree, yield of fruits in number, and quality along with their tolerance to various biotic and abiotic stresses.

(e) **Gene Insertion or Deletion**

The gene knockout/in on BGC has shown to affect the SM production. The conventional methods of gene editing take a lot of time but CRISPR-Cas9 approach is very efficient and rapid (Rubio et al. 2016). For modification of genes at many different places, a new technique is used, which is known as multiplexed site-specific genome engineering (MSGE). Using this method, multiple BGCs are edited (Li et al. 2017). This modification helps to enhance the yield of secondary metabolites.

To use the technique of metabolomics, many sophisticated instruments are required, which include colorimetric array detectors for the detection of redox compounds, fluorescent spectrometers for the detection of aromatic compounds, and ELSD for the detection of lipids. Newer technologies such as nuclear magnetic resonance (NMR) spectroscopy and mass spectrometry (MS) are utilized for the detection of various secondary metabolites and their related macromolecule (amino acids, lipids, sugars, organic acids, etc.). Some of the common MS-based analytical methods are GC-MS, CE-MS, and LC-MS (Emwas et al. 2019).

## 2.5 Omics-Based Technology and Enhancing Yields of Secondary Metabolites

Many areas of research come under omics. These are proteomics, transcriptomics, genomics, metabolomics, lipidomics, and epigenomics. These techniques are very helpful in the analysis of proteins, RNA, genes, metabolites, lipids, methylated DNA, histone proteins in chromosomes, and in increasing the yields of metabolites. These approaches have helped in the understanding of biological systems. The proteomic technique has been helpful in detecting kidney diseases, prostate cancer, and breast cancer recurrence as omic-based measurements give an accurate predictive model.

### 2.5.1 Genomics

The complete sequence of DNA in a cell or organism is termed as the genome present in the cell nucleus and in other organelles. This can be analyzed with single nucleotide polymorphism (SNP) chips and DNA sequencing technology. When arrays of many oligonucleotide probes hybridize with specific DNA sequences, they form SNP chips. When genomic analysis is made, it is able to detect insertions, deletions, and copy number variation. DNA can also be modified by methylation of cytosines as they are involved in health and disease (Honda and Littman 2011; Kinros et al. 2011; Tilg and Kaser 2011).

### 2.5.2 Transcriptomics

A set of RNA transcripts from DNA in a cell or tissue are known as transcriptome. It involves ribosomal RNA (rRNA), messenger RNA (mRNA), transfer RNA (tRNA), micro RNA (miRNA), and other non-coding RNA (ncRNA). The measurement of the transcriptome is made with microarrays and RNA sequencing (RNAseq) in which probes are not required.

### 2.5.3 Proteomics

The proteome is the entire set of proteins present in a cell at any given time. The study involves many challenges due to posttranslational modifications (glycosylation, phosphorylation, acetylation, ubiquitylation proteins). They have different configurations and localization in a cell. The analysis can be done using techniques such as mass spectrometry and protein microarrays (Ahrens et al. 2010; Wolf-Yadlin et al. 2009). Proteomics refers to the experimental analysis of proteins and proteomes, which often entails the protein purification and the mass spectrometry analysis. The amino acid sequences of many proteins differ between microorganisms, and many different analytical techniques are used to characterize and differentiate microorganisms based on the proteome analysis.

The gold standard in the identification and classification of bacteria is the sequencing of the gene encoding 16S rRNA (16S rDNA) due to its high conservation within species, inter-species variability, and the stable rate of evolution. The 16S rRNA gene is universal in bacteria so that relationships between all bacteria can be measured, and the comparison of gene sequences enables the differentiation between microorganisms on the genus level and the classification of whispers on the species and subspecies levels. The MALDI-TOF MS method detects proteins ranging in mass from 2 to 20 kDa, which mainly represents ribosomal proteins and essential metabolism proteins. Ribosomal proteins and fragments of the 16S rRNA gene use MALDI-TOF MS, which has been found to be better than 16S rDNA sequencing, as it analyzes the antibiotics used and the detection of antibiotic resistance.

MALDI-TOF MS analysis can identify the pathogens present in the sample. Jones et al. (2003) have shown that the MALDI-TOF MS can characterize specific ribosomal proteins directly from *Escherichia coli* cells.

### 2.5.4 Protein Chip Arrays

These array surfaces nurture proteins in a selective manner. Therefore, they are known as surface-enhanced laser/desorption ionization time of flight mass spectrometry (SELDI-TOF-MS).

### 2.5.5 Epigenomics

When a reversible chemical modification is made to the DNA or to the histones (binding the DNA), there is a change in gene expression. However, the base sequence is not changed. This usually happens due to any disease. It varies in different cell types of the same organism. It could be due to methylation of DNA cytosine residues (at CpG) and change in histone proteins in the chromosomes (Maas 2010).

### 2.5.6 Lipidomics

The complete lipid profile in a cell, tissue, or the entire body is known as lipidome.

Gram-negative bacteria have a membrane of lipopolysaccharide (LPS), an amphiphilic endotoxin made of lipid A, an oligosaccharide core, and an O-antigen. Maldonado et al. (2016) studied the lipopolysaccharide modification in Gram-negative bacteria during chronic infection. Gram-positive bacteria contain glycolipids, glucolipids, and lipoteichoic acid (LTA), which is a typical component of the cell membrane. The microbial lipids can be directly identified in body fluids as blood, urine, and serum without checking on agar plates or in a liquid medium.

### 2.5.7 Metabolomics

A small set of metabolites in any biological sample such as carbohydrate, lipid, amino acid, nucleic acid, and other biochemical pathways are called as metabolome. This can alter if there is a change in diet. Even stress, physical activity, or any disease in the body can change the metabolome. It can be analyzed quantitatively by using MS and NMR spectroscopic techniques (Zhang et al. 2009). Microbial metabolomics is becoming common in disease control. It has been effective in differentiating the drug susceptible phenotypes in *Leishmania donovani* (t'Kindt et al. 2010).

Bioinformatics and computational methods have been useful in analyzing omics data. In proteomics the development of selected reaction monitoring (SRM) technique is used (Picotti et al. 2010), which is available in public domain (Omenn 2009; Omenn et al. 2011). Untargeted "shotgun" mass spectrometry-based proteomics have generated peptide atlases for plasma, liver, and other organs and biofluids (Farrah et al. 2011). Aptamer chips help in the quantitative measurements of proteins (Gold et al. 2010). A combination of computational algorithms for protein structures in proteomics seems to be an effective preposition in many harmful diseases (Sreekumar et al. 2009). Scientists are working for omic measurements (such as transcriptomics, proteomics) on single cells (Tang et al. 2011).

The recent interest has focused on measuring multiple omics data types on a single set of samples to integrate different types of molecular measurements into an omics-based test. Such multidimensional datasets have the potential to provide deep insight into biological mechanisms and networks, allowing for the development of more powerful clinical diagnostics. An encouraging example of simultaneous measurement of multiple types of omics data is the DNA-encoded antibody libraries approach (Ryan et al. 2007), which can measure DNA, RNA, and protein from the same sample.

The approach has been tried to study neurodegeneration in mice (Hwang et al. 2009; Omenn 2009). In order to predict the drug targets, the study of genomics, transcriptomics, proteomics, and metabolomics combined together has been found effective (Committee on the Review of Omics-Based Tests for Predicting Patient Outcomes in Clinical Trials 2012). Scientists have been able to identify the disease-causal genes among the whole genome (Ng et al. 2010; Roach et al. 2010). Janiszewska et al. (2022) worked on the identification of antibiotic resistance using omic technique in bacteria.

### 2.5.8 Labeling of GEM-Produced Food Substances as GMO

Given the many regulations, it is not necessary that GEM-produced food substances should be considered as GMO for labeling. If the modified DNA is not in the detectable levels in the desired product, then the producer can be exempted of labeling the food as GMO. Also, exemption of GMO labeling is given if the modified food substance has been produced with some genetic techniques or if the genetic modification has happened naturally or has been obtained through traditional breeding. Also, if a food is produced using GEM-derived processing aids/incidental additives, it need not be labeled as GMO. Under the European Union (Regulation (EC) No. 1830/2003), the United States (7 CFR 66), and Canada (CAN/CGSB-32.315-2004), food substances that are produced with a GEM (GEM are essentially a processing aid or an additive) need not be labeled as GMO. However, the food substances that are derived from/produced from a GEM or other genetically engineered sources should be labeled as GMO. If DNA from the GEM is not detectable by any analytical method, then it need not be labeled as GMO. Usually, purification steps remove all the traces of DNA from the GEM in the finished food substance. Therefore, under these regulatory frameworks, such as the recently finalized United States National Bioengineered Food Disclosure Standard (7 CFR 66) as well as under Regulation (EC) No. 1830/2003 in the European Union, there is no need to label the GEM-produced foods as GMO. Moreover, each jurisdiction has its own labeling regulation. There are various regulatory interpretations when protein engineering is done for enzymes on the course of its interpretation to be considered as GMO or not. When the foods are produced via nontargeted mutagenetic techniques and are also different in phenotypes, it does not have the necessity to be labeled as GMO, as in ruby red grapefruit (Da Graca and Korsten 2004) and docosahexaenoic acid (DHA, from algal oil) (Fu et al. 2016).

Many private companies have made their own independent framework for the implementation of regulations.

### 2.5.9 Safety Evaluation of GEMs and Authorization of GEM-Produced Secondary Metabolites

Different countries expect similar regulations for novel foods produced with GEMs ingredient. However, if the food product, after genetic engineering, does not show much change in the composition or structure, then it does not come in the category of being novel. The evaluation of any novel food substance is made by assessing the entire manufacturing process, which starts with the organism used, fermentation media, equipment, filters used, processing aids, and the ingredients used in the formulation of the product (Redan 2020). Under a new European Union regulatory framework, the Food Improvement Agents Package or FIAP (European Commission (2011), European Commission (2012), food enzymes (including those produced with GEMs) must undergo an authorization that is different from the novel food authorization process. The safety evaluation of GEMs for the production of food enzymes depends on the European Food Safety Authority (EFSA Panel on Biological Hazards (BIOHAZ, (2017)) risk assessment guidance. In the United States, food enzymes (including those produced with GEMs) are categorized on the basis of "generally recognized as safe" (GRAS). The novel food ingredients are also categorized in the similar way (Hanlon et al. 2017). However, in the European Union, the food enzymes produced by GEMs are evaluated on the same principle as that for the other food substances. The safety assessment of the production strain and the evaluation of its pathogenic and toxigenic potential are very important. Pathogenic refers to the ability of a microorganism to produce an infection (pathogenic *E. coli* and *Salmonella* sp.), whereas toxigenic refers to the ability of a microorganism to produce toxins that would harm someone consuming them (*C. botulinum* toxin and *Staphylococcus aureus*). According to the European and United States agencies, the risk assessment differs in genetically engineered plants than of food and feed substances (EFSA 2011). The ASTM (American Society for Testing and Materials) classification is based on four genotype classes, which range from microbes without any intentional alteration, microorganisms that undergo genetic alteration without the introduction of any non-native DNA or microbial strains altered with addition of non-native DNA, and novel microbes that contains or produces chemical substances like non-natural amino acids or novel substances produced through engineered biochemical pathways. Using GEMs in foods have become an integral part of the food supply chain. As a result, regulatory agencies are evolving toward more efficient assessment of the foods. However, GMO labeling needs more clarity.

### 2.5.10 Enhancement of Product Yields Using Omic Sciences

Gonzáles-Jiménez et al. (2020) used the technique to study the protein content in wines produced by different strains of *Saccharomyces cerevisiae*. The volatile compounds are produced by the yeast used. The flor yeast strain is known to have higher concentration of proteins and it also imparts a specific organoleptic property. The characterization of specificities in wine, enzymes involved, and the study of volatile pathways have been possible by using the proteomic analysis.

Recent genetic techniques have made it possible to select a host in which the production of metabolite molecules can be done. This is very beneficial as the metabolite yields increase in a particular host organism in a laboratory and not on the land where the climatic conditions are a major concern. The production of β-caryophyllene has been increased in this manner (Hilgers et al. 2021). This compound can kill many pathogens. The yields of *N*-methylanthranilate (NMA) amino acid, which is a precursor of many bioactive compounds, could be increased by engineering the strain of *C. glutamicum C*1. The extraction of products from microorganisms needs expertise. The optimization of various parameters play a major role before extraction. Many newer techniques of extraction have come into the market.

Many antibiotics produced by microbial fermentation are able to treat bacterial infections. However, for maximum product, yields optimization of physical environment, nutrients, pH, time, temperature, oxygen, and chemical parameters are an absolute necessity. Production of steroids and vitamins by microbes are very common. A good strain selection helps in enhancing the product yield. Microbial fermentation helps in the production of better dairy products such as cheese and yogurt. In dairy products, the use of lactic acid bacteria converts lactose sugar to lactic acid. Microbial enzymes are another product that can be produced in large quantities, both by submerged and solid state fermentation. In submerged aerobic fermentations, the microorganisms come directly in contact with the media. This does not happen in solid state fermentation. However, in solid state fermentation, the end product is found to be more stable. The use of recombinant *E. coli*/yeast has the ability to produce human insulin called Humulin.

In agriculture, the use of chemical pesticides has been found to be very harmful as they spoil the soil and later they get discharged into the rivers. This kills the marine life and their intake by humans can cause fatal diseases. A study has been conducted to assess the bacterial community structure in soil (Gelsomino et al. 1999). It is therefore very beneficial to produce natural products by microorganisms. Natural biofertilizers and biopesticides produced by microorganisms are safe and are biodegradable. These are obtained from a living organism or natural-occurring substances. Microbial pesticides obtained from a virus, bacterium, or fungus can kill pests. According to De Oliveira (2021), there are many techniques by which these biopesticides can be applied (emulsion, electrospraying, fluidized bed, spray drying, extrusion, lyophilization, spray cooling, and coacervation). There are many bacteria and fungi that can prevent the entry of pathogens in the soil, which are the main causes of diseases. Biopesticides also help in controlling many diseases in plants (Kumar et al. 2021).

*Bacillus thuringiensis*, also known as Bt, is a Gram-positive bacterium, which is used for the production of microbial biopesticides. This spore-forming bacterium produces a delta endotoxin that destroys the lining of the digestive system of the insect or the pest. This makes the pest unable to feed on the crop or the plant. It is used commercially by the farmers for vegetables such as cabbage and potato (Ruiu

2018). Biopesticides are known to be safe with very less gas emissions and are specific. However, as they are produced from living cells, their storage conditions and the way of transportation are also specific, which might be a challenge.

Recently, with modern developments in nanosciences, scientists have come up with pesticides using nanotechnology. Encapsulation of nanomaterials (magnesium oxide, magnesium hydroxide, copper oxide, and zinc oxide) is done on a nanocarrier (usually less than 500 nm and is made up of antibodies that are agrochemical ingredients). It can be in the form of nano-gel, nano-encapsulation, nano-fibers, or nanosphere (Pan et al. 2023). Copper oxide nanoparticles and zinc oxide nanoparticles can control *Alternaria citri*, which causes citrus black rot disease. Copper nanoparticles can kill *Tribolium castaneum*, which infests grains (El-Saadony et al. 2020). The efficacy of nanobiopesticides can be made with its composition, surface charge, concentration, size, chemical and physical changes (Sofia et al. 2023). For knowing the sustainability and efficacy of nanopesticide, understanding of the molecular mechanism is important (Chaud et al. 2021).

Microbes also help in the biosynthesis of the essential amino acids such as L-methionine, L-lysine, L-tryptophan, as well as the nonessential amino acid such as L-glutamic acid. L-glutamic acid helps in the formation of monosodium glutamate (MSG), which is used as a food-flavoring agent. Microbes can produce organic solvents. There are many microbial products that are commonly eaten (cheese, bread, and yogurt). Wu et al. (2013) brought out the use of vinegar to restore the ovulation in women having polycystic ovary syndrome. In a paper (Margotta 1968), it is mentioned that Hippocrates (400 BC) used vinegar for medicinal purposes, which is a microbial product. For many years, human beings have made cheese with different species of fungi and bacteria, which produces different characteristic flavors. Microorganisms are a rich source of secondary metabolites that have significant pharmaceutical, biomedical, and food applications. Nowadays, the development and integration of gene editing tools, especially CRISPR-Cas9 (gene cloning, gene refactoring, and gene insertion or deletion) in metabolomics, provide the identification and detection of novel SMs and ways to enhance the yields of SMs although the screening costs are high (https://www.intechopen.com/chapters/73140).

### 2.5.11 Antibiotics: Biosynthetic Pathways for the Antibiotics and Modular Biosynthesis

These are products of secondary metabolism that inhibit the growth of disease-causing microorganisms. Some of them can kill many pathogens and have a wide spectrum. However, many pathogens are able to survive the treatment and are therefore called as antibiotic-resistant organisms, which could be either inherent or acquired.

Antibiotics are substances that kill bacteria and are known as bactericidal, whereas the ones that inhibit their growth are called as bacteriostatic. There are antiseptic substances that are used for the sterilization of surfaces with living tissues as they can cause infection, especially during surgeries. However, disinfectants kill many pathogenic microorganisms. These are nonselective antimicrobials and are used on nonliving surfaces.

Apart from bacteria there are many harmful fungi and viruses responsible for diseases in human beings. Microbes have the potential to produce compounds that can kill these harmful organisms. These are called as antifungals and antivirals. These antimicrobial compounds have enhanced the life of humans in the recent past due to the successful treatment of these infections. These compounds are also known to inhibit tumors. Lactic acid bacteria (LAB) produce antimicrobials known as bacteriocins, which are antimicrobial peptides and have the ability to preserve food for longer shelf storage. These compounds come under GRAS category, and therefore they are used as an alternative to antibiotic therapy. They are also found to be useful where the antibiotics do not work due to resistance. *Bacillus* strains produce peptides with antibiotic activity, which are used in the food industry. Nisin is produced by lactic acid bacteria, which has been approved by FDA, and is an important commercial bacteriocin. It is applied in pharmaceutical, veterinary, and microbiological industries. *Lactococcus lactis* subsp. *lactis* uses cheese whey as a substrate to produce nisin and lactic acid simultaneously. Pediocins obtained from *Pediococci* are used as food biopreservatives, providing effective measures to control pathogens.

The antimicrobial compounds like antibiotics produced by microorganisms are classified as follows:

Bacteriostatic compounds (inhibit protein synthesis):

(a) Glycylcyclines: Tigecycline
(b) Tetracyclines: Doxycycline, minocycline
(c) Lincosamides: Clindamycin
(d) Macrolides: Azithromycin, clarithromycin, erythromycin
(e) Oxazolidinones: Linezolid
(f) Sulfonamides (inhibits metabolic pathway): Sulfamethoxazole
(g) Quinolones (inhibits nucleic acid synthesis).

Bactericidal compounds (inhibit the cell wall synthesis):

(i) Aminoglycosides: Tobramycin, gentamicin, amikacin
(ii) β-Lactams (penicillins, cephalosporins, carbapenems): Amoxicillin, cefazolin, meropenem
(iii) Fluoroquinolones: Ciprofloxacin, levofloxacin, moxifloxacin
(iv) Glycopeptides: Vancomycin
(v) Cyclic lipopeptides: Daptomycin
(vi) Nitroimidazoles: Metronidazole

β-Lactam and cephalosporin are the most effective antibiotics. These antibiotics inhibit the cell wall (peptidoglycan) synthesis of the pathogen bacteria. Bacitracin is a peptide with antimicrobial activity. It is known to improve the growth in animal feeds. Most of the Gram-negative bacteria can be killed with amino glycosides.

Tuberculosis is caused due to pathogen bacteria *Salmonella* sp. The antibiotic, namely Streptomycin, is effective in inhibiting it. The antibiotic tetracycline inhibits the synthesis of proteins both in Gram-positive and Gram-negative bacteria. To enhance the yield of antibiotic, the technique like cell immobilization has been found to be very effective and is able to increase the yield by 108% (*Streptomyces clavuligerus*) after immobilization as compared to the free cells. Ellaiah et al. (2004) were able to increase the yields of neomycin in *Streptomyces marinensis* after immobilization of cells and optimization of parameters (using calcium alginate).

Different entrapment materials have been employed to enhance the yields of cephamycin C (sponge) and rifamycin (glass wool) from microorganisms.

Multidrug-resistant strains of bacteria are causing a lot of concern especially in health sector. These are the strains of *A. baumanii*, *C. difficile*, *Enterobacter* spp., vancomycin-resistant *enterococci* (VRE), *E. coli*, *Haemophilus influenzae*, *K. pneumoniae*, *P. aeruginosa*, *Staphylococcus epidermidis*, *S. pneumoniae*, extensively drug-resistant *M. tuberculosis* (XDR-TB), or MRSA. Carbapenem antibiotic resistance is very challenging as most of the infections take place because of carbapenem-resistant bacteria (Yohei and Henry 2015). Epsilometer testing (E-test) is the routine method of analysis. Recently, MALDI-TOF MS has been very useful to detect antibiotic resistance. Other effective techniques to identify antibiotic resistance are laser desorption/ionization (LDI), which is a nanotechnology-assisted laser desorption/ionization time-of-flight mass spectrometry (NALDI-TOF MS) and SELDI-TOF MS (proteins of interest in a sample become bound to a surface before MS analysis).

### 2.5.12 Biosynthetic Pathways for the Antibiotics

In order to enhance the yield of metabolites, it is important to understand the biosynthetic pathway. Omic sciences can then be applied easily. The production of the same compound/metabolite from different microorganisms usually follows a different pathway.

### 2.5.13 Penicillin, Tetracycline, Streptomycin, and Thiocillin

(A) *Penicillin*: This antibiotic is produced mostly from a filamentous fungus, *Penicillium chrysogenum*. In the process, three enzymes are involved, which get encoded with three separate genes namely: acvA, ipnA, and aatA. All these three genes are arranged in a cluster. The lateral chain of L-alpha-aminoadipate gets exchanged with a hydrophobic lateral chain. However, when penicillin is produced from the genus *Acremonium*, isopenicillin N, then it gets converted into penicillin N using enzymes acyl-CoA-synthetase and acyl-CoA-racemase (Brakhage and Caruso 2004).

(B) *Tetracyline*: These bacteriostatic killers are synthesized by *Streptomyces* sp. by submerged fermentation in a nutrient media. Many functional groups are attached to a tetracyclic nucleus. They bind to the 30S subunit of ribosome. They become different due to the presence of different groups such as chloro, methyl, or hydroxyl, which may impart a different pharmacological property. The core comes from malonamyl-CoA along with eight two-carbon fragments.

### 2.5.14 Modular Biosynthetic Pathways

When the antibiotics are synthesized in the laboratories, long reaction sequences are not used as that may reduce the yields. The role of enzymes and optimization of parameters to use them are very vital in enhancing the product yields. It is the modularity of pathways that decides the sequence of the reaction and also the molecular diversity. In penicillin pathway, three protein domains are present in each module, which ultimately forms the tripeptide core. The first domain selects the amino acids, the second one activates it, which is then inserted by the third protein domain. The sequence of the module decides the peptide chain.

For tetracycline, the basic building block is derived from acetate molecule, which resembles a lipid molecule. In streptomycin, sugars are the basic building material. The antibiotics have a similar modular assembly as that of the proteins. Many new antibiotics can be produced by using the classical genetic methods, which include mutation, duplication, deletion, and rearrangement. Scientists have shown that it is possible for antibiotics to modulate the transcriptional profiles of target bacteria.

### 2.5.15 Detection of Secondary Metabolites

Mass spectrometry (MS) measures the mass to charge ratio in a molecule. The principle of chromatography is to detect the retention time of the constituents that travel at different speeds under a specific condition. Nuclear magnetic resonance (NMR) spectroscopy is based on using the magnetic properties of atomic nuclei to determine the chemical and physical properties of atoms or molecules. In NMR mechanism, the electromagnetic radiation reemits at a specific resonance frequency. This depends on the magnetic properties of the isotope of atom and the strength of the magnetic field. NMR can be used to differentiate between the structural isomers. However, MS provides information on the formula of the molecule. In comparison to NMR, mass spectrometry (MS) is more sensitive and is able to detect a large number of metabolites. The initial data obtained from the NMR spectroscopy or MS is converted into computer formats, which is compatible with software packages. By using principal component analysis (PCA), the data can be simplified without losing the main features. Generally, the PCA principle is based on reducing the dimensionality of the dataset. Different univariant statistical tests can be used to analyze isolated metabolites such as ANOVA, nonparametric Wilcoxon signed-rank test, Kruskal–Wallis test, and the parametric Student's *t*-test.

Due to the development of various bioinformatics software, many metabolites can be identified by putative identification and definitive identification. In putative identification, one or two molecular properties are utilized for identification. However, in definitive identification, two properties such as the retention time and accurate mass and/or fragmentation mass spectrum and/or NMR spectrum can be used and spiked with the authentic chemical standard. The definitive identification gives more accurate results. There are many databases are available online that are based on spectra or chemical structure for metabolite identification. Some of the common databases used in nuclear magnetic resonance (NMR) spectroscopy are METLIN (http://metlin.scripps.edu), the Human Metabolome Database (HMDB, http://www.hmdb.ca), and Biological Magnetic Resonance Databank (http://www.bmrb.wisc.edu/metabolomics/), whereas commonly used databases for mass spectrometry are NIST (http://www.nist.gov/srd/nist1a.htm), MassBank (http://www.massbank.jp), the Golm Metabolite Database (GMD, http://csbdb.mpimp-golm.mpg.de/csbdb/gmd/gmd.html), METLI, and MMCD (http://mmcd.nmrfam.wisc.edu).

**FAQ's**

1. Which growth phase is best for primary metabolite production?
2. Which growth phase is best for secondary metabolite production?
3. What is the difference between primary and secondary metabolites?
4. Which parameters affect the secondary metabolite production?
5. Why do microorganisms produce secondary metabolites?
6. What is the function of microbial product in textiles?
7. What is the effect of stress on the production of secondary metabolites?
8. What are the advantages of CRISPER technology?
9. What is gene slicing?
10. What are the metabolites which that are formed from glucose?

## References

Ahrens CH, Brunne E, Qeli E, Basler K, Aebersold R (2010) Generating and navigating proteome maps using mass spectrometry. Nat Rev Mol Cell Biol 11(11):789–801

Alamillo JM, López CM, Martínez Rivas FJ, Torralbo F, Bulut M, Alseekh S (2023) Clustered regularly interspaced short palindromic repeats/CRISPR-associated protein and hairy roots: a perfect match for gene functional analysis and crop improvement. Curr Opin Biotechnol 79:102876

Armstrong DW, Schneiderheinz JM, Kullman JP, He L (2001) Rapid CE microbial assays for consumer products that contain active bacteria. FEMS Microbiol Lett 194:33–37

Baltz RH (1999) Mutagenesis. In: Flickinger MC, Drew SW (eds) Encyclopedia of bioprocess technology: fermentation, biocatalysis and separation. Wiley, New York, pp 1819–1822

Barrangou R, Fremaux C, Deveau H, Richards M, Boyaval P, Moineau S, Romero DA, Horvath P (2007) CRISPR provides acquired resistance against viruses in prokaryotes. Science 315(5819):1709–1712

Bérdy J (2005) Bioactive microbial metabolites. J Antibiot 58(1):1–26

Brakhage A, Caruso ML (2004) Genetics and biotechnology. In: Kück U (ed) Biotechnical genetics of antibiotic biosynthesis. Springer Science & Business Media, pp 317–345
Chaud M, Souto EB, Zielinska A, Severino P, Batain F, Oliveira-Junior J, Alves T (2021) Nanopesticides in agriculture: benefits and challenge in agricultural productivity, toxicological risks to human health and environment. Toxics 9(6):131
Committee on the Review of Omics-Based Tests for Predicting Patient Outcomes in Clinical Trials; Board on Health Care Services; Board on Health Sciences Policy (2012) Institute of Medicine. In: Micheel CM, Nass SJ, Omenn GS (eds) Evolution of translational omics: lessons learned and the path forward. National Academies Press (US), Washington (DC). 2: Omics-Based Clinical Discovery: Science, Technology, and Applications. Available from: https://www.ncbi.nlm.nih.gov/books/NBK202165/
Da Graca JV, Korsten L (2004) *Citrus huanglongbing*: review, present status and future strategies – diseases of fruits and vegetables, vol 1. Kluwer Academic Publishers Printed in Netherland, pp 229–245
De Oliveira JL (2021) In: Jogaiah S, Singh H, Fernandes L, Lima FR (eds) Nano-biopesticides: present concepts and future perspectives in integrated pest management. Advances in nano-fertilizers and nano-pesticides in agriculture. Woodhead Publishing, Cambridge, pp 1–27
Demain AL (1999) Pharmaceutically active secondary metabolites of microorganisms. Appl Microbiol Biotechnol 52(4):455–463
DeModena J, Gutiérrez S, Velasco J et al (1993) The production of cephalosporin C by *Acremonium chrysogenum* is improved by the intracellular expression of a bacterial hemoglobin. Nat Biotechnol 11:926–929
Edward CO, Rebecca W, Tom P, Eshwar M (2019) Understanding the challenges of non-food industrial product contamination. FEMS Microbiol Lett 366(23):fnaa010
EFSA (2011) Use of the EFSA comprehensive European food consumption. Database in Wiley Online Library
EFSA Panel on Biological Hazards (BIOHAZ), Ricci A, Allende A, Bolton D, Chemaly M, Davies R, Fernández Escámez PS (2017) Scientific opinion on the update of the list of QPS-recommended biological agents intentionally added to food or feed as notified to EFSA. EFSA J 15(3):e04664
Elander RP, Lowe DA (1992) In: Arora DK, Elander RP, Mukerji KG (eds) Handbook of Appl Mycol 4 fungal biotechnology: an over-view. Marcel Dekker, New York, pp 1–34
El-Saadony MT, Abd El-Hack ME, Taha AE, Fouda MMG, Ajarem JS, Maodaa NS, Allam AA, Elshae N (2020) Ecofriendly synthesis and insecticidal application of copper nanoparticles against the storage pest *Tribolium castaneum*. Nanomaterials (Basel) 10(3):587
Emwas AH, Roy R, McKay RT, Tenori L, Saccenti E, Gowda GAN, Raftery D, Alahmari F, Jaremko L, Jaremko M, Wishart DS (2019) NMR spectroscopy for metabolomics research. Metabolites 9(7):123
European Commission (2011) Communication from the commission to the European parliament, the council, the European economic and social committee and the committee of the regions a renewed EU strategy 2011–14 for corporate social responsibility
European Commission (2012) European food safety Authority. The EU summary report on antimicrobial resistance in zoonotic and indicator bacteria from humans, animals and food in 2012. Pub.2014. EFSA J Wiley Online Lib (2014) 12(3):3590
Farrah T, Deutsch EW, Omenn GS, Campbell DS, Sun Z, Bletz JA, Mallick P, Katz JE, Malmström J, Ossola R, Watts JD, Lin B, Zhang H, Moritz RL, Aebersold R (2011) A high-confidence human plasma proteome reference set with estimated concentrations in PeptideAtlas. Mol Cell Proteomics 10(9):M110.006353
Fu J et al (2016) Enhancement of docosahexaenoic acid production by low-energy ion implantation coupled with screening method based on Sudan black B staining in *Schizochytrium* sp. Bioresour Technol 221:405–411
Gelsomino A, Keijzer-Wolters AC, Cacco G, van Elsas JD (1999) Assessment of bacterial community structure in soil by polymerase chain reaction and denaturing gradient gel electrophoresis. J Microbiol Methods 38:1–15

Gold L, Ayers D, Bertino J, Bock C, Bock A, Brody EN, Carter J, Dalby AB et al (2010) Aptamer-based multiplexed proteomic technology for biomarker discovery. PLoS One 5(12):e15004. https://doi.org/10.1371/journal.pone.0015004

González J, María GM, Teresa M, Juan SL, Irene PA, Moreno J, Moreno-García J (2020) Comparative study of the proteins involved in the fermentation derived compounds in two strains of *Saccharomyces cerevisiae* during sparkling wine second fermentation. Microorganisms 8:1209. https://doi.org/10.3390/microorganisms8081209

Hamed HS, Tawwab MA (2021) Dietary pomegranate (*Punica granatum*) peel mitigated the adverse effects of silver nanoparticles on the performance, haemato-biochemical, antioxidant, and immune responses of Nile tilapia fingerlings. Aquaculture 540(736742):0044–8486

Hanlon P, Daines L, Campbell C, McKinstry B, Welle D, Pinnock H (2017) Telehealth interventions to support self-management of long-term conditions: a systematic metareview of diabetes, heart failure, asthma, chronic obstructive pulmonary disease, and cancer. J Med Internet Res 19(5):e172

Hilgers F, Habash SS, Loeschcke A, Ackermann YS, Neumann S, Heck A, Klaus O, Hage-Hülsmann J, Grundler FMW, Jaeger KE, Schleker ASS, Drepper T (2021) Heterologous production of β-caryophyllene and evaluation of its activity against plant pathogenic fungi. Microorganisms 9(1):168

Hiyoung K, Chang-Hun J, Hyun-Woo J, Jong-Pyung K, Hahk-Soo K (2020) mpCRISTAR: multiple plasmid approach for CRISPR/Cas9 and TAR-mediated multiplexed refactoring of natural product biosynthetic gene clusters. ACS Synthetic Biol 9(1):175–180

Honda K, Littman DR (2011) The microbiome in infectious disease and inflammation. Ann Rev Immun 30:759–795

Hwang D, Lee IY, Yoo H, Gehlenborg N, Cho JH, Petritis B, Baxter D, Pitstick R, Young R et al (2009) A systems approach to prion disease. Molecular Syst Biol 5:252

Janiszewska D, Szultka-Młyńska M, Pomastowski P, Buszewski B (2022) Omic approaches to bacteria and antibiotic resistance identification. Int J Mol Sci 23(17):9601. https://doi.org/10.3390/ijms23179601

Jiang W, Zhao X, Gabrieli T, Lou C, Ebenstein Y, Zhu TF (2015) Cas9-assisted targeting of chromosome segments CATCH enables one-step targeted cloning of large gene clusters. Nat Commun 6:8101

Jones JJ, Stump MJ, Fleming RC, Lay JO, Wilkins CL (2003) Investigation of MALDI-TOF and FT-MS techniques for analysis of *Escherichia coli* whole cells. Anal Chem 75(6):1340–1347. https://doi.org/10.1021/ac026213j

Joseph PW, Ana CP, Luca R, Amias AM, Vicki S, Sophie V, Vivien Y, David PM et al (2022) Multi-omic based production strain improvement (MOBpsi) for bio-manufacturing of toxic chemicals. Metabolic Eng 72:133–149, ISSN 1096-7176, https://doi.org/10.1016/j.ymben.2022.03.004

Kang HS, Charlop-Powers Z, Brady SF (2016) Multiplexed CRISPR/Cas9- and TAR-mediated promoter engineering of natural product biosynthetic gene clusters in yeast. ACS Synth Biol 5(9):1002–1010

Kinros JM, Darzi AW, Nicholson JK (2011) Gut microbiome-host interactions in health and disease. Genomic Med 3(3):14

Kumar J, Ramlal A, Mallick D, Mishra V (2021) An overview of some biopesticides and their importance in plant protection for commercial acceptance. Plants (Basel) 10(6):1185

Kusale S, Attar YC, Sayyed RZ, Malek RA, Ilyas N, Suriani NL, Khan N, El Enshasy HA (2021) Production of plant beneficial and antioxidants metabolites by *Klebsiella variicola* under salinity stress. Molecules 26(7):1894

Li L, Zheng G, Chen J, Ge M, Jiang W, Lu Y (2017) Multiplexed site-specific genome engineering for overproducing bioactive secondary metabolites in actinomycetes. Metab Eng 40:80–92

Lian J, McKenna R, Rover MR, Nielsen DR, Wen Z, Jarboe LR (2016) Production of biorenewable styrene: utilization of biomass-derived sugars and insights into toxicity. J Indust Microbiol Biotechnol 43:595–604. https://doi.org/10.1007/s10295-016-1734-x

Liu Z, Zhao Y, Huang C, Luo Y (2021) Recent advances in silent gene cluster activation in *Streptomyces*. Front Bioeng Biotechnol 9:632230

Maas S (2010) Gene regulation through RNA editing. Discov Med 10(54):379–386
Maldonado RF, Sá-Correia I, Valvano MA (2016) Lipopolysaccharide modification in gram-negative bacteria during chronic infection. FEMS Microbiol Rev 40(4):480–493
Margotta R (1968) The story of medicine. Golden Press, New York
Martín-Valmaseda M, Devin SR, Ortuño-Hernández G, Pérez-Caselles C, Mahdavi SME, Bujdoso G, Salazar JA, Martínez-Gómez P, Alburquerque N (2023) CRISPR/Cas as a genome-editing technique in fruit tree breeding. Int J Mol Sci 24:16656
Miranda J, Azocar F, Organista K, Dwyer E, Arean P (2003) Treatment of depression among impoverished primary care patients from ethnic minority groups disadvantaged medical patients. Psychiatr Serv 54(2):219–225
Monciardini P., Iorio M., Maffioli S., Sosio M., Donadio S. Discovering new bioactive molecules from microbial sources. Microb. Biotechnol. 2014;7:209–220. https://doi.org/10.1111/1751-7915.12123
Nawaz A, Chaudhary R, Shah Z, Dufossé L, Fouillaud M, Mukhta H, Haq IU (2020) An overview on industrial and medical applications of bio-pigments synthesized by marine bacteria. Microorganisms 9(1):11
Ng SB, Bigham AW, Buckingham KJ, Hannibal MC, McMillin MJ, Gildersleeve HI et al (2010) Exome sequencing identifies MLL2 mutations as a cause of Kabuki syndrome. Nat Genet 42(9):790–793
Noskov V, Kouprina N, Leem SH, Koriabine M, Barrett JC, Larionov V (2002) A genetic system for direct selection of gene-positive clones during recombinational cloning in yeast. Nucleic Acids Res 30(2):E8
Omenn GS (2009) A landmark systems analysis of prion disease of the brain. Mol Syst Biol 5:254
Omenn GS, Baker MS, Aebersold R (2011) Recent workshops of the HUPO Human Plasma Proteome Project (HPPP): a bridge with the HUPO cardio vascular initiative and the emergence of SRM targeted proteomics. Proteomics 11(17):3439–3443
Ough CS, Trioli G (1988) Urea removal from wine by an acid urease. Am J Enol Viticult 39:303–307
Pan X, Guo X, Zhai T, Zhang D, Rao W, Cao F et al (2023) Nanobiopesticides in sustainable agriculture: developments, challenges, and perspectives. Environ Sci Nano 10:41–61
Papaianni M, Cuomo P, Fulgione A, Albanese D, Gallo M, Paris D, Motta A, Iannelli D, Capparelli R (2020) Bacteriophages promote metabolic changes in bacteria biofilm. Microorganisms 8:480
Pérez-Llarena FRG, Antonio E, Francisco M, Juan L, Paloma (1998) The pcd gene encoding Piperideine-6-Carboxylate Dehydrogenase involved in biosynthesis of α-aminoadipic acid is located in the cephamycin cluster of *Streptomyces clavuligerus*. J Bacteriol 180:4753
Picotti P, Rinner O, Stallmach R, Dautel F, Farrah T, Domon B, Wenschuh H, Aebersold R (2010) High-throughput generation of selected reaction-monitoring assays for proteins and proteomes. Nat Methods 7(1):43–46. http://www.proteomexchange.org/ (accessed February 23, 2012)
Potts BC, Lam KS (2010) Generating a generation of proteasome inhibitors: from microbial fermentation to total synthesis of Salinosporamide A (Marizomib) and other Salinosporamides. Mar Drugs 8(4):835–880
Raymond KN, Dertz EA, Kim SS (2003) Enterobactin: an archetype for microbial iron transport. Proc Natl Acad Sci USA 100(7):3584–3588
Redan BW (2020) Processing aids in food and beverage manufacturing: potential source of elemental and trace metal contaminants. J Agric Food Chem 68(46):13001–13007. https://doi.org/10.1021/acs.jafc.9b08066
Roach JC, Glusman G, Smit AF, Huff CD, Hubley R, Shannon PT, Rowen L, Pant KP et al (2010) Analysis of genetic inheritance in a family quartet by whole-genome sequencing. Science 328(5978):636–639
Rubio A, Luoni M, Giannelli SG, Radice I, Iannielli A, Cancellieri C, Di Berardino C, Regalia G et al (2016) Rapid and efficient CRISPR/Cas9 gene inactivation in human neurons during human pluripotent stem cell differentiation and direct reprogramming. Sci Rep 18(6):37540
Ruiu L (2018) Microbial biopesticides in agro ecosystems. Agronomy 8:1–12
Ryan CB, Gabriel AK, Caius GR, Owen NW, James RH (2007) J Am Chem Soc 129(7):1959–1967. https://doi.org/10.1021/ja065930i

Sayed AM, Alhadrami HA, El-Gendy AO, Shamikh YI, Belbahri L, Hassan HM et al (2020) Microbial natural products as potential inhibitors of SARS-CoV-2 main protease (Mpro). Microorganisms 8:970

Schneider Y, Jenssen M, Isaksson J, Hansen KØ, Andersen JH, Hansen EH (2020) Bioactivity of serratiochelin A, a siderophore isolated from a co-culture of *Serratia* sp. and *Shewanella* sp. Microorganisms 8(7):1042

Schwab H (1988) Strain II, Strain improvement in industrial microorganisms by Recombinant DNA techniques. In: Bioprocesses including animal cell culture. https://api.semanticscholar.org/CorpusID:83597324

Shamikh YI, El Shamy AA, Gaber Y, Abdelmohsen UR, Madkour HA, Horn H, Hassan HM, Elmaidomy AH, Alkhalifah DHM, Hozzein WN (2020) Actinomycetes from the red sea sponge *Coscinoderma mathewsi*: isolation, diversity and potential for bioactive compounds discovery. Microorganisms 8(5):783

Sofia M, Ruth P, Rose M, OF S (2023) Nanobiopesticides: are they the future of phytosanitary treatments in modern agriculture? Sci Total Environ 896:66401. Issn:0048-9697

Springthorpe et al (2020). https://morf-db.org/projects/DETOX/styrene

Sreekumar A, Poisson LM, Rajendiran TM, Khan AP, Cao Q, Yu J et al (2009) Metabolomic profiles delineate potential role for sarcosine in prostate cancer progression. Nature 457(7231):910–914

Srinivasulu Ellaiah P, Kunamneni B, Adinarayana (2004) Optimisation studies on neomycin production by a mutant strain of *Streptomyces marinensis* in solid state fermentation. Process Biochem 39:529–534. https://doi.org/10.1016/S0032-9592(02)00059-6

t'Kindt R, Scheltema RA, Jankevics A, Brunker K, Rijal S, Dujardin JC et al (2010) Metabolomics to unveil and understand phenotypic diversity between pathogen populations. PLoS Negl Trop Dis 4(11):e904. https://doi.org/10.1371/journal.pntd.0000904

Tang F, Lao K, Surani MA (2011) Development and applications of single-cell transcriptome analysis. Nat Methods 8(4 Suppl):S6–S11

Tilg H, Kaser A (2011) Gut microbiome, obesity, and metabolic dysfunction. J Clin Invest 121(6):2126–2132

Vinci VA, Byng G (1999) Strain improvement by nonrecombinant methods. In: Demain AL, Davies JE (eds) Manual of industrial microbiology and biotechnology, 2nd edn. ASM Press, Washington, pp 103–113

Voitsekhovskaia I, Paulus C, Dahlem C, Rebets Y, Nadmid S, Zapp J, Axenov-Gribanov D, Rückert C, Timofeyev M, Kalinowski J et al (2020) Baikalomycins A-C, new aquayamycin-type angucyclines isolated from lake Baikal derived *Streptomyces* sp. IB201691-2A. Microorganisms 8:680

Wolf-Yadlin A, Sevecka M, MacBeath G (2009) Dissecting protein function and signalling using protein microarrays. Curr Opin Chem Biol 13(4):398–405

Wu D, Kimura F, Takashima A, Shimizu Y, Takebayashi A, Kita N, Zhang G, Murakami T (2013) Intake of vinegar beverage is associated with restoration of ovulatory function in women with polycystic ovary syndrome. Tohoku J Exp Med 230(1):17–23

Yohei D, Henry F (2015) Chambers, 2. In: John EB, Raphael D, Martin JB, Mandell D, Bennett's (eds) Other β-Lactam antibiotics, principles and practice of infectious diseases (Eighth edition). WB Saunders, pp 293–297.e2. ISBN 9781455748013. https://doi.org/10.1016/B978-1-4557-4801-3.00022-9

Zhang F, Gu W, Hurles ME, Lupski JR (2009) Copy number variation in human health, disease, and evolution. Annu Rev Genomics Hum Genet 10:451–481. https://doi.org/10.1146/annurev.genom.9.081307.164217

# Screening of Industrially Important Microorganisms

# 3

**Abstract**

Screening of microorganisms are very important for the strains that are to be used by the industries. The industries undertake a process that gives high yields of the product along with good quality and long shelf life storage. For this purpose, a high-yielding microbial strain has to be selected. Therefore, the strains of microorganisms that can produce the metabolite of interest should be screened to select the best strain. For basic screening of the culture strain, various techniques are available, as explained in this chapter. The best strain is then selected and is taken for scaling up and production process by the industry. Basically, the screening is done by classical methods, which is usually by enriching the culture medium. The strain can also be improved by classical mutagenesis. This chapter deals with the isolation, screening, and improvement of the industrially important microorganisms by classical, molecular engineering methods and also by modulating the metabolic pathway.

**Keywords**

Screening · Isolation · Metabolism · Protoplast fusion · Genetic engineering · Mutation · Strain improvement · Precursor

## 3.1 Introduction

In any fermentation process, it is important to use the right microorganism. The place of isolation and the weather conditions in which they are isolated play a major role in the product formation. These microorganisms can be improved for their efficiency by classical or genetic modification. There are many different methods to isolate microorganisms. For any metabolite production, the colonies of microbial strain of interest are isolated, then they are grown in a nutrient medium. All the



R. Agrawal, *Textbook of Industrial Microbiology*,
https://doi.org/10.1007/978-981-97-9582-6_3

colonies grown are then screened for metabolite production. The colony that produces the maximum metabolite is the high-yielding colony, which is selected for further use by the industry.

## 3.2 Initial/Primary Screening

The colonies that have the ability to produce the desired product are selected by many different methods. The commonly used methods include crowded plate technique, auxanography, and enrichment culture technique. A method that is fast and cheaper is preferable. Usually, initial screening takes a lot of time as each colony needs to be checked for metabolite production (Ezemba et al. 2022).

**Secondary Screening of Microorganisms for Metabolite Production** In the secondary screening, the optimization of media conditions and nutrient ingredients is done for the maximum production of the metabolite. Apart from this, if the microbial strain is improved in any way, then it is important to check the stability of the strain (Higgs et al. 2001).

Conditions for growing the culture in the initial flask as compared to the fermentor have been found to be very different. Therefore, it is important to optimize the conditions at all levels before applying them for industrial production. There are different pathways for the formation of various metabolites when the initial substrate is used as glucose (Fig. 3.1).

## 3.3 Improvement of Industrially Important Microorganisms

With the recent advances, newer methods are being adopted to improve the industrial strains after isolation. These include gene mining and supply of precursors in the biosynthetic pathway of metabolite production (Pickens et al. 2011).

**Addition of precursors for improving the secondary metabolites**
Malonic acid pathway is an essential source of phenolics in fungi and bacteria. Precursor is either added before or during the fermentation process (Fig. 3.2). The addition of enzymes to enhance the precursor reaction has been found effective. This gives higher yields of the product. However, it is difficult for the slow growing cultures. Aromatic amino acids are good precursors for many secondary metabolites, which either come from shikimate pathway or can be supplied by external means. Most of the phenolic compounds come from melonic acid pathway (Méndez-Hernández et al. 2023).

*Amycolatopsis* sp. produce vancomycin, which along with the precursor 4-(4-chlorophenyl) benzyl group produces oritavancin. This antibiotic is approved by FDA and can kill methicillin-resistant *Staphylococcus aureus* infections. The importance of mixed cultures has been worked out by the scientists. It has been

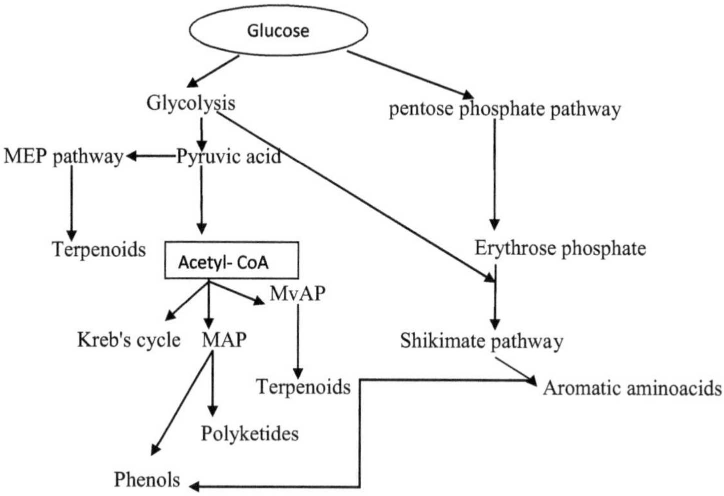


**Fig. 3.1** Formation of metabolites from glucose

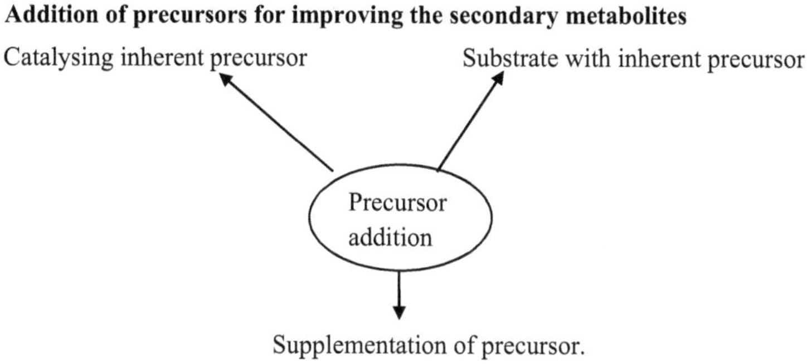


**Fig. 3.2** Precursor addition for high metabolite production

found that when *F. oxysporum* AB2 and *Epicoccum nigrum* TORT cells are fused, they produce a metabolite named as beauvericin, which is many folds higher in yields than the monoculture (Vásquez-Bonilla et al. 2022). This secondary metabolite has insecticidal, antimicrobial, and antitumor properties. The yields of isoamyl acetate (banana flavor) could be increased by adding the precursor (isoamyl alcohol).

Protoplast fusion is another way to enhance the yields of metabolites as it gives rise to a high-yielding colony when protoplasts of two desirable colonies are fused with the known attributes. The fusion helps in obtaining a better colony that can yield higher metabolite/biomass or whatever attribute is required for an economical industrial production (Beltrametti et al. 2007). Manipulation is also done at the level of ribosome that increases the productivity of the metabolite (Zhu et al. 2019). The yields of metabolite, 2-phenyl ethanol, has been found to increase with the addition of precursor, phenylalanine. Polyketide production has been enhanced when the precursor is added only on limitation and not otherwise. Optimization of substrate parameters and addition of precursor monosodium glutamate helped in increasing the yield of γ-polyglutamic acid in *Bacillus amyloliquefaciens* (Yong et al. 2011).

**Metabolic Pathway Engineering** This procedure has been successful with the addition or deletion of genes in order to produce higher yields of metabolites. Scientists have observed that this procedure allows the precursors to carry on the reaction at higher levels as seen in engineered *Streptomyces platensis* strains that overproduce antibiotics platensimycin and platencin (Smanski et al. 2009). The construction of a protease-deficient strain for the fission yeast *Schizosaccharomyces pombe* was found useful for effective production of protease-sensitive heterologous proteins (Idiris et al. 2006). The removal of genes, which are not needed for metabolite production, has shown to increase the yields of the metabolites. Success has been achieved to improve the secretion of recombinant proteins using this technique. This has been applied for human insulin precursor and α-amylase from *A. oryzae* and for vaccine against the hookworm infection.

A combination of engineering methods at the levels of metabolic pathway and proteins have shown to increase the production of levopimaradiene many folds. The yields of γ-aminobutyric acid, which is known to prevent nerve problems, can be increased when glutamate is used as a precursor in the process. 2-Deoxystreptamine-containing aminoglycoside used as an antibiotic can be manipulated in their biosynthetic pathways to obtain higher yield of the metabolite (Park et al. 2013). This approach has helped in increasing the secondary metabolite yields and an advancement toward successful commercial preposition for the production of novel and useful compounds. Biological activities of functional groups can be increased by making structural changes on them with the help of oxidase genes present in the polyketide pathways. This has been well worked out (Zhang et al. 2012) to prevent tumor formation by the tumor-associated macrophages in solid tumor.

**Mutasynthesis** Change in the structure of the metabolite can also be done by disrupting the genes and simultaneously supplementing the precursor. In mutasynthesis, new analogues of products are formed. Barbuto et al. (2021) demonstrated the enhanced production of drugs in *Streptomycetes* using this technique. Scientists (Idrees et al. 2021) have worked on the morphology, genetics, and pathogenesis in *Staphylococcus aureus*. They also worked on the removal of biofilms. A review has been written (Lim et al. 2022), detailing the design and various probes needed in different bioreactors along with heat and mass transfer process. Their work also provides the validation for manufacturing strategies in tissue engineering. This will help the researchers and industrialists to design a perfect bioreactor for a specific product.

Antibiotics are used as therapeutic agents against highly resistant pathogenic bacteria. However, they do not work on the pathogens that have become (multi) drug-resistant. Therefore, the scientists have come up with new strategies to make a derivative of the antibiotic into a new compound. As derivatization is difficult and time-consuming, scientists have innovated another strategy known as mutasynthesis. Mutasynthesis is a method of metabolic engineering that combines chemical synthesis along with genetic engineering techniques. A study has been done on the Streptogramins, which are the last line of defense antimicrobials with pristinamycin (Hennrich et al. 2023). Using mutasynthesis, when 4-fluorophenylglycine was supplemented into the biotransformation fermentation broth and utilizing *E. coli*, the scientists were able to produce high levels of 6-fluoropristinamycin I.

## 3.4 Isolation of Pure Cultures

There are many bacterial culture methods and the selection is based on the yield and other required qualities for industrial production.

**Plating Method Sequence**

(a) Streak plate/pour plate
(b) Serial dilution
(c) Enrichment culture

(a) Streak /Pour plate from bacterial suspension (Fig. 3.3):

**Different methods to streak inoculums on an agar plate (Fig. 3.4)**

(a) **Streak Plate/Pour Plate**

Initially, to isolate a pure colony, the inoculums can be streaked in a petri plate with nutrient media (Fig. 3.4) in four directions, from top to bottom, starting at one point and streaking in different directions, vertical and horizontal streaking, streaking in a clockwise direction or in three directions as shown in Fig. 3.4. After isolating and selecting a good yielding culture, strain can be grown for further use.

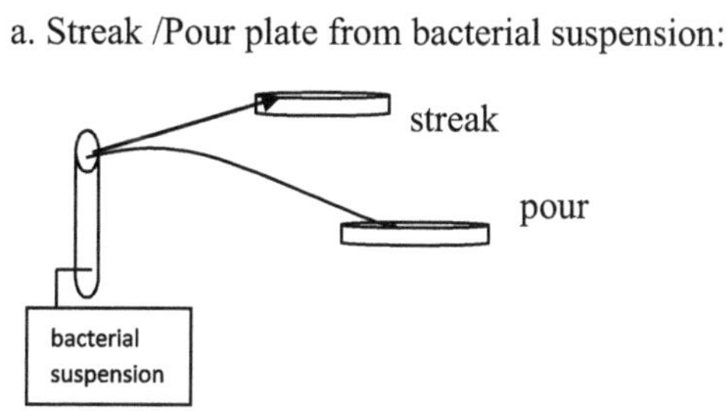


**Fig. 3.3** Streak and pour plating of bacterial suspension

**Fig. 3.4** Streaking in different directions

**Agar plates** Microbes can be grown in petri dishes of different sizes containing a thin layer of agar-based growth medium. The desired bacterial strain is inoculated in the media and then the plates are incubated at the optimum temperature of the selected bacteria (usually at 37 °C, for cultures from humans or animals, or lower temperatures for environmental cultures). For preservation, the agar plates are usually kept in a refrigerator in an upside-down position. In case the bacteria in question needs any additives for growth, then that particular additive is/are added into the media. In case a strain of bacteria needs to be tested for antibiotic resistance, the selected antibiotic can be added into the agar. This helps to grow the antibiotic-resistant colonies.

(b) **Serial Dilution**

In this procedure the concentration of the microorganism that needs to be plated is diluted serially so that it is reduced in a systemic method. This is done by taking an aliquot from the initial tube and resuspended in a fixed volume of the diluents (buffer/saline). The volume of diluents is taken in such a way that it can help in the logarithmic reduction of the organism. Therefore, usually multiple of 10 are used. For the procedure, 1 mL of the starting sample is added to 9 mL of Dilution Blank Tube 1. This is then followed by the same procedure, where 1 mL from Tube 1 is added to 9 mL of Tube 2. Then, 1 mL from Tube 2 is added to 9 mL of Tube 3, and so on until the desired concentration is reached.

(c) **Enrichment Technique**

The growth of the microorganism can be increased by enriching/supplementing the nutrient medium with nutrients/precursors of metabolite or substrate/s whichever helps. This usually increases the biomass/the metabolite of interest.

Agar-based dipsticks are miniature versions of agar plates in the form of dipstick formats, such as Dip Slide and Digital Dipstick, that can be used for diagnostic purposes. These are cost-effective and it does not require expertise or any laboratory.

**Stab cultures** Stab cultures are similar to agar plates; however, these are prepared by solid agar in a test tube. In this method the bacteria are introduced with the help of an inoculation needle, which is straight and not with a loop. The inoculum is stabbed into the center of the agar. After the growth of the bacteria, they can be seen in the punctured area. Motile and nonmotile bacteria can be differentiated along the stab lines. Motility of the culture strain can also be checked easily as the motile bacteria grows out of the stab line, whereas the nonmotile bacteria grow only along the stab line. Stab cultures are very helpful for short-term storage/shipment of the cultures.

**Broth cultures** In this method the desired bacteria are suspended in a liquid nutrient medium, which are taken in a conical flask in an upright position. This helps to grow large amounts of bacteria for a variety of downstream applications. Liquid cultures can be grown for the optimized time period to obtain maximum product (using an incubator shaker for uniform growth). After the growth, aliquots of the sample are tested for various functional activities. However, incubation is also done statically. In static conditions, the liquid cultures are not shaken as it provides an oxygen gradient to the microbes.

**Solid plate culture for thermophilic microorganisms** Low acyl clarified gellan gum are good gelling agents for solid plate cultures of thermopiles such as *Bacillus acidocaldarius*, *Bacillus stearothermophilus*, *Thermus aquaticus*, and *Thermus thermophilus*.

**Virus and phage cultures** Virus and phage cultures need a host where they can multiply. For bacteriophages, cultures are grown by infecting the bacterial cells. The phage can then be isolated from the resulting plaques on an agar plate. Streak plate method is used to isolate any particular microbial colony. This is done by using an inoculating loop over the solid agar plate by spreading the inoculate with the loop moving back and forth. Each single colony can be isolated and grown. Once a microorganism has been isolated in pure culture, it is necessary to preserve it in a viable state for further study and use. Stock cultures should be maintained in a way to preserve their biological, immunological, and cultural characters.

## 3.5 Culture Collection Centers

This aspect of preserving cultures in a collection center plays a vital role in metabolite microbiology. Globally, there are many microbial culture collection centers. The major objective is to keep the select cultures with the functional properties intact. Starting with their acquisition, the cultures need proper authentication. The various steps involved for the metabolite production should be written in a log book. Under the optimized conditions (as worked out by the researcher), each culture is preserved in these culture collection centers, which are catalogued properly for easy accession. The vial of viable cultures/cell lines can be obtained from these culture collection centers as and when required by the researchers/industrialists, as these are repositories of type strains.

These microbial cultures can be obtained by the researchers for further research work, for industrial production of metabolites, and also for teaching students to show particular metabolite production at the laboratory level. Therefore, it is vital to maintain each deposition of culture in these centers. Modern molecular engineering work is also done to improve the strains for higher productivity of the metabolite of interest by obtaining the basic culture strain from such centers.

The first culture collection center named as NCIM (National Collection of Industrial Microorganisms) was established in the twentieth century (1951). This is a center that has most of the industrially important microbial strains. Established in 1951, the NCIM is one of the oldest biological resource centers holding especially industrially important microorganisms in India. This was started by the Indian government at Pune in the National Chemical Laboratory (NCL). The center maintains more than 5000 strains of various bacteria, fungi, actinomycetes, yeasts, and algae, which can be obtained just via an e-mail. The cultures are categorized as per the necessity, which could be for a specific activity such as antimicrobial, antitumor, immune enhancement, industrially important enzymes, useful in agriculture, health improvement, and many more.

Among the culture collection centers, the Biological Resource Centre (BRC) preserves many biological materials with a view to conserve the genetic resources. The information is provided as and when there is a need. Presently, alternatives to natural resources are being looked into, with best one being the microbial cultures. Production of useful metabolites by microbial means it will boost the economy. In this context, these biological resource centers play a vital role. Recently, the microbial cultures are being used for the production of biofuels, antibiotics, and other medicinal drugs. They are also used in various foods as nutraceuticals. Using this facility, the countries can become self-sufficient, as they can purchase the culture strain for industrial production of any particular metabolite. This will improve the country's economy and more innovative start-ups can be initiated in the near future. It will also help in bilateral relations with other countries by signing up the MoUs for many efficient and economical bioproducts. The World Data Centre for Microorganisms (WDCM) is very user friendly to the scientists of all communities. It has a total of 152 culture collection centers that are spread around 51 countries in the world. It has a total of 530,267 strains. Among the various strains, 68,534 are plant associated, 63,929 are host associated, 61,799 are soil associated, 36,328 are engineered, 19,377 are from the environment, and 16,872 have been obtained from water. Preservation methods for all these microorganisms is a challenge as all the properties of the culture should be maintained. Therefore, authentic long-term preservation strategies need to be followed. It has various databases. CCINFO is a database that provides worldwide information of cultures. ABC database is an analyzer of bioresource. GCM database provides a reliable catalogue of microorganisms. They also have FN, which is a global data bank of fungal cultures. It was the initiative of Prof. Karl, working at Prague, of the Czech Republic to have culture collection centers. It was his idea to preserve the cultures and make them available to others by paying a fee. However, after his demise the collection was taken up by the University of Vienna.

Important culture collection centers around the world with each one consisting of particular number of strains has been studied (Table 3.1).

**Culture Deposition for Patent Filing under Budapest Treaty**

Many biological products are produced from microbes. Therefore, the governments around the globe have decided that the patented cultures must be deposited in a culture collection center. To finalize this, a meeting was held in Budapest, Hungary, in 1977 in which an International Treaty was signed by 78 countries including India. This came into effect from the year 1980 and the administration was given to the World Intellectual Property Organization (WIPO), Geneva. Accordingly, whichever culture was deposited at the International Depository Authority (IDA), it had the recognition for patenting.

As it is the responsibility of the IDA to maintain the cultures, it is important to recruit staff with good skills. The administration has set up regulations for IDA, which should be followed by the staff effectively and uniformly around the globe among the participating countries. The culture maintenance should be to keep the culture in the pure and viable form at all times. However, it does not give the right to check the culture(s) for the details of the properties at the depository for which the culture has been patented. This aspect needs to be checked thoroughly by the depositor him/herself. It is always not easy to show that an invention is either new or useful as many discoveries are of a scientific nature and are not necessarily inventions for the purpose of patent laws. Many times, it becomes necessary to approach the court to decide if the invention is patentable or not. Before applying for a patent, it is important to take advice from an attorney who is well aware in the patent laws. All over the world the general ideas for patentability are the same. The current Patent Law in the United States is the United States Code Title 35—Patents (Revised August 3, 2005) and is administered by the Patents and Trademarks Office, while the equivalent UK Patent Law is the Patent Act 1977. Patents are valid in the United Kingdom and some other countries for a period of 20 years, whereas they are valid in the United States for 17 years. International laws have helped to bridge many differences among the patent practices in many countries. Protection of industrial property made in "The Paris Convention" was signed by many countries. Accordingly, each country guarantees to the citizens of other countries the same

**Table 3.1** Some of the important culture collections around the globe with number of strains

| Country | Approximate number of strains |
|---|---|
| Japan | 24,040 |
| USA | 242,500 |
| Brazil | 177,000 |
| India | 152,849 |
| China | 146,170 |
| Korea | 145,000 |
| Germany | 93,400 |

rights in patent matters as their own citizens. It also gives the right of priority if there is any dispute. In case of an applicant filing a patent, he/she can apply for protection in all the other member countries and it is regarded as filed on the same day as initially applied.

It was in Washington, DC (1968) that another international treaty was signed. This is known as the Patent Cooperation Treaty, which facilitates the filing of patent applications in different countries by providing standard formats among other things. According to the European Economic Community, the following can be patented: the creation of new plasmid vectors, isolation of new DNA restriction enzymes, isolation of new DNA-joining enzymes or ligases, creation of new recombinant DNA, creation of new genetically modified cells, means of introducing recombinant DNA into a host cell, creation of new transformed host cells containing recombinant DNA, a process for preparing new or known useful products with the aid of transformed cells, and novel cloning processes. These patents are regarded as a process and not as a substance. If a microbiological process is patented, then the patentee must deposit the culture in a recognized culture collection center. This helps the culture to be maintained permanently and is readily available. The patentee must provide a pure culture in lyophilized vials. The major culture collection centers around the globe are given below.

**Culture Collection Centers in India**

Under the WDCM, India has around 27 culture collection centers. However, all of them are not active in maintaining the cultures for scientific needs. The main centers that are active are Microbial Culture Collection (MCC, Pune), Microbial Type Culture Collection (IMTECH, Chandigarh), National Fungal Culture Collection of India (Pune), National Collection of Industrial Microorganisms (Pune), agriculturally important National Bureau of Microorganisms and National Collection of Dairy Cultures (Karnal) (http://www.wfcc.info/ccinfo/collection/col_by_country/i/91).

After many discussions and meetings among the top scientists of the country, it was decided to maintain the active cultures at one place. Thus, the MCC at National Centre for Cell Science (NCCS), Pune, was decided the place for all the cultures to be maintained and kept. This is recognized by the WIPO, Geneva, Switzerland under Budapest Treaty and is also recognized as Designated National Repository for Microorganisms by the Ministry of Environment and Forests (MoEF), New Delhi, India under Biological Diversity Act 2002. The MCC accepts microorganisms for deposition under general, safe, patent, and IDA deposit. The center has the facility and skilled staff to maintain aerobic, anaerobic, and phototrophic cultures. The cultures received by the depositors are preserved either in the lyophilized form or in liquid nitrogen or at low temperatures (−)80 °C. The maintained cultures are given to the clients in lyophilized ampoules or on an agar slant. The cultures are supplied not only to the educational and research institutions but also to the industries for various metabolite production. The center also caters to molecular and biochemical characterization of the culture.

**Shipping of Cultures and Guidelines for Shipping and for Receiving Biological Materials**

There are many government agencies that are responsible for shipping of the biological materials. The skilled person who does the shipping has the responsibility toward correct identification, classification, packaging, labeling, marking, and for the documentation for all shipments of infectious substances. However, advance arrangements need to be made between the consignee and the operator. It is important and compulsory to acquire all permits, especially if any infectious substance is being imported from foreign countries. In case the federal and international regulations are not followed, the airline can refuse the shipment or it may amount to paying penalty, jail, or both. In no case, any infectious substance can be carried personally by air. A website can be looked into for the regulations. http://www.fedex.com/us/services/packageshipment/index.html?CMP=PAC-GTSVURL008•DOT regulates the shipment of Etiological Agents, http://www.federalregister.gov/articles/2008/01/23/E8-1050/interstate-shipment-of etiologic-agents • IATA—Dangerous Goods Regulations (DEHS has copies of the regulations) • WHO—Guidelines for the Safe Transport of Infectious Substances and Diagnostic Specimens, 1997 • Postal Service Hazardous Materials Domestic Mail Regulation, http://www.ups.com/content/us/en/resources/ship/hazardous/index.html.

It is very important to obtain all permits needed before getting any biological material by ship. It is also important to send all copies to the shipper so that they are aware of all details of the cultures being transported. In case the shipment is from a foreign country, then the shipping regulations should be followed as per the MoU made in the United States, which needs to be signed before the shipment takes place. The material must have the correct labels. These should be sent to the foreign shipper. It is worthwhile for each institution to have an Institutional Biosafety Committee (IBC). This helps in taking the approval to work with the material that requires a Biosafety Level 2 or higher precaution or is a biologically derived toxin. An institutional staff should be deputed at the site where the material will arrive. Approvals should be taken from the academic department and also the dean to receive any material. After the biological material is received, it should be notified to the Biosafety Officer to have Biosafety Level 2 and a copy of the Biological Safety must be obtained. For any pathogen possession/use from the receiver's institution, an official approval must be sent to the Safety and Emergency Management. It should be determined if a Material Transfer Agreement is required or not. It is important to know and understand the material's classification and the Classification Flow Chart. In case the material is being classified under category A or B of infectious substance or shipped on dry ice or a toxin, the Department of Biological Science Lab Manager should be contacted who would supervise and sign on all the shipments. In the absence of the lab manger, the safety and emergency management director must be contacted. It is important to get the correct type of packaging and labeling to classify the material. In case a permit is required, then a copy can be obtained from them before shipping the material. When shipment goes to another country, a custom's courtesy letter is needed, which has to be obtained from the country where the shipment has to reach. A person should be assigned to receive the package when it reaches the particular country. In case any clarification is needed, a government agency should be approached for clarification and updates regarding the permit requirements. An updated training should be given to the concerned personnel to fulfil the requirements as per the US Department of Transportation and the shipping companies. Shipping training must be updated every 2 years or more frequently if regulations change. If shipping is of nondangerous biological materials, then it does not require training. Federal permits are needed for importing/exporting disease causing agents for humans and animals, vectors for those agents, animal products, plants, plant products, and plant pests. Sometimes permits are required for domestic transport. The US receiver (importer) is responsible for the package being sent to them from a foreign country. The receiver must assure that the foreign shipper has packed and labeled the material as per the requisite of the US Public Health Service and the International Air Transport Association (IATA) regulations. This is done with proper shipping labels and a copy of their import permit to the shipper. When the foreign import regulations are complied, it does not deny entry at the customs.

**USDA/APHIS Transport Permits**

This regulates the transport of materials that could harm US agricultural products such as the livestock, poultry, and crops. For such materials, permits are required from the APHIS for any import/export/interstate transport of animal or plant pathogen, pathogen vector, animal, animal product, plant, plant product, and introduction of genetically modified organism into the environment.

## 3.6 Methods to Identify Microorganisms

Industries are looking for rapid and cost-effective methods for the identification of microorganisms rather than conventional methods, which are time-consuming and need a lot of manpower. Techniques such as growing the microorganism on an agar plate or any biochemical method are not preferred as they take a lot of time. Molecular techniques of identification as real-time PCR and 16S RNA have become common for the identification.

Taxonomy involves a major part for the identification of microorganisms. This consists of the description, classification, and inventory of the organism. Aristotle (384–322 B.C.) was the first scientist to propose classification of living organisms that were based on the movement. He made two categories as plants that are immovable and animals that are movable. Later, it was Linnaeus (1707–1778) who introduced the binomial classification, which is used till date. Initially, it was only the phenotypic characterization for the classification. However, it had many analytical difficulties. Therefore, he soon developed a binomial system that was simple and had a combination of two Latin names (as one identifies humans with a name and a surname). He named it as genus and species. This helped in making a family tree and lineage. Later, classification was done based on the pH-based reactions, enzyme profiling, carbon source utilization, acid detection, and cellular fatty acids (Horikoshi 1999). He categorized the organisms into three categories as acidophile organisms being acid loving can grow at low pH (0.0–5.5), neutrophiles grow best at pH between 5.5 and 8.5, and alkalophiles, being alkali loving organisms, grow at high pH (8.5–11.5) conditions. Microbial metabolism gets affected by the pH. During fermentation, the microbes use organic material and after metabolic reactions, they produce a number of products. Looking into this phenomenon, many biochemical tests were used for their identification. Gram staining was the first test to be done for the identification of microorganisms. Later, many biochemical tests came into being, which included catalase and oxidase tests. Basically, decomposition of carbohydrate was evaluated through the production of acid (pH) and end products ($CO_2$, acetate, and lactate). Many other tests were conducted for various enzymes such as glucuronidase, glucosidase, galactosidase, and fucosidase as these were involved in the cell metabolism. With the passage of scientific studies, researchers added many other tests for the identification of microorganisms in a precise manner. These included the products formed during protein and amino acid metabolism, tests like indole and $H_2S$ production, gelatin and casein digestion, decarboxylation of lysine, ornithine, arginine, arginine dihydrolase, phenylalanine deaminase, and urease. The breakdown of lipids and the metabolic products have been studied by testing the activity of lipase and lecithinase. Digestion of tween has been also taken as a test for checking the breakdown of lipids. Cell wall receptors like optochin are also being checked. Other tests to check lipid breakdown are lysozyme susceptibility and the bile solubility tests. These are used for the identification of Gram-positive cocci. With scientific advances in recent times, scientists have invented ready-to-use commercial kits that can identify the microorganisms as per the phenotypic system. In the late twentieth century, many genetic techniques and sequence databases have been developed that can identify any microbial species easily with the help of PCR (polymerase chain reaction). Further advances have given rise to many other methods that are both culture dependent and also culture independent. Molecular DNA sequencing, 16S rRNA sequencing, and hybridization have further helped in rapid identification of a microbe. Techniques such as omics, metagenomics, proteomics, lipidomics, transcriptomics, and metabolomics have made characterization and quantification of biological molecules to translate into the structure, function, and dynamics of an organism. They are used in microbial ecology, phylogeny, functional genome analysis, and transcriptional profiling. To classify the genetically dissimilar species, only phenotypic characters are not enough and other tests need to be done, which are based on the morphological, physiological, and biochemical features along with the genetic profile. Franco-Duarte et al. (2019) discussed the advances in chemical and biological methods to identify microorganisms from the past to the present. Among microorganisms, bacteria, yeasts, and molds produce many metabolites. They are important in biotechnology, food sciences, medicine, and genetic engineering as these can prevent a large number of diseases. The disease-causing microorganisms are also bacteria and fungi. Among these the major bacteria include *Campylobacter jejuni*, *Escherichia coli*, *Shigella* spp., *Vibrio cholerae*, *Aeromonas* spp., enterotoxigenic *Bacteroides fragilis*, and *Clostridium difficile*. Therefore, correct identification of the microorganism is extremely important before taking up any commercial/industrial production of metabolite/s from the microorganisms. Also, to obtain the optimum levels of metabolites and for the treatment of diseases, it is vital to initially identify the organism with all available techniques, including morphology, biochemical tests, and molecular tests. After proper identification, the media conditions and other parameters must be optimized in the process line to obtain maximum yield of the product.

### 3.6.1 Identification Methods Using Chromogenic Media

Earlier, the microorganisms were identified based only on the cultivation method. A pure culture of a single type of microorganism is obtained by enrichment cultures. The isolation methods include seeding by dewatering, deep seeding in solid media, and liquid dilution. For organisms that form colonies on the solid medium plates, the seeding technique is rapid. If repeated seeding of an isolated colony is done, then a pure culture can be obtained easily. Purification of both aerobic and anaerobic organisms on solid medium plates is possible by using the seeding method by depletion. Many times, the identification can be made based on the metabolite production by the microorganism, which is released into the medium. To obtain a pure culture, different isolation methods are utilized. These include seeding by dewatering, deep seeding in solid media, and liquid dilution. When seeding is done by exhaustion, the isolation of pure colonies is very rapid and easy. Purification of aerobic and anaerobic organisms can be done on solid media plates using incubation devices. When the identification of microorganisms is done only on the basis of cultivation technique and not by using depletion technique, the results are inaccurate and incomplete. Also, the study of biodiversity cannot be done if it is based only on cultivation.

### 3.6.2 Microscopic Techniques

Microscope is an instrument that helps in the analysis of shape, motility of a culture, and staining details. These aspects help in the identification of a microorganism. It can differentiate among cocci, tetrad, or a bacilli shape of a culture. Therefore, it becomes very helpful in clinical as well as in research work (Bedrossian et al. 2017). Under microscopy, other techniques are contrast microscopy and fluorescence microscopy which are very helpful. The principle in contrast microscopy is based on a phase shift of light that passes a clear specimen and makes the image bright. This helps to show cellular details and the different phases of cell division. The cellular components can be quantified by this method in cases where the use of fluorescence is not possible.

In fluorescent microscopy, the 3D effect is very clear. During the procedure, fluorescent tags are attached to the antibodies that attach to the target. Light of a specific wavelength is supplied to the target specimen that gets absorbed by the fluorophores and then they emit light of longer wavelength as some energy gets utilized by the fluorophores. The color of the emitted light is different. This depends on the wavelength of light emitted. It helps to visualize the cellular organelles. There are wild-field microscopy, confocal microscopy, and total internal reflection fluorescence microscopy. Basically, the sample is supplied with a particular wave length of light. The emitted light and the excitation lights are separated using optical filters. The fluorescent dyes are used, which help in visualizing the microbial cells very clearly. Added tools in electron microscopy help in the identification of microorganisms. These include transmission electron microscopy (TEM), scanning electron microscopy (SEM), confocal microscopy (CLSM), and atomic force microscopy (ATM). The microscopic technique is very effective in the identification of microorganisms that are present in a biofilm. In a capillary tube, even a single cell can be studied. Flow cytometry, which has an electronic detector, has been found to be very helpful in counting and evaluating the microorganisms in a liquid. However, it can evaluate only select microorganisms on the basis of size, shape, or fluorescent properties of individual cells. Microscopic images of microorganisms help in analyzing their shape and motility. The morphological differences in bacteria can be visualized as in *Streptococci*, *Staphylococci*, bacilli (*Listeria monocytogenes*, *E. coli*, or *Salmonella* spp.), and *Vibrio*, which are useful both in clinical and research work. However, microscopy alone cannot identify the organism completely as the cells are small. There is variation in the sizes of prokaryotes and also to differentiate the living or dead cells becomes difficult. The phylogenetic diversity of the microorganisms cannot be known with microscopy.

### 3.6.3 Biochemical Analytical Methods to Detect Microorganisms

**Traditional Biochemical Methods** These are mainly based on the biochemical tests done after the isolation and dilution of microorganisms and then growing them on cultivating media containing agar. These are very time-consuming and gives false positives. Therefore, the method becomes unreliable. In some laboratories, traditional identification methods are still being used. ELISA (enzyme-linked immunosorbent assay) technique is an immunoassay method, which is expensive, but it is a very efficient method for some bacterial species. Another standard method is wherein the microorganism is integrated into the miniaturized reaction couples and is known as the analytical profile index (API). Earlier (till 1992), API 20E (analytical profile index 20E) testing apparatus was used very commonly. For rapid testing, the BBL Crystal (Becton Dickinson, NJ, USA) method was used earlier, which did not have the reading of test reactions. In this method, the identification is made by one single step to inoculate the organism. As it is a closed system, the handling is very easy and efficient. The designing in this method is made in a way not to add any reagent. Studies have shown that the results are very accurate. Later, Biolog Omni Log System was invented, which is rapid, standardized, and can determine the oxidation reaction (tetrazolium redox dye) of different and simultaneous carbon sources. However, many more biochemical methods with more accuracy have been evolved, which are used for the identification.

Traditionally, the increase in turbidity has been used as a parameter of growth. There is a Wickerham card (2 × 3″) that has white and black lines to read the turbidity. However, it is not a standard method and the incubation period is very long (Procop and Koneman 2016). Immunoassays, such as the enzyme-linked immunosorbent assay (ELISA) (Mandy et al. 2023), are efficient but the method cannot be performed on all bacteria and the cost of analysis is very high. There is an analytical profile index (API) that can classify bacteria based on the biochemical tests. However, only known bacteria can be identified by this method and has been known as the gold standard in clinical microbiology. Although this method has a huge database but it is very time-consuming. The plastic strip has 20 small chambers that contain dehydrated media with defined composition for each test. It can usually analyze most of the enzymatic activities undergoing in the cell. When the bacterial suspension is infused into the well, the media gets rehydrated. As the reaction takes place, a color change can be seen. With all the reactions in different wells the microorganism can be identified.

Another method is the Vitek system (bioMérieux, Craponne, France) (Patrick et al. 2020). By this method, the bacteria can be identified easily and the study of antimicrobial susceptibility can also be done simultaneously. An automated broth is used that applies attenuation of light measured by an optical scanner for growth. The improvement on Vitek system is the Biolog Omni Log method, which can analyze the oxidation of tetrazolium dye of carbon sources along with pH, salt tolerance, and chemical sensitivity.

### 3.6.4 Mass Spectrometry Based Methods for the Identification of Microorganisms

This method saves time due to the semi-automatic and automatic systems, and it uses biochemical methods for analysis. For microbial typing, this method is very useful as it is rapid and involves less cost. It is simple and can be used for bacteria, archaea, and fungi. Many ionization and separation techniques can be coupled with MS such as gas chromatography (GC), matrix-assisted laser desorption ionization time-of-flight mode (MALDI-TOF), electromigration techniques, and electrospray ionization (ESI).

The various chromatographic techniques are as follows:

A. **Liquid Chromatography (LC)-Based Methods**

   The combination of liquid chromatography (LC) with MS (LC-MS) can determine metabolome and can identify microorganisms by analyzing nonvolatile or thermally labile high molecular compounds in cases where GC-MS approaches do not work. It works under low temperatures and the volatility of the sample is not required. This method is useful in clinical applications to identify the microorganism. The sample can be injected into the solvent, which is divided within the column where the stationary phase is bound chemically.

Advances in technology have given rise to fast LC-MS, LC-MALDI-MS, LC-ESI-MS-MS, LC-Nuclear Magnetic Resonance (NMR)-MS, hydrophilic interaction liquid chromatography (HILIC)-MS, reverse phase LC-MS, and ion mobility spectrometry. The sample mixture can be injected into the sampler that goes into the column. The flow is adjusted with the pumps. The method of HPLC is expensive and the equipment needs a lot of maintenance. Denaturing high-performance liquid chromatography (DHPLC) is a new and promising approach for bacterial identification and monitoring. It also helps in the separation of PCR products with the use of ion-pair, reversed-phase, high-performance liquid chromatography (IP RP HPLC).

B. **Gas Chromatography–Mass Spectrometry**

Complex biological mixtures can be identified with GC coupled with MS. A constant flow or pressure is used by a flow program. GC separates efficiently, but MS delivers individual mass spectra that distinguishes various chemically different metabolites. The method is sensitive, robust, easy-to-use, cost-effective, and has accessibility to commercial and public libraries. Nonpolar molecules can be analyzed easily as lipid components and lipid elements of microorganisms help in their classification. Curie-point pyrolysis-MS (Py-GC-MS) is applied for the distinction between Gram-positive and Gram-negative bacteria. In whole cells thermally assisted hydrolysis, methylation-GC and MALDI-MS with on-probe sample pretreatment has been utilized to analyze *E. coli* K-12 phospholipids and also in yeasts. However, in GC-MS, the analytes are required in a volatile form. It needs to be derivatized before injecting into the system, extra steps are needed to be worked out which takes time. There is a conjugation technique that is very useful. It has two different GC columns that are conjugated together. This increases the metabolite detection coverage but increases the cost. The speed of scanning rate (TOF-MS) is also increased, providing extra sensitivity for detection. The flame ionization detector (FID)-GC-FID gives rapid, very sensitive results and does not cost much. For rapid identification and classification of microorganisms, MALDI-TOF MS (matrix-assisted laser desorption/ionization (MALDI)-time-of-flight (TOF)) method is used, which is based on the ionization of the microbial cells with short laser pulses and then accelerating the particles in a vacuum system using an electric field. A molecular fingerprint is obtained in the form of a specific spectra profile for each microorganism. This is compared with the database and identification is made by an automated program. Another technique is called as electrospray ionization (ESI)-MS, which analyzes samples in a liquid state and the ionization is done at atmospheric pressure.

C. **Spectroscopic Methods**

Recently, simple technique known as fiber optics spectroscopy is very reproducible. The spectra depend on the molecular groups. The chemical composition of proteins, lipids, carbohydrates, membranes, pharmaceuticals, and human tissues are taken into consideration. For the identification of microorganisms, these methods are very important as no sample is destroyed.

D. **Fourier Transform Infrared Spectroscopy (FTIR)**

Fourier transform infrared spectroscopy (FTIR) is an easy, rapid, noninvasive method and a cheap method of analysis. It analyzes both the chemical composition and the physical state of the sample. However, the biological samples should be preprocessed before putting it into the analyzer. The analyzer is fitted with an IR energy source. This emits light in longer wavelength region. This specific light is (IR beam) absorbed by the sample and results are obtained when spectral data is coupled with other standard methods. Several compounds can be analyzed at the same time and can be used for real-time process monitoring, as well as the accomplishment of high-throughput screenings.

E. **Raman Spectroscopy–Vibrational Spectroscopy**

In case of noninvasive samples, this method is good for rapid characterization and identification of microbes. It is low cost and rapid. It utilizes vibrational, rotational, and other low-frequency modes in the system to produce a structural fingerprint. Species and strains can be identified within a few hours. Due to the vibration of atoms, the polarizability of functional groups gets changed. It evaluates the inelastic scattering of radiation of monochromatic light, with a spectral shift, known as Raman shift.

F. **Nuclear Magnetic Resonance (NMR) Spectroscopy**

In this method, strong magnetic fields and radio frequency pulses to the nuclei of the atoms are applied and in the case of atoms such as $^{1}H$ or $^{13}C$, the magnetic field causes a nuclear spin, absorbing the radio frequency energy (low-energy to high-energy spin states) and the emission of radiation is detected.

G. **Electrokinetic Separation Methods**

Differences in microbial composition have different migration patterns that separate different microbial species. In capillary electrophoresis (CE)-MS, the separation process is through electrophoresis along with MS detection. The separation efficiencies are very good and a very low amount of sample volume is required. The cations, anions, and uncharged molecules can be separated simultaneously. CE can be combined with capillary isoelectric focusing (CIEF) to separate and identify seven microbial species with a wide range of sizes and shapes (Armstrong et al. 1999). CE with fluorescence can monitor the operational conditions and the microbial dynamics. The method can mainly differentiate between size, shape, and charges and help in microbial identification. Electrical field-flow fractionation (ElFFF) technique uses the ability of microorganisms to migrate in an electric field on the separation of sample components in a channel as a result of different layers (fractionation) of each group of components under the influence of various electrical fields.

H. **Microfluidic Chips**

Separation and detection of sample constituents are made by controlling the movement of fluids within the microfluidic chips, without the need for special sample preparation or reaction. These are small portable chips with microchannels of tens to hundreds of micrometers, pressure systems, and detection systems. The method is very sensitive as the pathogenic bacteria and viruses can be detected using a very little sample volumes as just a nanoliter or even less is sufficient to give the desired result. The method is commonly used in food safety control, environmental monitoring, and clinical diagnosis.

I. **Fluorescence In Situ Hybridization (FISH)**

Recently, the use of combination of analytical standard techniques inside the microfluidic chips has become common. This does not need labeling procedures. Scientists have worked on combinations of microfluidic devices with PCR, MS approaches, spectrometry, and electrochemistry. The technique FISH (fluorescence in situ hybridization) involves the use of phylogenetic dyes, which are fluorescent oligonucleotides whose base sequences are complementary to the ribosomal RNA sequences (16S or 23S RNA in prokaryotes, or 18S or 28S RNAs in eukaryotes). The dyes penetrate into the cells and form hybrids with the microbial ribosomal RNA without lysing them. As they are specific, they react with one specific species. Phylogenetic probes can also be used in this technique. Each probe is designed in a way to react with a particular organism or group of organisms, with a specific fluorescent dye. With the association of FISH and CLSM, it is possible to study the microbial populations in greater detail and use it in biofilm study. Gene expression of the organisms can also be measured.

J. **Molecular Methods Used to Detect Bacteria**

The genotypic methods have enabled the identification of a diverse unknown taxa. These are based on variations in DNA analysis, which could be either amplification or sequence-based such as PCR, real-time PCR, and RAPD-PCR. Other complex methods include restriction fragment analysis, targeted gene and whole-genome sequencing, and mass spectrometry. Matrix-assisted laser desorption/ionization time-of-flight mass spectrometry (MALDI-TOF-MS) have also been studied.

16S rRNA PCR-sequencing a very sensitive technique to detect bacteria. As the 16S rRNA gene is specific to each bacterial species, it helps in the correct identification. However, the PCR inhibitors should be removed, which will help in maximum extraction from the bacteria without contamination.

K. **Real-Time PCR**

Real-time PCR (RT-PCR) provides more accuracy. Through fluorescent intensity, the DNA amplification can be monitored in real-time. Using the Cq value (cycle number at which fluorescence intensity rises above the detectable level), the amount of DNA can be quantified. The primers have been designed on the basis of alignment sequence of 962,279 bacterial 16S rRNA gene sequences. It helps to identify or quantify bacterial loads in clinical infections and gastrointestinal mucosal biopsies.

Random amplification of polymorphic DNA-RAPD-PCR short primers (8–12 nucleotides long) with arbitrary sequences bind nonspecifically to template bacterial DNA for bacterial identification. RAPD-PCR reactions of isolated DNA are amplified in the presence of a RAPD primer and at low levels of magnesium. Unique RAPD fingerprints are obtained after agarose gel electrophoresis. The designed primers bind randomly to the template DNA. Both Gram-positive and Gram-negative species can be identified.

L. **Restriction Fragment Length Polymorphism (RFLP)**

Bacteria strains can be identified using unique fingerprints on the basis of variations (polymorphisms) in homologous DNA sequences. The different fragments are separated by agarose gel electrophoresis to generate a unique pattern of bands for each bacterial strain. Differences in banding patterns identifies the diversity in bacterial strains. Among few patients, this technique could identify differences in a cluster of MRSA strains.

M. **Amplified Fragment Length Polymorphism (AFLP)**

The method utilizes restriction enzymes (usually a pair) to fragment genomic DNA. With the use of ligated adaptors, it amplifies a subset of restriction fragments by using complementary primers that have unique nucleotides. The AFLP fingerprints are analyzed by gel electrophoresis. Using fluorescently labeled PCR primers, the PCR fragments are amplified and are analyzed after separation by gel electrophoresis.

N. **Pulsed-Field Gel Electrophoresis (PFGE)**

This can separate large fragments of DNA and can characterize bacteria for epidemiological studies. The bacterial strains in agarose are treated with enzymes and detergents (proteases and SDS). This releases the chromosomal DNA, which are then incubated with restriction enzymes. It generates DNA fragments at specific sites. An electric current is passed with alternate rotations in a magnetic field. This enhances the movement of large DNA fragments that separate them.

O. **Ribotyping**

This employs rRNA based on phylogenetic analysis. As rRNA genes are highly conserved within a bacterial species, identification of 16S rRNA gene polymorphism helps in bacterial classification, taxonomy, epidemiology, and population biology. It is a multistep process with restriction enzymes targeting the genomic sequence, followed by a Southern blot transfer and hybridization with probes. Then the ribotype RFLP bands are analyzed.

P. **Whole-Genome Sequencing (WGS)**

This method is very accessible and affordable tool for bacterial genotyping, especially for pathogen identification and antimicrobial resistance typing.

Q. **MALDI-TOF-MS in Bacteria**

The spectra patterns by the bacterial cells have the information that identifies and characterizes the bacterial species. The sample is mixed with organic matrices and ionized by a laser beam. When the ions reach the mass analyzer, the mass:charge ratio forms a spectral pattern, which is compared with fingerprints in the library.

R. **Molecular Methods Used to Detect Yeasts**

The fungal species may differ in virulence and drug resistance. In any probe-based identification, a double-stranded hybrid is obtained after the binding of the single-stranded DNA or RNA of the organism to a complementary sequence. DNA polymorphism of fungi can be analyzed by RFLP and Southern hybridization using appropriate DNA probes, DNA RAPD, PCR-based methods, electrophoretic karyotyping by pulsed field gel electrophoresis (PFGE), and sequencing-based methods. PCR-based methods are highly sensitive, rapid, specific, and permit both intraspecies differentiation and species identification of yeast isolates. As the fungal cell walls are strong and difficult to disrupt, a lot of effort is needed for DNA isolation. Therefore, the use of glass beads, sonication, and phenol–chloroform helps in enzymatic digestion. The selection of primers plays an important role. Multiplex PCR helps in the detection of many fungi at a time. The coding regions are 18S, 5.8S, and 28S among fungi.

S. **Quantitative Real-Time PCR (RT-PCR)**

RT-PCR assay is very accurate and specific. SYBR Green and TaqMan-based assays that use primers and probes for the 43-kDa glycoprotein (gp43) gene are able to detect small gene copies (about 2 fg of DNA) per reaction. The qPCR assays have a positive linear correlation between CFU and qPCR but are unable to distinguish between living and dead yeasts.

T. **DNA Fingerprinting Methods**

Pulsed field gel electrophoresis (PFGE) method helps in the separation of fungal chromosomal DNAs on the basis of size in agarose gels that help in the identification of species and strain specific profiles.

U. **Restriction Fragment Length Polymorphisms (RFLP)**

Same strain with variations in DNA structure can be identified by RFLP. By using specific DNA restriction enzymes by gel electrophoresis, the fragments get divided based on the molecular size. Staining of the gel is done with ethidium bromide and are then visualized under UV light. PCR-amplified ITS2 region DNA is a rapid method to identify clinical yeasts. ITS-rDNA region (ITS1–5.8SrDNA-ITS2) can also be amplified by PCR along with universal primers, which can be digested with *Msp*I to identify a huge number of clinical yeast strains. Closely related strains can be identified using mitochondrial DNA (mtDNA) by variation in the cleavage outline, which is specific for each strain.

V. **Random Amplified Polymorphic DNA (RAPD)**

DNA fragments after PCR amplification of random segments of genomic DNA having a single primer of nucleotide sequence helps in the identification. By the use of OPE-18, OPE-04, and OPA-18, a characteristic molecular fingerprint can be obtained. Primers such as M2 and P4 show different profiles.

W. **Amplified Fragment Length Polymorphism (AFLP)**

The genomic DNA is digested using two restriction enzymes (*Eco*RI and *Mse*I). Double-stranded oligonucleotide adaptors are then linked to the fragments that are the targets for the primers. This is labeled with a fluorescent dye during PCR amplification. This is then separated and scrutinized using a software.

## FAQs

1. What are the factors that influence the growth of microorganisms for primary and secondary metabolites?
2. In which category do anthocyanins come?
3. Which secondary metabolites are useful for human welfare?

4. How long a microbial culture maintains its purity through subculturing cycles?
5. What is the biosafety level of the microorganism from culture collection center?
6. How can the microorganisms purchased from microbiologists can be disposed?
7. What is the purpose of microbial screening?
8. Why is it important to write an expiry date on the product?
9. How should a lyophilized culture be used?
10. From where should a certificate of quality be obtained?

## References


Armstrong DW, Schulte G, Schneiderheinze JM, Westenberg DJ (1999) Separating microbes in the manner of molecules. 1. Capillary electrokinetic approaches. Anal Chem 71:5465–5469

Barbuto FS, Cammarota M, Schiraldi C, Restaino OF (2021) *Streptomycetes* as platform for biotechnological production processes of drugs. Appl Microbiol Biotechnol 105(2):551–568

Bedrossian M, Lindensmith C, Nadeau JL (2017) Digital holographic microscopy, a method for detection of microorganisms in plume samples from *Enceladus* and other icy worlds. Astrobiology 17(9):913–925. https://doi.org/10.1089/ast.2016.1616

Beltrametti FD, Barucco R, Rossi SE (2007) Vicuron pharmaceuticals. protoplast fusion and gene recombination in the uncommon actinomycete *Planobispora rosea* producing GE2270. J Antibiot 60(7):447–454

Ezemba CO, Oluchi A, Vivian E, Arinze OP et al (2022) Insights in industrial microbiology 1

Franco-Duarte R, Černáková L, Kadam S, Kaushik KS et al (2019) Advances in chemical and biological methods to identify microorganisms-from past to present. Microorganisms 7(5):130. https://doi.org/10.3390/microorganisms7050130

Hennrich O, Weinmann L, Kulik A, Harms K, Klahn P, Youn JW, Surup F, Mast Y (2023) Biotransformation-coupled mutasynthesis for the generation of novel pristinamycin derivatives by engineering the phenylglycine residue. RSC Chem Biol 4(12):1050–1063

Higgs RE, Zahn JA, Gygi JD, Hilton MD (2001) Rapid method to estimate the presence of secondary metabolites in microbial extracts. Appl Environ Microbiol 67(1):371–376

Horikoshi K (1999) Alkaliphiles: some applications of their products for biotechnology. Microbiol Mol Biol Rev 63:735–750

Idiris A, Bi K, Tohda H, Kumagai H, Giga-Hama Y (2006) Construction of a protease-deficient strain set for the fission yeast *Schizosaccharomyces pombe*, useful for effective production of protease-sensitive heterologous proteins. Yeast 23:83–99

Idrees M, Sawant S, Karodia N, Rahman A (2021) *Staphylococcus aureus* Biofilm: morphology, genetics, pathogenesis and treatment strategies. Int J Environ Res Public Health 18(14):7602

Lim D, Renteria ES, Sime DS, Ju YM et al (2022) Bioreactor design and validation for manufacturing strategies in tissue engineering. Biodes Manuf 5(1):43–63

Mandy A, Muhammad Z, Aisha F (2023) Enzyme linked immunosorbent assay. NCBI Bookshelf. A service of the national library of medicine, national institutes of health. In: Stat Pearls [Internet]. Stat Pearls Publishing, Treasure Island (FL)

Méndez H, Jazmín RD, Luis VPL, Jesús SC, Soto-Cruz NO (2023) Strategies for supplying precursors to enhance the production of secondary metabolites in solid-state fermentation. Fermentation 9:804. https://doi.org/10.3390/fermentation9090804

Park SR, Park JW, Ban YH, Sohng JK, Yoon YJ (2013) 2-Deoxystreptamine-containing aminoglycoside antibiotics: recent advances in the characterization and manipulation of their biosynthetic pathways. Nat Prod Rep 30:11–20

Patrick G, Emilie R, Myriam F, Karama R, Jean LM, Isabelle P (2020) Comparative evaluation of the QMAC-dRAST V2.0 system for rapid antibiotic susceptibility testing of Gram-negative blood culture isolates. J Microbiol Meth 172:105902. ISSN 0167-7012

Pickens LB, Tang Y, Chooi YH (2011) Metabolic engineering for the production of natural products. Annu Rev Chem Biomol Eng 2:211–236

Procop GW, Koneman EW (2016) In: Procop GW, Church DL, Hall GS, Janda WM, Koneman EW, Schreckenberger PC, Woods GL (eds) Koneman's color atlas and textbook of diagnostic microbiology, 7th edn. Wolters Kluwer

Smanski MJ, Peterson RM, Rajski SR, Shen B (2009) Engineered *Streptomyces platensis* strains that overproduce antibiotics platensimycin and platencin. Antimicrob Agents Chemother 53:1299–1304

Vásquez-Bonilla JN, Barranco-Florido JE, Ponce-Alquicira E, Rincón-Guevara MA, Loera O (2022) Improvement of beauvericin production by *Fusarium oxysporum* AB2 under solid-state fermentation using an optimised liquid medium and co-cultures. Mycotoxin Res 38(3):175–183

Yong X, Raza W, Yu G, Ran W, Shen Q, Yang X (2011) Optimization of the production of poly-γ-glutamic acid by *Bacillus amyloliquefaciens* C1 in solid state fermentation using dairy manure compost and monosodium glutamate production residues as basic substrates. Bioresour Technol 102(16):7548–7554

Zhang QW, Liu L, Gong CY, Shi HS, Zeng YH, Wang XZ, Zhao YW, Wei YQ (2012) Prognostic significance of tumor-associated macrophages in solid tumor: a meta-analysis of the literature. PLoS One 7:e50946

Zhu S, Duan Y, Huang Y (2019) The application of ribosome engineering to natural product discovery and yield improvement in *Streptomyces*. Antibiotics (Basel) 8(3):133

# Preservation of Microorganisms

# 4

**Abstract**

Microorganisms are very important for metabolite production. More so, given the microbial diversity and each strain being highly specific. To achieve this, it is important to preserve the best genome sequence for maximum yield of the metabolite. For research work and various projects, the microorganisms need preservation in the viable form. Starting from the isolation and growing the pure isolate strains, the preservation should be done in such a way that does not change the morphological, physiological, and genetic characteristic of the culture strain. This chapter covers the different methods of preservation of microorganisms. The preservation through cryopreservation and lyophilization has been explained. The importance of culture collection centers has been covered in this book.

**Keywords**

Culture · Preservation · Immobilization · Encapsulation · Biocontrol agents · Liquid nitrogen

## 4.1 Introduction

In any environmental process, the importance of microbes is vital. Among the genus, there are many species and among the species each microbe has many strains. Each strain has a particular and a different metabolic pathway that results in the production of a specific metabolite. Therefore, for the production of a particular metabolite, a specific strain needs to be preserved along with the viability and functional properties. It is also necessary that the specific culture strain has a long shelf storage. This helps the scientists/industrialists to obtain a specific viable culture strain from any culture collection centers that enables them to produce the desired metabolite.



R. Agrawal, *Textbook of Industrial Microbiology*,
https://doi.org/10.1007/978-981-97-9582-6_4

Before the preservation, a thorough study of the structure, functions, growth phases, and other physical and chemical parameters of the microbe should be done. Many a times due to climate change and other changes in the environment, the microbiome can get mutated or can get lost. This loses the credibility of a microbial process and makes it difficult for any industrialist to venture into microbial processes.

However, after the preservation of any microbe, a thorough study should be done at regular intervals of the preserved culture strain at least once in 6 months. This will help the client to know if there have been any changes or not. With recent researches, scientists have been able to study the microbiome using omic sciences. A lot of work is still under progress till date in this direction.

Sometimes, for a particular product formation, a mixed culture consortium is required, which involves two or three cultures together. This makes the preservation more complex. In such cases of mixed cultures, the study of the cultures is done at the macromolecular level involving the study by omic sciences. Most of the human diseases are related to gut- microbiome. Therefore, to find a cure for such diseases, it is important to study their microbiome.

## 4.2 Techniques for Culture Preservation

Apart from microbiome, it is important to preserve the culture strains under optimized conditions for maximum product formation whenever the culture is taken out for the purpose. The culture should be preserved under optimum growth conditions and phase of culture growth. The use of microbes for metabolite production with higher yields is much more economical and beneficial as compared to obtaining the metabolite from the plants. Any microbial process can be conducted in a small shed and does not require a big land as it is required when plants are to be used. Even the weather conditions do not affect the microbial product formation as they are produced in a laboratory. Microbial products are commonly used in clinical work, agriculture, pharmaceuticals, medicines, and in the prevention of serious diseases (Lancini and Demain 2013).

To study the microbes in a particular soil, which directly affects the plant growth, it is necessary to study the conditions of the soil, as well as the interactions between the plant and the microbiome in the environment. All these are responsible for plant health and the environment health, which ultimately affect the health of human beings.

(a) **Metagenomics.** Recently, with the study of metagenomics, the biodiversity of the microbial strains can be studied. This has brought forward the importance of a particular strain for the production of a specific metabolite and for their preservation. It is also vital to preserve the culture in the pure form, which should be viable for a long storage time and should maintain the genetic stability and functionality. The culture strain should be able to reproduce the same yield of the product even after a long shelf life, as it was reported initially, by the scientist when it was deposited at the culture collection center for preservation. The culture collection centers are placed at different locations around the globe. The scientists/industrialists obtain a particular culture of interest from these culture collection centers. This makes the culture very vulnerable. Therefore, the methods used for the preservation of microbial cultures should sustain the transportation and must maintain the qualitative and quantitative properties of the culture strain.

In recent times, the preservation of the microbiome has become important in order to keep the stability of the culture for a long storage time. The method of preservation is specific to each microorganism. They are named as bacteria microbiome, fungal microbiome, plant microbiome, and animal microbiome. With the understanding of microbiomes, it becomes easy to understand the health of the soil, plants, and animals. In case the structure of the microbiome gets disrupted or damaged in any way, then it is known as dysbiosis in which important components could be lost. Due to the disruption, many diseases can appear, as there can be lack of immunity. Therefore, the microbiota must be maintained in a manner that does not allow any disruption of the microbiota. It is important to conserve microbial diversity as microbes are responsible for most of the reactions all-around the globe. This could be soil, plants, animals, or environment. And to have the right microbiota, the preservation of each culture strain is a necessity, which needs to be done under the right conditions. The properties, functions, and the genetic sequence of the culture strain should be maintained in the collection center when the culture is preserved. This will give the same results time and again, whenever taken by any scientist/industry, as claimed by the researcher when the culture was initially deposited at the culture collection center. It has been found that the microbiome can get changed if the storage conditions are not proper. Therefore, the storage conditions should be appropriate to keep the cultures in their original state (Fig. 4.1).

Proper storage conditions, the media of the culture, the temperature inside the tube and outside, all the parameters play a major role in the preservation of the

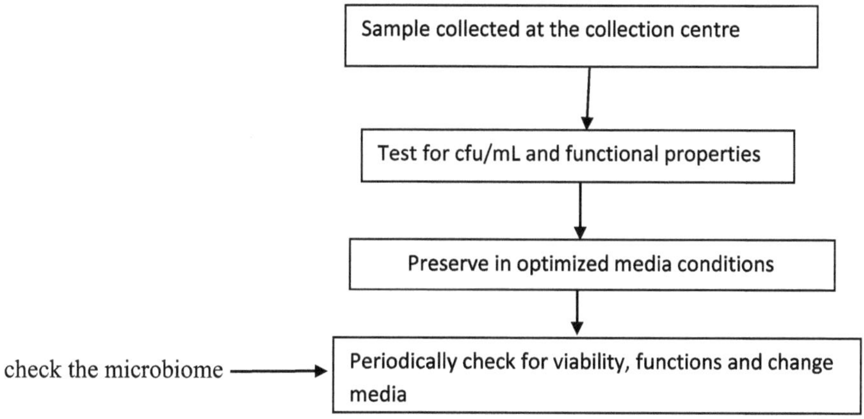


**Fig. 4.1** Flow diagram showing different steps for culture preservation

cultures. Parameters such as temperature and pH contribute mostly during the preservation. It has been found out that when the culture strains are kept at room temperatures, especially during summers wherein the temperatures go very high, the microbiome of the culture can get altered and the desired results cannot be obtained.

(b) **Cryoprotectants.** Sudden freezing and then thawing can also damage the microbiome. Many times, preservation along with cryoprotectants has shown beneficial results as they help in protecting the microbiome intact (Meryman 2007) (Fig. 4.2).
(c) **CAS (cell alive system).** Scientists have invented a new technique known as CAS (cell alive system). In this system, water is spinned around the culture tube during the procedure of preservation of the culture, which helps in avoiding the formation of crystals and preserves the microbiome (Morono et al. 2015). Prakash et al. (2020) have written an excellent review article on the preservation techniques of microbial microbiome.
(d) **Liquid nitrogen.** Many methods are used for the preservation of microorganisms. The storage in liquid nitrogen is one of the best methods for preserving most of the microorganisms. However, microbial contamination has been observed in storage containers among different culture collections. The risk of cross-contamination has been found to be low in liquid nitrogen storage tanks. This could be possible if sealed devices are used and then they are stored at (−) 150 °C freezers. Therefore, in a survey, it has been observed that when such procedures are followed in biobanks and standard operating procedures (SOPs) are followed, the contamination can be kept to a minimum (Bajerski et al. 2021). Viruses can also be preserved in a better way when they are frozen along with a cryoprotective agent. Storing them at (−) 70 °C has been found to be effective and equivalent as preserving in liquid nitrogen.

The preservation for short time (3–12 months) can be done on agar slants or as stab cultures in tightly capped tubes. However, for long-term storage and for the organism to withstand the dehydration, a method of lyophilization is more reliable. The culture tubes should be always refrigerated to avoid dehydration, which can be

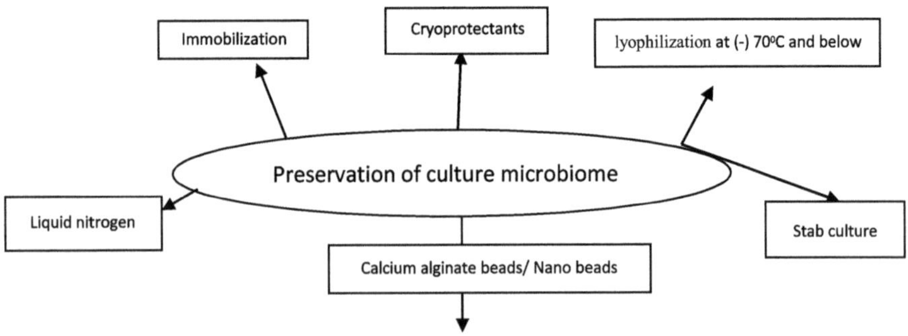


**Fig. 4.2** Different methods to protect the microbiome of microbial culture strains

otherwise mutagenic due to the changes in the microbiome. Most of the cultures are refrigerated for preservation in collection centers for future propagation. Many food and pharmaceutical industries use drying technologies to be the preferred methods for preservation. The industries conduct microbiological testing and quality control using positive control samples as reference microbial strains. The drying techniques make the cultures stable and transportable at ambient temperatures. The emerging fields such as probiotics and biocontrol agents also require preservation of certain microorganisms. According to Bircher et al. (2018), survival of the original cell population is good in freeze-dried samples and allows continuation and maintenance of the culture. The growth of bacteria has four distinct phases: lag, log, stationary, and death phases when grown in batch culture. Studies have shown that the growth phase and the growth media have a significant effect on the freeze-dried survival of microorganisms (Carvalho et al. 2003, 2004). A study has been conducted (Palmfeldt and Hahn-Hagerdal 2000) to understand the effects of lowering the pH during freeze-drying (*Lactobacillus reuteri* to pH 5.0). The use of protective agents during the growth of the microorganism or prior to freezing or drying has been found to enhance preservation. The type of protectant may vary with each organism used. The protectants could be nonfat milk solids, serum, trehalose, glycerol, betaine, adonitol, sucrose, glucose, lactose, and polymers such as dextran and polyethylene glycol (Hubalek 2003). The cryoprotectants help in maintaining the cell matrix during the preservation time. The use of glycerol and dimethyl sulfoxide (DMSO) is very common. As the storage medium contains water molecules, the addition of cryoprotectants has been found to be beneficial. When glycerol is added as a cryoprotectant, it forms hydrogen bonds with them and, in turn, enhances the storage life of the cells with more stability. The preservation methods may be different from genus to genus, and also among species and strains, they may vary depending on their growth conditions. The preservation methods, media, and the physical and chemical parameters are different for bacteria, fungi, and yeasts. Nyanga et al. (2012) studied various preservation methods in yeast. A prebiotic, galacto-oligosaccharides (GOS), has been found to be a good preservative agent for yeasts and also for the mixed cultures. The same cryoprotectant cannot be used for all microorganisms. Studies have been done for the preservation of a probiotic, *Lactobacillus* spp., in which addition of amino acids has shown to enhance the shelf life of live cells (Mokoena 2017). These bacteria contain acyl fatty acids in their cell membrane. With the addition of amino acid, the fatty acids become more mobile. This results in keeping the cell viability for a longer time. The composition of lipids changes, which help in maintaining the cell membrane at different temperatures.

World famous collection centers such as the American Type Culture Collection (ATCC) and the National Collection of Type Cultures (NCTC) use the freeze-drying methods for preservation. The freeze-dried material can also be easily shipped and is cost-effective. It is generally recommended to freeze-dry concentrated cultures ($>10^7$ cells) to ensure that there are sufficient cells remaining after the freeze-drying process, long-term storage, and rehydration. It is important that the culture maintains viability during transportation. Lyophilization is the best method for the safe transportation, which can keep the culture preserved. However, many times the equipment cost becomes too high for smaller laboratories. For this purpose, the use of gelatine discs has proved beneficial. In this process the cells are frozen to low temperatures and later preserved in gelatin discs. Storage up till the period of 1 year has been found good (Obara et al. 1981). Spores of thermophiles and many pathogenic cultures can be preserved by this process for a year. However, to keep them for a longer shelf life, it has shown many challenges. Moreover, no study has been done on the preservation of total microbiome using this method.

Rehydration is a very critical step for the revival of cells after drying. There are many rehydration media, but they have given varying results. Morgan et al. (2006) have written an excellent review article on the drying methods for preserving microorganisms.

(e) **Immobilization techniques.** The cells can be preserved for a long time when immobilized in a gelling substrate such as alginate and acacia gum. Small beads are made of known diameter entrapping a known number of cells (cfu/mL). These cellular beads have been found good to preserve the cell viability along with the functional properties for a long duration. However, the physical and chemical nature of the gelling agent plays a major role in the shelf storage of the culture. Hence, it is advised to optimize all the parameters before preserving the cells. Also, addition of cryoprotectants and antioxidants in gelling matrix have shown better preservation results especially in mixed culture preservation. It has been noticed that in probiotic industries the cells are preserved by immobilization technique and are then frozen. However, this may bring osmotic disturbances that reduce the viability and functionality of the cells. Good results can be obtained when the immobilized cells are later dried using fluidized-bed method. Alonso (2016) has written a good article on the choice of polymers for cell immobilization.

(f) **Microencapsulation.** This method includes electrospinning and electrospraying for cells, which are very sensitive to the environmental conditions. Both are different based on the viscosity of the substrate material used. In electrospinning, the substrate should be more viscous as compared to electrospraying. As the substrate is viscous in electrospinning, the resultant comes out in the form of solid fibers, whereas in electrospraying, in which the substrate is free, the resultant comes out in the form of droplets. By adjusting the parameters (viscosity of the substrate, the pressure, and rate of flow), it is possible to quantify the number of cells in one solid fiber/droplet. This makes the method very precise and quantitative. In these methods, the temperature is not involved. However, the electromagnetic fields can influence the formation of ice in non-thermal procedures. This helps to keep the protein intact, which, in turn, keeps up the viability and functionality of the culture. The viability is usually calculated by checking the most probable number (MPN) method. With an instrument, namely fluorescence activated cell sorter (FACS), live and dead cells can be measured using propidium iodide (PI) and SYTO9. López-Rubio et al. (2004) used the method for the preservation of probiotic cells by this method and found it very effective. In a study (Nogal et al. 2021), it was found that the

**Table 4.1** Methods for the preservation of microorganisms

| Culture | Method used | Shelf life | Reference |
|---|---|---|---|
| *Bifidobacteria* sp. | Heat shock | Postprocess stability | Nguyen et al. (2014) |
| *Lactobacillus acidophilus* | Nutrient starvation | Increased tolerance to low temperature | Wang et al. (2005) |
| Yeasts, *Saccharomyces cerevisiae* | Acid shock | Higher viability after fermentation | Chu-Ky et al. (2013) |

production of short chain fatty acid can be used as a functional marker. The preservation of microorganisms that produced butyrate can be done using glycerol and inulin. According to Vekeman and Heylen (2015), most of the cultures could be preserved well in deep freezers along with 5% DMSO. Similar results have been found encouraging for the preservation of gut-microbiota (Gaci et al. 2017). In another study (Lee et al. 2019), it has been found that a combination of DMSO, ethylenediaminetetraacetic acid, saturated salt (DESS), 15% glycerol in phosphate-buffered saline (PBS) was the best for the preservation of fungal cultures with excellent viability and functionality of the cultures.

(g) **Preservation of anaerobic cultures.** It is found that the anaerobic cultures, especially the nitrification/denitrification cultures, present in the biofilms and human fecal material can be best preserved when cryoprotectants are supplemented in the substrate matrix during preservation. The use of glycerol has been found good (Yu et al. 2015). The preservation of cultures such as *Bifidobacterium* and *Enterococcus* has been worked out using this method (Bircher et al. 2018). The optimization of cryopreservation of pathogenic microbial strains has been studied by Ning et al. (2020). Different methods of preservation for specific microorganisms have been given in Table 4.1.

## 4.3 Storage and Packaging of the Preserved Culture

During preservation, it is important that the viability and the functionality is maintained as that of the initial culture strain. The deterioration in any of these qualities makes the storage useless and unfit for use by the researchers and the industrialists. The cultures to be preserved by any method should be protected from any environment changes such as oxygen, moisture, light, contamination of microbes from the environment, and temperature. Only then the cultures can have a better shelf life. Therefore, packaging plays a major role. The packaging material taken should be able to protect the culture from any change. Storing the cultures in glass ampoules or glass vials have been found to give very effective results.

Usually, the microbial culture is incubated in an agar solid medium. Agar is obtained from an algae. After storing, it is not advisable to leave the culture as such in the same vial for very long periods. Therefore, it needs proper maintenance by subculturing and other methods. This would prevent any loss biologically, immunologically, and in cultural characters such as viability and functionality.

## 4.4 Maintenance of Cultures for Long Duration

To keep the microbiome intact for a longer time, the culture strain needs to be maintained in a culture collection center or in an industry. Usually, there are three methods to keep the culture viable and functionally safe for long periods. These include:

1. Microorganism to be dried in soil and then stored on agar slants.
2. Using lyophilization to remove the water from the cells or spores and then storing them.
3. Thorough cleanliness is required to avoid any type of contamination.
   This requires the following:
   (a) Clean surfaces while subculturing
   (b) Using pure culture and continuously maintaining them
   (c) Properly sterilized slants and broths

## 4.5 Culture Maintenance Methods

I. **Subculturing**

Any microbial culture goes through different growth phases. These are lag phase, log phase, stationary phase, and cell decline phase. It is important to maintain the culture in the correct growth phase so that when the culture is used by the researchers/industrialists it is viable and produces maximum metabolite of interest. Usually, it is the log phase in which the culture should be preserved. For maintenance, the culture in the log phase should be transferred onto a fresh medium. Periodically, the culture should be checked for mutation, if any, by the gene sequencing technique. It is vital for the culture to maintain the gene sequence for the functionality required in studies or by the industry.

The method of subculturing has many advantages:

(a) It is low cost.
(b) No special equipment are needed.
(c) Small collection centers/research labs/industries can maintain the culture easily.
(d) A single colony can also be retrieved.

In long duration, changes in physiological and genetic characters needs monitoring. The duration of the method is long.

II. **Freezing**

In this method, the microorganisms of culture are rapidly frozen in liquid nitrogen at (−) 196 °C in the presence of stabilizing agents such as glycerol or dimethyl sulfoxide (DMSO) that prevents the cell damage due to the formation of ice crystals and promotes the cell survival.

III. **Storage under Liquid Nitrogen**

A cryoprotective agent is used in this method and therefore it is also called as cryogenic storage method. Commercially important microorganisms are stored under very low temperature, which ranges from (−) 150 °C to (−) 196 °C by using liquid nitrogen. At this low temperature, all metabolic activities of the microorganisms are reduced to a large extent. This method has been found beneficial to preserve: fungi, bacteriophages, viruses, algae, yeasts, animal cells, plant cells, and tissue cultures. In the log growth phase of the culture, it is suspended in 10% glycerol. This suspension is transferred (0.5 mL) into small ampoules made up of borosilicate glass. The cultures are stored in the stationary phase. The ampoules with the cultures are frozen under nitrogen and sealed. Any fermentation process is basically dependent on the characters of the culture strain. Therefore, it is important to isolate a specific microorganism with utmost care. Usually, the microorganisms are isolated from their natural habitats such as soil, lakes, river mud/unusual habitats/environments such as extreme cold, high altitude, deserts, deep sea, and petroleum fields. These are tested for the particular product of interest. If it is found favorable, then it is isolated. In this way, scientists have isolated many microbes from various locations that have the desired functionality. After isolation, the selection of the correct colony is very important. This is done by primary and secondary screening. The method has many advantages:

- Preservation can be done for a long time.
- With the pure culture strain, the fermentation processes have been found to be economical.
- The fermentation processes are much easier as they can be carried out in a laboratory and do not require large land area.
- The culture strain can remain viable for 10–30 years if maintained under proper conditions.
- The transportation is very easy and they can be sent to any part of the globe easily by following the regulations.

However, in case the liquid nitrogen is used, then it needs to be replaced periodically.

IV. **Employment of Dried Cultures in Soil**

The storage of fungi and actinomycetes is usually done by this method. The preserving medium is soil. The soil is sterilized, cooled, and then inoculated with the particular culture. This is incubated for growth and then stored at room temperature or in a refrigerator. Sometimes, in place of soil, silica gel or porcelain beads can also be used. Good viability of cultures (20 years) has been reported by this method.

V. **Mineral Oil Preservation**

This is a very economical and easy method. The oil is placed (1 cm) above the agar slants. Generally, paraffin oil is used. Before placing the oil, it is sterilized in McCartney bottles for 15 min at $103.41 \times 10^3$ N/m$^2$ pressure or autoclaved for 2 h. Then the bottles are dried in an oven at 170 °C for 1–2 h before use. By this method, the viability of cultures can stay intact, between 10 and 20 years, depending on the culture. This method has an added advantage as, at the time of use, a little portion of the culture can be taken out and utilized. To the rest of the tube, some more mineral oil can be added. This is done to keep the culture immersed in the oil. This keeps the shelf life, viability, and functionality of the culture safe. The addition of oil should be done under sterilized conditions, in a laminar hood with HEPA filters. This avoids air-borne contamination.

VI. **Silica Gel Storage**

To store cultures by this method, Pyrex glass test tubes (13 × 100 mm) are taken with cotton wool stoppers. The test tubes are half-filled with silica gel (grade as specified by Perkins 1962) and are dry sterilized at 175 °C for 1.5–2 h. Precooled microbial suspension (0–5 mL) is added dropwise to each chilled tube of sterilized anhydrous silica gel. The tubes are then kept at 0 °C for a period of 10–15 min. Then the tubes are kept for 1 week in a desiccator containing activated silica gel prior to a viability test. Finally, each tube is sealed with Parafilm and stored over self-indicating silica gel in a sealed jar. The reactivation of the dehydrated culture is done by incubating several granules of the silica gel stock in the appropriate liquid medium. The remaining dehydrated material can be resealed and stored again. Therefore, each tube of dehydrated culture can be used many times. The medium used to reactivate each culture is the same as that in which the culture was grown initially. Then the spores/culture tubes can be preserved further by storing them in sterilized air-tight container at 4 °C. The viability of the culture by this method has been found to be from 2–3 years. This method has many advantages as it is simple and retains the stability of the culture. It has been found suitable for fungi and *Neurospora*. However, it needs repeated retrieval and is not favored much for bacteria.

## FAQs

1. What is the best technique for preservation?
2. By which method subculturing can be done only once in a year?
3. What is the temperature of liquid nitrogen?
4. What are the cryoprotective agents?
5. What criteria must be met in order to accept the new medium for use in microbial enumeration tests?
6. What should be done to hydrate the lyophilized pellets?
7. What are the basic principles of preservation?
8. Why is it important to preserve microbial cultures?
9. What are the short-term methods of preservation?

10. What are the long-term methods of preservation?

## References


Alonso S (2016) Novel preservation techniques for microbial cultures. In: Ojha K, Tiwari S, Brijesh K (eds) Novel food fermentation technologies.1. Springer, Cham, pp 7–33

Bajerski F, Nagel M, Overmann J (2021) Microbial occurrence in liquid nitrogen storage tanks: a challenge for cryobanking? Appl Microbiol Biotechnol 105(20):7635–7650

Bircher L, Geirnaert A, Hammes F, Lacroix C, Schwab C (2018) Effect of cryopreservation and lyophilization on viability and growth of strict anaerobic human gut microbes. Microb Biotechnol 11(4):721–733. https://doi.org/10.1111/1751-7915.13265

Carvalho AS, Silva JH, Teixera P, Malcata FX, Gibbs P (2003) Impedimetric method for estimating the residual activity of freeze-dried *Lactobacillus delbrueckii ssp. bulgaricus*. Int Dairy J 13(6):463–468

Carvalho AS, Joana S, Ho P, Paula T, Xavier FM, Paul G (2004) Relevant factors for the preparation of freeze-dried lactic acid bacteria. Int Dairy J 14(10):835–847. ISSN 0958-6946. https://doi.org/10.1016/j.idairyj.2004.02.001

Chu-Ky SV, Laurent L, Siriluck S, Klanarong TM (2013) Acid adaptation for improvement of viability of *Saccharomyces cerevisiae* during freeze-drying. Int J Food Sci Technol 48. https://doi.org/10.1111/ijfs.12114

Gaci N, Chaudhary PP, Tottey W, Alric M, Brugère JF (2017) Functional amplification and preservation of human gut microbiota. Microb Ecol Health Dis 28(1):1308070. https://doi.org/10.1080/16512235.2017.1308070

Hubalek Z (2003) Protectants used in the cryopreservation of microorganisms. Cryobiology 46(3):205–229. ISSN0011-2240. https://doi.org/10.1016/S0011-2240(03)

Lancini G, Demain AL (2013) Bacterial pharmaceutical products. In: Rosenberg E, DeLong EF, Lory S, Stackebrandt E, Thompson F (eds) The Prokaryotes. Springer, Berlin, Heidelberg. https://doi.org/10.1007/978-3-642-31331-8_28

Lee KM, Adams M, Klassen JL (2019) Evaluation of DESS as a storage medium for microbial community analysis. Peer J 5(7):e6414. https://doi.org/10.7717/peerj.6414

López-Rubio AA, Eva HM, Pilar L, Jose MC, Ramon GR (2004) Overview of active polymer based packaging technologies for food applications. Food Rev Int 20:357–387. https://doi.org/10.1081/FRI-200033462

Meryman H (2007) Cryopreservation of living cells: principles and practice. Transfusion 47:935–945. https://doi.org/10.1111/j.1537-2995.2007.01212.x

Mokoena MP (2017) Lactic acid bacteria and their bacteriocins: classification, biosynthesis and applications against uropathogens: a mini-review. Molecules 22(8):1255. https://doi.org/10.3390/molecules22081255

Morgan CA, Herman N, White PA, Vesey G (2006) Preservation of microorganisms by drying-A review. J Microbiol Methods 66(2):183–193. ISSN0167-7012. https://doi.org/10.1016/j.mimet

Morono Y, Terada T, Yamamoto Y, Xiao N, Hirose T, Sugeno M, Ohwada N, Inagaki F (2015) Intact preservation of environmental samples by freezing under an alternating magnetic field. Environ Microbiol Rep 7:243–251. https://doi.org/10.1111/1758-2229.12238

Nguyen HT , Razafindralambo H, Blecker C, N'Yapo C, Thonart P, Delvigne F (2014) Stochastic exposure to sub-lethal high temperature enhances exopolysaccharides (EPS) excretion and improves Bifidobacterium bifidum cell survival to freeze–drying. Biochemical Eng J 88: 85–94

Ning G, Qiang W, Yi X (2020) Optimization of cryopreservation of pathogenic microbial strains. J Biosaf Biosecurity 2:266–270. ISSN 2588-9338

Nogal A, Valdes AM, Menni C (2021) The role of short-chain fatty acids in the interplay between gut microbiota and diet in cardio-metabolic health. Gut Microbes 13(1):1–24. https://doi.org/10.1080/19490976.2021.1897212

Nyanga LK, Noutm MJ, Smid EJ, Boekhout T, Zwietering MH (2012) Yeasts preservation: alternatives for lyophilisation. World J Microbiol Biotechnol 28:3239–3244. https://doi.org/10.1007/s11274-012-1118-y

Obara YA, Yamai SH, Nikkawa TA, Shimoda YU, Miyamoto YA (1981) Preservation and transportation of bacteria by a simple gelatin disk method. J Clin Microbiol 14:61–66. https://doi.org/10.1128/JCM.14.1.61-66.1981

Palmfeldt J, Hahn-Hagerdal B (2000) Influence of culture pH on survival of *Lactobacillus reuteri* subjected to freeze drying. Int J Food Microbiol 55:235–238

Perkins DD (1962) Preservation of Neurospora stock cultures in anhydrous silica gel. Canadian J Microbiol 8:591–594

Prakash O, Nimonkar Y, Desai D (2020) A recent overview of microbes and microbiome preservation. Indian J Microbiol 60(3):297–309

Vekeman B, Heylen K (2015) Preservation of microbial pure cultures and mixed communities. https://doi.org/10.1007/8623_2015_51

Wang Y, Corrieu G, Béal C (2005) Fermentation pH and temperature influence the cryotolerance of *Lactobacillus acidophilus* RD758. J Dairy Sci 88:21–29

Yu C, Reddy AP, Simmons CW, Simmons BA, Singer SW, Vander GJS (2015) Preservation of microbial communities enriched on lignocellulose under thermophilic and high-solid conditions. Biotechnol Biofuels 2(8):206. https://doi.org/10.1186/s13068-015-0392-y

# Organizations Involved in Microbiological Work

# 5

**Abstract**

Microbiological work is very important for the production of various metabolites and to cure many deadly diseases caused by pathogens. In this regard, many organizations have started working on different important aspects of microbiology. The national bodies also have collaborative research work with other organizations involved with microbiological work around the globe. This promotes international exchange of ideas among the microbiologists working in the discovery, research, and microbial industries in the development of newer and innovative products. This chapter deals with most of the microbiological organizations working around the globe in the field of microbiology.

**Keywords**

Organizations · World · Microbiological · Globe · Institute · Research · Science · Education · Scientific · Molecular biology

## 5.1 Introduction

There are many organizations around the world that are involved with microbiological work. As in the recent times, the importance of microorganisms is being known to the common man and it has become important to study and understand the microbiology subject. Many colleges around the globe have started with undergraduate courses. In postgraduation, microbiology is taught in depth. There are many universities that are providing various specialized courses in the field of microbiology. There are many useful and beneficial microbial industrial processes. These have helped the mankind to a large extent. Therefore, microbial processes are being taken up by start-ups, entrepreneurs, and industrialists.

In India, following are the major institutions working in microbiology:



R. Agrawal, *Textbook of Industrial Microbiology*,
https://doi.org/10.1007/978-981-97-9582-6_5

I. **Academy of Scientific and Innovative Research (AcSIR)**

The Academy of Scientific and Innovative Research is an Indian institute with a good standing at the national level. The headquarter is situated at the Council of Scientific and Industrial Research—Human Resource Development Centre Campus in the city of Ghaziabad in Uttar Pradesh state of India (Wikipedia 2011).

The main objectives of ACSIR are as follows:

It provides teaching and research in science at high levels with innovations. They have varied number of courses that are provided many different fields. These are basic sciences, life sciences, mathematical sciences, computer sciences. Studies are provided in the field of medicine, engineering, arts in applied field, humanities, social sciences, and law. It also provides scholarship in advanced science and technology based on merit. Scientific technologies are taught and at the same time, the students are encouraged to establish net working with the industries present in India and even abroad. Collaboration is encouraged with reputed universities and institutions from all around the globe for good scientific work. The organization also helps in the promotion of scientific work and takes up extension services.

II. **All India Institute of Medical Sciences (AIIMS)**

The AIIMS Act (1956) governs these institutes. They have an autonomous status and they report to the Ministry of Health and Family Welfare. Special health care services are provided to the patients (Rajkumari Amrit Kaur 2020).

III. **Agharkar Research Institute (ARI)**

This institute is autonomous in their status and comes under the Department of Science and Technology (DST), Government of India. When it was founded in 1946, it was named as Maharashtra Association for the Cultivation of Science (MACS). The registration of the association has been done under the Societies Registration Act, which was recognized in 1850. However, as the number of activities were very high, a separate identity was given to it. The founder Director was Professor Agharkar. Therefore, it was in his honor in the year 1992 that the institute was named again under his name as Agharkar Research Institute (ARI). The institute works on different themes. The various themes in the study course are biodiversity and paleobiology, bioenergy, bioprospecting, developmental biology, genetics, plant breeding, and nanobioscience (Agharkar Research Institute 1946).

The students are encouraged to write research projects to get grants from various agencies such as DST, CSIR, ICAR, ICMR, DBT, MoEF, and MNES. The ARI encourages funded research work and technology transfers.

IV. **Indian Institute of Science (IISc)**

This is an excellent academic institution for research and development. The technology transfers have enhanced India's wealth creation and betterment of human beings in the society. The main focus is on the fundamental and applied research in science and engineering, which has a high-impact and has innovative knowledge. With the deep knowledge, technical know-hows are created, which can be utilized by the industries and society (Indian Institute of Science 2020). It is situated in Bangalore, Karnataka, India.

V. **Indian Institute of Science Education and Research (IISER)**

The Indian Institute of Science Education and Research is situated in Pune (IISER Pune 2023). It is an autonomous university, which was established in 2006. Excellent work is done in science and technology. The research work is undertaken under many themes and students send their project proposals to many funding agencies for supporting their work ideas. The institute has been getting funds from different agencies such as Centre for Integrative Studies (CIS), DBT Centre of Excellence in Epigenetics, DST Unit on Nanoscience, Max-Planck Partner Group in Quantum Field Theory, Max-Planck Partner Group in Glyco-nanotechnology, National Facility for Gene Function in Health and Disease, Centre of Excellence in Science and Mathematics Education (CoESME), and Centre for Energy Science.

VI. **Institute of Science Nirma University (ISNU)**

The Institute provides excellent facilities for research activities. It encourages collaboration with national and international universities, research institutes, and biopharma companies. The Nirma University, Ahmedabad, was initially founded in the year 2003 and today it has a place of a renowned national university (Wikipedia 2021).

## 5.2 International Organizations

(a) **European Molecular Biology Organization (EMBO)**

The organization works without any profit. Therefore, it is classified as a nonprofit organization. The research work is mainly carried out in the field of life sciences and enables international exchange between the scientists. Around 1800 life scientists work in this institution. It was founded in 1964 (History: fifty years of EMBO). In the beginning, it created scientific interactions between researchers in Europe. The founding chairman was Dr. Max Perutz and the secretary general was Dr. John Kendrew (Anon 2020).

(b) **American Society for Microbiology (ASM)**

The society is a major contributor toward the advancement of microbiology. Organization of various conferences, publications, and certifications are plenty by this society. This provides a lot of opportunity to young scientists to showcase their work. With a lot of awareness on viruses and viral diseases, it has become a special field. Therefore, for virology, the American Society for Virology has been construed, which contributes in exchanging information and discussion among virologists. The American Society for Microbiology (ASM), originally the Society of American Bacteriologists, is a professional organization for scientists who study viruses, bacteria, fungi, algae, and protozoa, as well as other aspects of microbiology. It was founded in 1899. Many scientific journals, books for various courses, educational material on the latest microbiology work, and infectious diseases are published regularly by the society.

Every year the society holds annual meetings, and workshops. It also comes out with the professional opportunities for the development of the members (Timeline of the Society. Reference Documents. Center for the History of Microbiology/ASM Archive 2019). These activities help in promoting and understanding the subject of microbiology and to know the microbes in a better way.

(c) **Mycological Society of America (MSA)**

In order to discuss the various scientific research work being done on fungi was founded in 1932. The society works untiringly for the advancement of science of mycology (the study of fungi). The studies can be in the fields of mushrooms, molds, truffles, yeasts, lichens, plant pathogens, and medically important fungi. It promotes the advancement in mycology and research work in this direction. The annual meeting has a number of talks to understand the research and exchange information in fungi. Fitzpatrick (1944) have written about the Mycological Society of America. It is a scientific society that works toward the advancement of science in fungi of all kinds. These are mushrooms, molds, truffles, yeasts, lichens, plant pathogens, and medically important fungi. The MSA also promotes research work and education in mycology. The mycologists from the United States and Canada are mainly involved in this society. Every year an annual general body meeting is conducted in which the scientists exchange their research information on fungi (Fitzpatrick 1944).

(d) **Society for Industrial Microbiology and Biotechnology (SIMB)**

This international association runs without any profit. Basically, it deals with the advancement of microbiological sciences, industrial products through microbes, microbial processes, and technology transfers. The exchange of scientific information is usually done by conducting meetings and by publications that covers the specialized fields and recent trends in microbiology. Initially, it was founded in 1949 by Walter, Charles Thom, and Charles Porter. The details have been given in The History of SIMB, which was later retrieved (2013-04-27).

(e) **World Health Organization (WHO)**

The United Nations has given the responsibility for taking care of international public health to a specialized agency, which is known as The World Health Organization (WHO). The headquarter of WHO is in Geneva, Switzerland. Around the world, it has about 150 field offices. It was founded in 1948. Among the documents, it has the International Classification of Diseases (ICD). It is the WHO that provides the news for any Disease Outbreak News (DONs) anywhere around the world. They have many programs with the federal agencies. The main mandate is to provide safety worldwide and protect the people from the epidemic. In this regard, it gives assistance to different countries by setting up international health standards. They also have the responsibility to make a survey on the health issues at the global level. This serves as evidence for scientific discussions related to health and in further policymaking. They regularly publish an official publication, the World Health Report, which provides an assessments of the worldwide health topics (The U.S. Government and the World Health Organization 2020). The WHO has played a leading role in many difficult public health works, such as the eradication of smallpox, the near-eradication of polio, and the development of an Ebola vaccine. Recently, they are working on the communicable diseases, such as HIV/AIDS, malaria, and tuberculosis, as well as on noncommunicable diseases such as the heart diseases and cancer. They are also working on the healthy foods, nutrition and food security, health of working persons, and substance abuse. The main job of the WHO is to improve the health of people around the globe, monitor the major public health risks, and make people aware of the risk consequences. In case of any health emergency, they come forward and coordinate in order to improve the health and well-being of human beings. The World Health Assembly (WHA) governs the body. This has 194 elected members and is therefore an elected body. This assembly (WHA) then elects an executive board composed of 34 health specialists. The executive board selects the WHO's chief administrator and the director-general (DG) (currently Tedros Adhanom Ghebreyesus of Ethiopia). The DG is responsible for setting up the goals and looking into the priorities. The DG has the powers to approve the budget and take up activities of concern around the world. The funds to the WHO usually come from member states. However, private donors are always welcome to contribute. The total approved budget for 2020–2021 was more than US$ 7.2 billion, which grew to more than US$ 6.2 billion in 2022–2023. However, the budget for 2024–2025 has been enhanced to US$ 6.83 billion.

(f) **Centers for Disease Control and Prevention (CDC)**

It is a health protection agency in the United States that saves lives and protects the people from health threats. They conduct experiments on the critical health issues in the population and ultimately provides them with the health information. This results in protecting the United States against expensive and dangerous health threats. The CDC is a national public health agency of the United States that works under the United States federal agency in the Department of Health and Human Services. The main office/headquarter is situated in Atlanta, Georgia. The main aim of the agency is to protect the public, improve their health and safety by controlling and preventing the diseases, injuries, and the disabilities in the United States or anywhere in the world. They play a major role in developing newer methods for not only controlling the diseases but also ways to prevent them. The work is being done in the field of the infectious diseases, how foods can be protected from the food-borne pathogens, the ways to improve the environment, safety techniques, and health for the workers in any food industry. They also work on the prevention of injuries. Many educational activities are developed to help in improving the health of the citizens of the United States. The target of the CDC is also on the noninfectious diseases. They conduct a lot of research work on these fields as well. The information is gathered after the experiments, which is dissipated to the public for their use. During the recent times, they have collected very useful information on obesity and diabetes. They have founded the International Association of National Public Health Institutes and Mandy Cohen took office as the Director in 2023 (Centers for Disease Control and Prevention under the United States Department of Health and Human Services 2019).

(g) **The United States Food and Drug Administration (FDA or US FDA)**

The FDA has a Coordinated Outbreak Response and Evaluation (CORE) network in the United States, which was created to manage not just the outbreak response but also surveillance and post-response activities related to incidents involving multiple illnesses linked to FDA-regulated human and animal food and cosmetic products.

It is a federal agency that works under the Department of Health and Human Services. Basically, it helps in protecting and promoting the public health by controlling and supervising the safety of food, tobacco products, caffeine products, dietary supplements, prescription, as well as pharmaceutical drugs that are available over the counter. They also supervise various vaccines, transfusions of blood, medical devices, electromagnetic radiation emitting devices (ERED), cosmetics, animal foods and feeds, and other veterinary products. The major focus of the FDA is on the Federal Food, Drug, and Cosmetic Act (FD&C). Many other laws are also covered, as provided in Section 361 of the Public Health Service Act and associated regulations. The work done under the regulations involve lasers, cellular phones, and condoms. They look into the controlling of diseases in pets. They also construct sperm banks where the sperms can be donated for use in assisted reproduction. A commissioner is the leader of the FDA who gets appointed by the President. For this purpose, the advice of the Senate is taken. Robert Califf was made the Commissioner in 2022. The FDA is led by the Commissioner's report to the Secretary of Health and Human Services. The HQ is situated in White Oak, Maryland under which there are 13 laboratories across the states of the United States. They have employees around the globe, including countries such as China, India, Costa Rica, Chile, Belgium, and the United Kingdom (FDA Centennial 1906–2006).

(h) **NIH/NIAID has Set Up a Division of Microbiology and Infectious Diseases (DMID)**

The National Institutes of Health(NIH) has researchers who actively experiment to prevent infectious diseases. They provide funding for research work against all human infectious agents. Basic research, preclinical development, and clinical trials are supported by providing funds. The National Science Foundation (NSF) was set up in the United States, which is an independent federal agency. It promotes work for the national health and welfare. The NIH has set up an institute, namely National Institute of Allergy and Infectious Diseases (NIAID). The institute addresses the research work done in the fields of various infections, diseases caused due to changes in immunological functions, and due to various allergies. The research encourages studies for their treatment and prevention (NIAID; National Institute of Allergy and Infectious Diseases 2012). In total, there are 27 institutes and different centers that come under the umbrella of the National Institutes of Health (NIH). This is an agency of the United States Department of Health and Human Services (HHS).

The major laboratories of NIAID are situated in Maryland and Hamilton, Montana. NIAID is known to work in coordination with the scientists in different academic institutions and in various industries. They also have a strong networking with the government and nongovernmental organizations. It has multidisciplinary activities and has been addressing all emerging health challenges. In the past the institute has tackled the pandemics caused due to H1N1/09 and the COVID-19.

They are working on the cure for HIV-infected individuals and are constantly engaged in developing newer strategies that can prevent the disease. These include vaccines and therapeutic strategies. Coinfections due to TB and hepatitis C in HIV-infected individuals are also being studied and treatments are sought to prevent them.

Emerging Infectious Diseases (BioD) are being studied in depth to understand the immune system and the underlying mechanisms.

In the Infectious and Immunologic Diseases (IID), scientists mainly study the diseases associated with asthma, allergies, response of the immune system in various autoimmune diseases, and the reasons of transplant rejection (NIAID History. National Institute of Allergy and Infectious Diseases 2012).

(i) **The U.S. Geological Survey (USGS)**

There are many natural calamities that are life threatening. This survey emphasizes and carries the studies to overcome these natural hazards. There are many areas in nature that come under this survey. It mainly involves the livelihoods, the essentials that are required every day by human beings. These are water, energy, minerals, and other natural resources. They also study the ecosystem and environment, as well as the impact of climate and land-use change. Newer methods are developed by the scientists to overcome the disasters on the earth. This was the main cause for forming this agency, the Geological Survey, in the year 1879. The initiation was from the government of the United States. In order to combat the effects of natural disaster, the survey is made in many fields. These are the fields of biology, geography, geology, and hydrology. In the beginning, the survey was made to study the landscape of the United States, the natural resources in different parts of the United States, and to understand the types of natural hazards happening and the reasons behind it. These studies, in turn, have helped in preventing many disasters in later times. From the data obtained from the US space probes, the scientists prepare different maps of the planets and moons. The head quarter is in Reston, Virginia and the main offices are in Colorado and California. The slogan is "Earth Science in the Public Service" (Who We Are. U.S. Geological Survey. Archived from the original on August 4, 2022. Retrieved August 3, 2022).

(j) **American Society for Virology (ASV)**

This agency discusses the research done in the field of viruses. Virology is a subdiscipline of microbiology. It started when the virologist found that the discussions on viruses were very scant in microbiological associations. The discussion on viruses were very poor even in the International Association of Microbiological Societies and in the American Society for Microbiology.

Therefore, the idea came up to form a separate society for viral diseases and the researches done in the field. In 1982, the society had a membership of about 1000 scientists. Wolfgang Joklik, was made the founding president and he served from 1981 to 1983. Annually the general body meetings are held in either in the United States or in Canada. They receive grant from the Alfred Sloan Foundation. There is a website that is maintained by former ASV president, Sondra Schlesinger. From 2021 onward, Dr. Colin Parrish of Cornell University has been serving as the president (Joklik and Grossberg 2006).

(k) **National Science Foundation (NSF)**

This foundation is an independent federal agency. It was created by Congress in 1950 with the goal of promotion and progress of science. It helps in improving the national health and welfare to secure the national defense.

**FAQs**

1. Where can an industrial microorganism be obtained from?
2. Which microorganisms are useful to industries?
3. Which microorganisms can be taken in treating industrial waste?
4. How do research institutions learn to better engage with Indigenous and other marginalized communities in microbiome research?
5. How does pollution affect the human microbiome and how the impact is distributed?
6. How does the climate change and related ecosystem degradation affect the environmental microbiome, what is the impact on health?
7. Why is microbiology course useful?
8. In which way microbiology can be useful for nurses?
9. What is the importance of medical microbiology?
10. What is the role of American Society for Virology?

## References

Agharkar Research Institute (Est. 1946) Maharashtra association for the cultivation of sciences, DST Pune. India

Amrit Kaur R (2020) The princess who built AIIMS. The Indian Express. 27 August

Anon (2020) EMBO facts & figures 2019 (PDF). Retrieved 2021-04-28

Centers for Disease Control and Prevention (2019) United States Department of Health and Human Services. October 4. Archived from the original on April 10, 2020. Retrieved May 15, 2020

FDA Centennial 1906–2006 (2016) Food and drug administration. Archived from the original on May 24. Retrieved September 13, 2008

Ferry G (2014) History: fifty years of EMBO. Nature 511(7508):150–151. https://doi.org/10.1038/511150a. PMID 25013879

Fitzpatrick HM (1944) The first twelve years of the Mycological Society of America. Mycologia 44(1):1–17

Indian Institute of Science. Archived from the original on 26 November 2020

Indian Institute of Science Education and Research, Pune. Retrieved 29 May 2023

Joklik WK, Grossberg SE (2006) How the American Society for Virology was founded. Virology 344(1):250–257
NIAID History. National Institute of Allergy and Infectious Diseases. (2012)
The History of SIMB (2013) Soc Industr Microbiol Biotechnol. Retrieved 2013-04-27
The U.S. Government and the World Health Organization (2019) The Henry J. Kaiser Family Foundation. Archived from the original on 18 March 2020. Retrieved 18 March 2020
Timeline of the Society (2019) Reference Documents. Center for the History of Microbiology/ASM Archives. Retrieved 2019-05-17
Who We Are. U.S. Geological Survey (2022) Archived from the original on August 4, 2022. Retrieved August 3, 2022
Wikipedia (2011) Parliament passes Academy of Scientific and Innovative Research Bill. December
Wikipedia (2021) 7 private universities in Gujarat granted Centre of Excellence by state govt. mint. Retrieved 20 November

# Fermentation Media and Industrial Sterilization

# 6

**Abstract**

Many useful metabolites, antibiotics, chemicals, and pharmaceuticals are being produced by industrial fermentation processes. Production of these compounds by microbes has proved very economical and it comes under GRAS category, which is generally regarded as safe. The term industrial sterilization covers the fermentation of bacteria, fungi, and also yeasts. Many organic acids are being successfully produced by utilizing filamentous fungi: for example, the production of citric acid by *Aspergillus niger* and Penicillin from *Penicillium chrysogenum*. With the recent inventions, the production of recombinant proteins such as insulin (helpful for diabetic patients) is being produced by fermentation utilizing *Escherichia coli* microbial cells. However, for industrial fermentations, the knowledge to use the fermentor, especially the engineering tools, the process control systems, and optimization of the parameters are important steps for the efficiency of the processes. This chapter covers the sterilization charts, designing of sterilizing equipment with deterministic and probabilistic approaches. Different tools in fermentation process development and in their operation have been discussed in detail.

**Keywords**

Fermentation · Media · Sterilization · Aerobic · Anaerobic · Design · Equipment · Heat · Fermentor · Optimization · Deterministic · Probabilistic

## 6.1 Fermentation

In this method the energy is generated by the carbon source using substrate level phosphorylation. The electron is finally accepted by an organic molecule, which arises after the breakdown of carbon containing compounds or by catabolic



R. Agrawal, *Textbook of Industrial Microbiology*,
https://doi.org/10.1007/978-981-97-9582-6_6

reactions (Berg et al. 2005). It can either be aerobic or anaerobic. Respiration is aerobic, but fermentation is anaerobic when it uses sulphate or nitrate as the carbon source (Müller et al. 2012). In industrial microbiology, fermentation is done when the electron is not accepted by any organic compound and the microorganism is grown at a bigger scale. In these reactions the growth could be aerobic as in the production of penicillin and the yeasts. According to a study (Maicas 2020), fermentation includes all alcoholic products. Microorganisms determine the nature of the food by producing different flavor compounds. Common fermented foods are cheese, bread, and yogurt (Jay 2000). In any fermentation process, the nutrients from the various raw materials go into the medium. Initially, the microorganism is taken in a shake flask before taking it into a fermentor. Then, it is transferred into a pilot fermentor. The methods and conditions usually vary from flask to fermentor depending on the organism. Sterility is a very important parameter that needs to be maintained. It has been found that autoclaving causes the fit of the respirator to fail as it is not dry heat (Yuen et al. 2022). The wastes from microbial fermentations are disposed of in an environmentally friendly manner. Quality control of the product material should be done in every batch. This ensures in obtaining a homogenous product in all the batches. Fermentation techniques have been explained by Bacame-Valenzuela et al. (2020) for the production of pyocyanin and Bansod (2021) by fermentation technology.

## 6.2 Thermal Death Time and Media Heat Sterilization

The cell number of microorganisms getting lysed (cfu/mL) at a particular temperature is known as the thermal death time. This is the time to give heat to the fermentor in order to obtain perfect sterilization. To obtain the value, the temperature is kept constant and the time necessary to kill the number of cells is measured. Thermal death time is the heating time required to give commercial sterility.

### 6.2.1 Batch Sterilization Principle

For liquid media sterilization, batchwise sterilization method is a common process (Sharma and Arindam 2018). Direct heating of vessel or passing steam into the medium is used for sterilization in this process. A condensate is used for the dilution of the medium. This may increase the volume up to 25%. The steam can also be introduced into the vessel either with the help of a double jacket or coils in order to sterilize the medium. The use of good quality steam helps in the prevention of any uptake of metal ions (Giulia 2017). The amount of heat required depends on the temperature and time of the process. Kennedy and Krouse (1999) have written an article on the various strategies for the improvement of fermentation medium performance. In order to keep the culture in log phase for a longer time, it is important that the sugar gets caramelized, the proteins are denatured, the vitamins are inactivated, and the reaction of sugars are prevented. However, when the sterilization of each component of the medium is done individually, these steps can be avoided. This method is followed to make polymers of unsaturated aldehydes. During these bioprocesses, the nutrients are sterilized earlier and supplied later as and when required. Heaters are used that are designed with similar dimensions using steam at high temperature but for shorter duration. In any batch sterilization, there are three major steps: heating, holding, and cooling period. Holding time is important to remove all the contamination. This can be reduced if cell destruction is accounted during these periods. Sterilization kinetics, determination of temperature, time for killing, the effects of heating, maintenance, and cooling down in a batch sterilization cycle have been well explained by Giulia (2017). The facts about steam and chemical indicators have also been given. High temperature during heating and cooling has been attributed to the decomposition of the vitamins, proteins, and sugars in nutrient solutions. Therefore, many times the large-scale sterilization is replaced by shorter and more effective continuous sterilization. Heat supplied can be either dry or wet. However, the wet heat usually kills the microbes more efficiently. With increased humidity, moisture, or water activity, the resistance to heat can be decreased. The presence of salts have both positive and negative effects. Carbohydrates and proteins lower the activity of water in the cells, increasing resistance to heat. pH values must be optimized as it can affect the resistance of heat. Biofilms excrete exopolysaccharides, resulting in increasing the heat resistance. During the log phase, the microorganisms are more heat sensitive. Heat resistance of strains and the temperature at which the culture is grown affects the fermentation process. The addition of $SO_2$ has a compounding effect on heat sensitivity. Thermophiles have the maximum heat resistance. Any spore formers usually have high resistance to heat. Among the bacteria, the Gram-positive and cocci can survive at higher temperature. A similar pattern has been observed in yeasts and also in fungi. Rutala (1948) provided guidelines for disinfection and sterilization. The analysis of genome-scale metabolic models have been explained (Aurich et al. 2016).

Fermentation media and metabolite profiling has been studied by Bolling and Fiehn (2005). Kozlovskii et al. (2013) worked on the fungi of the genus *Penicillium* and the fermentation conditions to produce physiologically active compounds. Work has been done (Lange et al. 2001) to improve the sampling for in vivo kinetics of intracellular metabolites in *Saccharomyces cerevisiae*. Larsson and Tornkvist (1996) studied the reasons for the cell inactivation of the low extracellular glucose concentrations during fed-batch cultivation. Leder (1972) brought forward the interrelation between cold shock and osmotic pressure on the permeability of the *Escherichia coli* membrane. Peter et al. (1995) explained the principles of fermentation technology. Schaefer et al. (1999) worked on an automated sampling device for monitoring the intracellular metabolite dynamics. Shitut et al. (2020) used the permanent wall-deficient cells to discover new metabolites. Theobald et al. (1993) used a rapid sampling technique in yeasts. Villas-Bôas and Bruheim (2007) used a cold glycerol-saline method to analyze intracellular metabolites in microbial cells accurately. Marcel (2010) worked on the characteristics and techniques of fermentation systems. The book on modern industrial microbiology and biotechnology contains good information on fermentation techniques used in microbial production of metabolites (Nduka 2007).

### 6.2.2 Fermentation and Sterilization for Products

**Application in Wine** In wine, heat has an adverse effect on the flavor profile and the color. Earlier, it has been shown that yeasts can tolerate more heat. According to the work of Couto et al. (2005), if heating is done above 32 °C, the cells of *Dekkera/Brettanomyces* get inactivated. Boulton et al. (1996) studied the principles in wine making. Heat is mainly used to denature the proteins. The bottling line is most of the times sterilized using heat. The filters and their holders are used to direct hot water/steam into the filler, filler bowl, and filler spouts. It is important that the hot water should enter at 82 °C and exit at 72 °C and the live steam (with 2 cm of invisible vapor) should be flown out of the filler spouts, minimum for 20 min time period. In addition to the bottling line, heat sterilization has been found very successful for cleaning the various parts of the processing unit, preventing contamination by pathogens (Pilone 1953). Christoph et al. (2004) showed the impact of the cold shock phenomenon on the quantification of intracellular metabolites in bacteria. Chung et al. (2023) worked on the fuzzy logic for the performance in food sterilization. Chuppa et al. (2020) evaluated the biodegradation of waste water from sugarcane industries. Costa et al. (2002) studied the effect of nitrogen and carbon sources on growth of the biocontrol agent. De Koning and Van Dam (1992) studied a method to determine changes in the glycolytic metabolites in yeast on a subsecond time scale using extraction at neutral pH. Denggang Wang et al. (2021) worked on a green process with an improved semicontinuous fermentation process of *pichia pastoris*, which was based on the principle of vitality of cell separation.

## 6.3 Advantages of Continuous Sterilization

For the production of many metabolites, the system of continuous sterilization has been found to be advantageous (Junker et al. 2006). The utilization of sanitary heat exchanger design, a cooling system, field I/O modules, and computerized calculations of various parameters helps in better result performance and makes the process very cost-effective.

At higher levels, the method of continuous sterilization utilizes around 60–80% less steam and also less energy for cooling of water. It ultimately helps in the conservation of energy. In batch sterilization, heating and cooling takes longer time due to the constant hold temperature, which brings down the heat transfer coefficients. Scaling-up using batch system has been found to bring down the efficiency. The sterilized medium is put into the empty fermentor that has been sterilized earlier. Some amount of water is put to cover all the probes. The time taken is very less, which reduces the overall turnaround time. However, treatment of medium is quite gentle in continuous sterilization as compared to the batch method of sterilization.

The heat sensitive medium components do not get degraded. To degrade the nutrients, an energy of usually 50–150 kJ/mol is required. This is much less than the energy needed to lyse the pathogenic microorganisms (250–350 kJ/mol). When compounds like amino carbonyl are produced, they give undesirable color and taste. This adversely affects the medium quality as they destroy the growth factors. Proteins and carbohydrates are sterilized separately using different mix tanks to which a sterile water flush is supplied. There is a lot of flexibility as the equipment design allows a wide range of time/temperature combinations. Scaling-up is linear with media flow rates in HTST (high temperature short time) pasteurization process. In this process, the heating is done at 71.5 °C and the process is continuous. This is then passed via a holding tube under the optimized holding time. Heat exchangers are designed in such a fashion that there is no fouling and it should have a cleaning ability so that the maintenance is less. It is vital that the process works under the optimized parameters. In case the media is not sterilized properly, steps should be taken to rectify the parameter immediately before taking up the process. Devasya (2017) studied the batch and fed batch cultivation and harvesting of nano *Chloropsis gaditana* for environmental applications. Fouillaud and Dufossé (2022) studied the microbial secondary metabolism and biotechnology. William (2008) has given guidelines for disinfection and sterilization healthcare facilities.

## 6.4 Design of Sterilization

In a continuous system the time period for the media to be heated and the time required for heating/cooling needs to be optimized. The rate at which the medium should be supplied into the system and the size of the coil should also be optimized before starting the fermentation process. The cooling is usually done by using two heat exchangers that are arranged one after the other by using the incoming media. In case of double spiral type heat exchanger, two sheets of stainless steel plates are used. These are wrapped with an axis in the center. Then both the sides are covered for sealing. From one spiral sheet, the medium is supplied and from the other sheet the steam is passed. The incoming medium, which is not sterile, acts as a coolant. The complete unit should be sterilized with steam before taking up the process (Blakebrough 1968). The method is beneficial as the contamination does not take place due to the two separate coil system. The steam is circulated by these plates, which are arranged in an alternating manner. These have gaskets in between to separate them and the steam can be circulated. The capacity can be increased by adding extra plates.

Unsterile broth is blown with steam, which is later cooled. For any fermentation process, clean steam is required, which should be free from any anticorrosion additives.

The benefits of continuous sterilization system are as follows:

I. The broth goes to high temperature to complete sterilization.
II. Alternate high and low temperatures are provided.
III. Relative humidity does not increase.

## 6.5 Probabilistic and Deterministic Approaches in Designing of the Sterilizing Equipment

Designing of sterilizing equipment with deterministic and probabilistic approaches is as follows:

**Probabilistic design** This design method deals with the effects of random variability during the design phase, which are related to quality and reliability and a small probability of failure is assumed in place of the safety factor. Each variable is a probability distribution that predicts the flow of variability and focuses on the prediction of random variability (Loucks and vanBeek 2017). Most of the quality problems can be predicted and rectified during the early design stages at minimum cost. Identification of the smallest effect of random variability is possible. Due to the pressure inside the pressure vessel, a stress is developed. This needs to be measured and analyzed. To do this, ASME codes are designed using a deterministic approach. The ASME is an American Society of Mechanical Engineers standard, which has the power to regulate the design and construct the boilers and also pressure vessels as per the need and quality specifications. There is a document written in this regard from where the expertise can be taken. This involves the use of safety factors for various input parameters. It is a very critical and an expensive manufacturing process. The probabilistic approach is usually taken to design various buildings and aeronautical. Typically, the pressure vessels are differentiated on the type of head used and the wall of the vessel. The vessels vary in the making material, shape, and other parameters needed to operate the process. This is different from deterministic design in which there is only one single pattern, which is specific for the material to be added. The ASME reference code is specific for each requirement such as the components for bearing the pressure, designing, and fabrication. Even the testing, inspection, and issuing of certificates is covered. This system is used all around the world. By using ASME codes and analytical methods, it is possible to calculate different types of stresses. These include hoop (circumferential) stress, axial (longitudinal) stress, or thermal stress in case of thermal loading.

**Deterministic designs** are based on the facts and lowers any type of risk. Therefore, this approach is commonly used. The method is very simple and economical. On the other hand, the probabilistic approach is based on the variability using statistical techniques (Zhu et al. 2024). Giulia (2017) studied the facts on the steam chemical indicators. Harrison and Maitra (1968) studied the role of ATP in the energy metabolism of growing bacteria. Jha (2021) worked on the methods of sterilization media and air.

## 6.6 Sterilization Chart

There are various methods for sterilization given in Fig. 6.1. Each one is explained as follows:

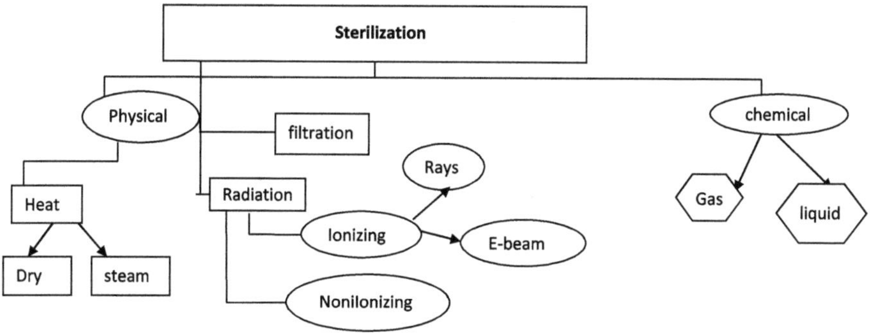


**Fig. 6.1** Different methods of sterilization

**Sterilization with steam.** In this method an autoclave is used in which the material is directly subjected to steam at 121 °C at 20-pound pressure for 20 min. The various important parameters for steam sterilization include the pressure, temperature, and time of the steam (Centers for Disease Control and Prevention, National Center for Emerging and Zoonotic Infectious Diseases (NCEZID), Division of Healthcare Quality Promotion (DHQP) 2016).

**Flash sterilization.** In this method the material is sterilized with the steam, which is penetrated either in open trays or in a closed vessel.

**Low temperature sterilization technologies.** They use gaseous or chemical sterilant to sterilize medical devices and instruments at temperatures less than 60 °C (152 °F).

**Ethylene oxide gas sterilization.** This agent helps in the alkylation and penetrates the material to decontaminate it. However, high doses of the chemical can cause cancer.

**Hydrogen peroxide gas plasma.** The vapors of the gas are exposed to the surface of the material for approximately 1 h. This inactivates the pathogenic microorganisms. The gas plasma is created by using a radio frequency that produces an electric current.

**Peracetic acid sterilization.** This is usually done by inserting a tube with per acetic acid using instruments such as endoscope and arthroscope, which help to decontaminate surface pathogens.

**Microbicidal activity with ethylene oxide.** Ethylene oxide (ETO) can sterilize at low temperatures. It is commonly used by healthcare institutions in the United States to sterilize medical devices that are temperature and moisture sensitive.

**Bioburden of surgical devices.** It is very difficult to get the actual number of cells on a medical device, container, or component. However, it can be obtained by plating out on specific media for total plate count, yeasts, and molds and can be measured as cfu/mL. According to the FDA, 0–150 cfu/mL is allowed beyond which the contamination should be removed. When Gram-negative bacteria are killed, they produce endotoxins, which are usually harmful to the human body. Therefore, it is vital to remove the bioburden before any process/procedure is taken up with the device.

## FAQs

1. Which method is most commonly used for media sterilization in the fermentation industry and why?
2. Which elements are frequently added in the fermentation media?
3. Which instruments work on the principle of batch sterilization?
4. What is the highest feasible temperature for batch sterilization?
5. How does pH affect fermentation?
6. Why is continuous sterilization preferred in industries?
7. Why should the media be sterilized before fermentation process?
8. Why should the fermentor be sterilized?
9. How can the substrates be sterilized and why is it necessary to sterilize the substrate?
10. What material should be used for a fermentor vessel and why?

## References

Aurich MK, Fleming RM, Thiele I (2016) MetaboTools: a comprehensive tool box for analysis of genome-scale metabolic models. Front Physiol 3(7):327. https://doi.org/10.3389/fphys.2016.00327

Bacame-Valenzuela FJ, Pérez-Garcia JA, Figueroa-Magallón ML, Espejel-Ayala F, Ortiz-Frade LA, Reyes-Vidal Y (2020) Optimised production of a redox metabolite (pyocyanin) by *Pseudomonas aeruginosa* NEJ01R using a maize by-product. Microorganisms 8:1559

Bansod TA (2021) Fermentation technology. World J Pharma Res 10(3):260–299. ISSN 2277–7105

Berg IA, Keppen OI, Krasil'nikova EN, Ugol'kova NV, Ivanovskii RN (2005) Carbon metabolism of filamentous anoxygenic phototrophic bacteria of the family *Oscillochloridaceae*. Microbiology 74:258–264. https://doi.org/10.1007/s11021-005-0060-5

Blakebrough N (1968) Preservation of biological materials especially by heat treatment. Chap 12. In: Blakebrough N (ed) Academic Press. NY. Biochem Biol Eng Sci 2:29–63

Bolling C, Fiehn O (2005) Metabolite profiling of *Chlamydomonas reinhardtii* under nutrient deprivation. Plant Physiol 139:1995–2005. https://doi.org/10.1104/pp.105.071589

Boulton RB, Singleton VL, Bisson IF, Kunkee RE (1996) Principles and practices of wine making. Chapman and hall, New York

Centers for Disease Control and Prevention, National Center for Emerging and Zoonotic Infectious Diseases (NCEZID) (2016) Division of Healthcare Quality Promotion (DHQP)

Christoph W, Jens OK, Patrick K, Tina B, Elmar H (2004) Impact of the cold shock phenomenon on quantification of intracellular metabolites in bacteria. Anal Biochem 327(1):135–139. ISSN 0003-2697. https://doi.org/10.1016/j.ab.2004.01.002

Chung CCF, Chao LC, Ho HT, Ching H (2023) Fuzzy logic for performance promotion and function enhancement of food sterilization. Int J Netw Secur 25(2). ISSN 1816-353X

Chuppa TG, Tan M, Adelard L, Shum CSA, François JM, Caro Y, Petit T (2020) Evaluation of filamentous fungi and yeasts for the biodegradation of sugarcane distillery wastewater. Microorganisms 8:1588

Costa E, Teixidó N, Usall J, Atarés E, Viñas I (2002) The effect of nitrogen and carbon sources on growth of the biocontrol agent *Pantoea agglomerans* strain CPA-2. Lett Appl Microbiol 35(2):117–120

Couto JA, Neves F, Campos F, Hogg T (2005) Thermal inactivation of the wine spoilage yeasts *Dekkera/Brettanomyces*. Int J Food Microbiol 104(3):337–344. https://doi.org/10.1016/j.ijfoodmicro.2005.03.014

De Koning W, Van Dam K (1992) A method for the determination of changes of glycolytic metabolites in yeast on a subsecond time scale using extraction at neutral pH. Anal Biochem 204:118–123

Denggang W, Wenjie L, Xinying Z, Shuli L, Ying L (2021) Green process: improved semi-continuous fermentation of *pichia pastoris* based on the principle of vitality cell separation. Front Bioeng Biotechnol 9. https://doi.org/10.3389/fbioe.2021.777774

Devasya RP (2017) Batch and fed batch cultivation and harvesting of nano *Chloropsis gaditana* for environmental applications Electronic thesis and Dissertation repository. p 4670

Fouillaud M, Dufossé L (2022) Microbial secondary metabolism and biotechnology. Microorganisms 10(1):123

Giulia G (2017) The facts about steam chemical indicators. Milestones 3:06–39

Harrison DE, Maitra P (1968) Role of ATP in control of energy metabolism in growing bacteria. J Gen Microbiol 53:S7–S8

Jay J (2000) Fermentation and fermented dairy products. https://doi.org/10.1007/978-1-4615-4427-2_7

Jha N (2021) Methods for sterilization of media and air. https://www.biologydiscussion.com/biotechnology/bioprocess-technology/methods-for-sterilization-of-media-and-air-with-diagram/10102

Junker B, Lester M, Brix T, Wong D, Nuechterlein J (2006) A next generation, pilot-scale continuous sterilization system for fermentation media. Bioprocess Biosyst Eng 28(6):351–378

Kennedy M, Krouse D (1999) Strategies for improving fermentation medium performance: a review. J Industr Microbiol Biotechnol 23:456–475

Kozlovskii AG, Zhelifonova VP, Antipova TV (2013) Fungi of the genus *Penicillium* as producers of physiologically active compounds (review). Appl Biochem Microbiol 49(1):1–10

Lange HCE, Zuijlen M, Visser G, Dam D, Frank JC, Mattos J, Heijnen JS (2001) Improved rapid sampling for in vivo kinetics of intracellular metabolites in *Saccharomyces cerevisiae*. Biotechnol Bioeng 75:406–415. https://doi.org/10.1002/bit.10048

Larsson G, Tornkvist M (1996) Rapid sampling, cell inactivation and evaluation of low extracellular glucose concentrations during fed-batch cultivation. J Biotechnol 49:69–82. https://doi.org/10.1016/0168-1656(96)01534-9

Leder IG (1972) Interrelated effects of cold shock and osmotic pressure on the permeability of the *Escherichia coli* membrane to permease accumulated substrates. J Bacteriol 111(1):211–219. https://doi.org/10.1128/jb.111.1.211-219.1972

Loucks DP, vanBeek E (2017) An introduction to probability, statistics, and uncertainty. In: Water resource systems planning and management. Springer, Cham. https://doi.org/10.1007/978-3-319-44234-1_6

Maicas S (2020) The role of yeasts in fermentation processes. Microorganisms 8(8):1142. https://doi.org/10.3390/microorganisms8081142

Marcel GA (2010) Characteristics and techniques of fermentation systems. In: Comprehensive food fermentation and biotechnology, Chap.7, vol 1. Asia Tech Pub., pp 183–227

Müller M, Mentel M, van Hellemond JJ, Henze K, Woehle C, Gould SB et al (2012) Biochemistry and evolution of anaerobic energy metabolism in eukaryotes. Microbiol Mol Biol Rev 76(2):444–495. https://doi.org/10.1128/MMBR.05024-11

Nduka O (2007) Modern industrial microbiology and biotechnology. Science Publishers, Enfield, New Hampshire

Peter FS, Allan W, Stephen JH (1995) Principles of fermentation technology, 2nd edn. Elsevier Science Ltd., Burlington

Pilone FJ (1953) The role of pasteurization in the stabilization and clarification of wines. Am J Enol Viticult 4:77–83

Rutala, William A. (William Anthony), 1948- and Weber, David J. (David Jay), 1951- "Guideline for disinfection and sterilization in healthcare facilities, 2008"

Schaefer U, Boos W, Takors R, Weuster-Botz D (1999) Automated sampling device for monitoring intracellular metabolite dynamics. Anal Biochem 270:88–96. https://doi.org/10.1006/abio.1999.4048

Sharma SK, Arindam SV (2018) In: Arindam K, Vinay S (eds.) Sterilization techniques used in fermentation processes. Chap 3. https://doi.org/10.1002/9781119460381.ch3

Shitut S, Bergman GÖ, Kros A, Rozen DE, Claessen D (2020) Use of permanent wall-deficient cells as a system for the discovery of new-to-nature metabolites. Microorganisms 8:1897

Theobald U, Mailinger W, Reuss M, Rizzi M (1993) In vivo analysis of glucose induced fast changes in yeast adenine-nucleotide pool applying a rapid sampling technique. Anal Biochem 214:31–37. https://doi.org/10.1006/abio.1993.1452

Villas-Bôas SG, Bruheim P (2007) Cold glycerol-saline: the promising quenching solution for accurate intracellular metabolite analysis of microbial cells. Anal Biochem 370(1):87–97. https://doi.org/10.1016/j.ab.2007.06.028

William AR (2008) Guideline for disinfection and sterilization healthcare facilities. U.S.A: Department of health and human services. https://www.cdc.gov/infectioncontrol/guidelines/disinfection/

Yuen JG, Marshilok AC, Benziger PT, Yan S, Cello J (2022) Dry heat sterilization as a method to recycle N95 respirator masks: the importance of fit. PLoS One 17(1):e0257963. https://doi.org/10.1371/journal.pone.0257963

Zhu Y, Zhao S, Zhang Y, Zhang C, Wu J (2024) A review of statistical-based fault detection and diagnosis with probabilistic models. Symmetry 16(4):455. https://doi.org/10.3390/sym16040455

# Fermentor

# 7

**Abstract**

Fermentors are specific vessels in which a chemical process takes place between microorganism and the nutrients to form a particular metabolite. The reaction may require air or may not need air. Depending on the requirement of the air, it is called as aerobic or anaerobic. Most of the time the vessel is cylindrical in shape so that it can accommodate maximum substrate. The size can vary, but the material used should be rust resistant and therefore stainless steel is used. The vessel and the various probes are arranged in such a manner that it gives a controlled and contamination-free environment. Fermentor and bioreactor are the same. The only difference is that when any mammalian, plant/stem cells are used, it is termed as a bioreactor, whereas when bacteria, yeast, or fungi are used, it is called as a fermentor. During the process, large nutrient molecules are utilized by the microorganisms with the help of many enzymes found in the microbial cellular system. The complex nutrient molecules get converted into simpler forms or into useful metabolites.

**Keywords**

Bioreactor · Fermentor · Aerobic · Anaerobic · Vessel · Bacteria · Fungi · Yeast · Microbial · Sensor · Yield

## 7.1 Introduction

Microbial fermentation processes are conducted using a fermentor, which is known as a biofermentor for mammalian cells. Industrial products such as antibiotics, enzymes, beverages, and other useful microbial products can be produced using a fermentor. It is designed in such a way that the maximum yield of the product is obtained. During the process, the microbial biomass is grown, which is then



R. Agrawal, *Textbook of Industrial Microbiology*,
https://doi.org/10.1007/978-981-97-9582-6_7

processed to obtain the desired metabolite under optimized conditions for maximum yield of the product. The surface inside the vessel is made up of a rust-proof material. Therefore, usually glass or stainless steel material are utilized for making the fermentor vessel. The vessel is mostly cylindrical in shape. This helps to take maximum volume of the substrate. The vessel is double jacketed to maintain the temperature and other parameters during the process. The air is introduced with the help of a sparger, which is placed at the bottom of the vessel.

There are many stages in any fermentation process. These include the preservation of the inoculum, building up of inoculum, and pre-fermentor culture, followed by production fermentation (Bansod 2021). Giulia (2017) looked into the facts about steam chemical indicators. Jay (2000) discussed about the fermentation processes in different fermented dairy products. Kennedy and Krouse (1999) wrote an excellent review article on the strategies for improving the fermentation medium performance.

## 7.2 Components of a Fermentor

It is made up of a large cylindrical tank in which air is introduced at the base through the perforated pipes. In order to increase the rate of mixing, most of the times, mechanical impellers are used. This also breaks and disperses the air bubbles. Compressed air is needed for this purpose.

To construct a fermentor, the basic components required are described as follows:

(a) **Top plate** The cover is known as the top plate, which is mostly made up of stainless steel. An inoculation pipe is fixed there, which helps to port the inoculum inside the fermentor. The impeller shaft is driven by the drive motor.
(b) **Impeller shaft** This is a shaft that holds the agitator in a central position. It has the ability to rotate the ball and roller bearings. It is usually connected with the help of a coupling so that it can be removed easily. In the selection of an impeller shaft, usually it is seen that the shaft should be able to move with high speed. If the viscosity is low, then it allows fast stirring. This is helpful in many different applications. Usually, a blade, which is made up of stainless steel, is very effective.
(c) **Impeller** This is an agitating device that is used for mixing the nutrients and the microorganisms uniformly. It helps in the mass transfer between the microorganism and the nutrients in a very homogeneous way during the fermentation process. Then, a centrifugal force is developed, which moves the fluid out. This is away from the center of the impeller. As the culture viscosity is less in the impeller area, it increases the rate of oxygen getting transferred into the liquid phase. However, it has been noticed that the culture present near the walls of the bioreactor/fermentor receives less aeration. Therefore, the growth of the culture is usually low near the walls.
(d) **Stirrer** This helps in mixing the gas bubbles with the liquid culture medium. In a fermentor, in which a stirrer is used, the transfer rate of the oxygen depends

on the speed of stirring. When a vortex is used, the impeller rotates with a very fast speed. This allows the air bubbles to enter into the liquid, which later gets dispersed with the help of the impeller.

(e) **Baffle** This helps in breaking down the gas bubbles and, in turn, improves the aeration efficiency and prevents the counter flow. It is used inside the fermentor. This enhances the aeration; however, it causes a lot of turbulence. To increase aeration in laboratory experiments, baffles can be used in conical flasks by using different types of metal coils.

(f) **Sparger** This supplies oxygen into the culture medium through the perforated tubes. Due to the application of sparger, the usage of vortex is not required. Sparger provides better aeration to the culture. Sparger is placed immediately below the impeller. This helps to break down the air into smaller bubbles as soon as the air is received.

(g) **Drain point** It withdraws either the cells or the medium to allow continuous fermentation.

(h) **Cooling jacket** This jacket is fitted externally to the fermentation vessel. In this jacket, steam and cold water are run continuously, which balance the heat generation during the process.

(i) **Controlling elements** Monitoring of the parameters such as temperature, pH, acid, bases, oxygen supply, and pressure are done with the help of controlling elements. These are needed during the formation of products.

## 7.3 Properties of Different Fermentor Parts

It is important that the fermentor should be designed in such a way that the production of metabolite is obtained under sterile conditions and also it should avoid any cross-contamination. The fermentor must allow homogeneous mixing of the substrate and the media under optimized conditions of aeration, which is vital for the proper growth of the culture.

### Properties of Various Parts in a Fermentor

**Pt-100** This monitors the temperature in the culture vessel.
**Foam probe** The probe senses the foam formation.
**pH electrode** This electrode monitors the pH in the culture vessel.
**Oxygen sensor** It maintains the dissolved oxygen content level.
**Heating pad** This provides the heat to the medium.
**Cold finger** There is a pipe through which cold water is passed inside the vessel. This helps in cooling the contents.
**Rotameter** This provides a variable airflow into the culture vessel.
**Pressure valve** It helps in maintaining the pressure in the fermentation system.
**Air pump** The pump supplies air throughout the medium.
**Peristaltic pump** It helps in pumping acid, base, and antifoam into the medium.

It is vital that the fermentor should be reliable for a long-time operation.

The crucial factors to be observed during a fermentation process are as follows:

- It should have the ability for aseptic operation, providing sterile conditions.
- There should be enough aeration and agitation for uniform mixing of the contents in the vessel.
- Power consumption must be low.
- Controller for probes must be provided in the fermentor, which can regulate parameters such as temperature, pH, and oxygen level.
- There should be an inlet for the inoculum and another for the medium to get into the vessel.
- It should be designed in such a way that there is no excessive evaporation loss.
- There should be minimum cost of labor input for the operation, harvesting, cleaning, and maintenance.

## 7.4 Phases of Growth of the Microbial Culture and their Role in Fermentation

Any microbial cell grows in a predictable pattern. This results in a growth curve that is composed of five distinct phases of growth: the lag phase, the time required before going into exponential/log phase, the exponential or log phase, followed by the stationary phase and then the death or decline phase. It is the exponential/log phase when the cells divide constantly. In the stationary phase, the microorganism stops replicating as the conditions become unfavorable for growth. Environment, plays a major role in the growth of microorganisms. Every microorganism needs nutrients and energy to grow. The nutrients include carbohydrates, fats, proteins, metals, and vitamins. These are usually supplied into the medium in which the microorganism is grown. During the lag phase the microorganism gets adjusted to the conditions and then in the log phase the culture starts growing in an exponential manner by dividing continuously via a process called as binary fission (Mason et al. 2013). Characterization of lag phase of any microbial growth curve is vital as it has importance in food sciences, environmental sciences, bioremediation, and in understanding the basic cellular processes.

When all the microbial cells are similar in physiology and are at the same stage of cell division, the growth is known as synchronous growth of a microbial population.

Looking into the supply of carbohydrates into the medium, if in case the media is supplied with two sugars, the culture may grow in two different growth phases simultaneously. This is known as diauxic growth, diauxic or diphasic growth. Diauxic growth means a double growth in which the organism is able to metabolize one of the sugars easily at a faster rate and the other sugar at a slower rate. This results in bringing the culture in two growth phases. However, when the microbial growth takes place at a constant rate in a standard set of environments (which means the growth is taking place under the steady state conditions), then it is called as a continuous culture of microorganisms (Ghasem 2007).

## 7.5 Bacterial Growth Curve/Kinetic Curve

When an organism is inoculated into the growth medium, it starts growing after the lag phase is over. The rate of growth of the organism increases steadily after the lag phase and this period is known as the log or the exponential phase. After this period, the growth slows down as the concentration of nutrients get depleted and the formation of toxic substances start, which slowly get accumulated. This phase is known as the decline phase. After this, the growth starts ceasing and the culture enters into a stationary phase/steady state. When the nutrients get totally depleted, the cells may become senescent and start lysing/dying. However, the total amount of biomass does not decrease, but the number of viable cells come down.

The time taken by any microorganism to divide and multiply is known as the generation time, which could be from few seconds to days. The growth rate can be calculated by dividing a factor 0.301 with the generation time. Generation time varies for each microorganism and helps to understand the evolution of microorganisms. It is very difficult to calculate the doubling time in nature. The doubling time takes place in an exponential/geometric manner. With each division, one cell gives rise to two cells, which then produces four and so on. All the details should be looked critically before taking up the process in the fermentor.

## 7.6 Fermentation Medium

The nutritional requirements for the growth of any microorganism are provided by the fermentation medium (Frost and Moss 1987). This contains carbon, nitrogen, and phosphorus as nutrient sources along with minerals, amino acids, vitamins, water, salts, other micronutrients, and some growth factors (Volesky and Luong 1985).

Before selecting the various nutritional sources for a fermentation medium, it is vital to keep a track of many factors. It is very important to check the nutritional requirements for the desired metabolite to be produced. The media should have enough nutritional components during the fermentation process, which will help in the growth of the particular microorganism being used. The nutritional requirements are different for primary and secondary metabolites. The physical and chemical parameters are specific for each product, and therefore it is important to optimize them for the production of a particular product/process. This felicitates the microorganism to grow at a faster rate. While doing the experiments at the laboratory level, pure grade chemicals are utilized. However, at the large-scale production the process needs to be cost-effective and therefore many times cheaper chemicals are used. For this purpose, initially, the optimization needs to be done at the pilot plant level, using cheaper constituents. All these steps are vital and necessary to optimize the conditions of the fermentation media before any industry starts a microbial process unit.

The optimization of physical and chemical parameters play a very important role in a microbial process. It involves the development of a fermentation medium that can produce a quality product with maximum yield. The optimization using central composite rotatory design (CCRD) has been found very useful as three to five parameters can be optimized at a time. Strain improvement by mutagenesis and optimum conditions for culture parameters by response surface methodology for lactose tolerance in a novel native culture isolate *Leuconostoc mesenteroides* sub sp. have been worked out by Shobharani et al. (2006).

The major nutrition sources are as follows:

**Carbon Source**

These mainly include sugars and carbohydrates. However, for the production of vinegar, alcohols are also used. Commonly used pure carbon sources are glucose, sucrose, and glycerol, whereas for industrial fermentation, inexpensive sources used are whey, malt extract, molasses, corn steep liquor, or sugarcane juice. Costa et al. (2002) studied the effects of nitrogen and carbon sources on the growth of the biocontrol agent *Pantoea agglomerans*.

**Nitrogen Source**

These are usually in the form of organic or inorganic compounds. Inorganic sources used are ammonium salts or the free form of ammonia and the cheaper sources are tryptone, peptone, soy meal, corn steep liquor, and yeast extract.

**Growth Factors**

There are some growth factors, such as yeast extract, that can provide vitamins and macronutrients. Trace elements such as copper, zinc, iron, cobalt, molybdenum, and manganese are present in the unrefined nitrogen sources and can be added to the medium.

**Miscellaneous**

The gas produced during fermentation can lower the process, as it forms a layer of foam. Therefore, an antifoaming agent can be added to the fermentation medium. For the stabilization of media pH, mineral buffering salts are used. These include phosphates and carbonates. For metals, addition of chelating agents to the media have been found good.

In case of wine production, grape must be used as the medium. In the production of bioethanol, a cheap carbon source is added to the medium. Sugars or other carbohydrates are used as carbon sources. For any large-scale fermentation to be successful at the commercial scale, it is important to use inexpensive sources of carbohydrates such as molasses, corn steep liquor, sugarcane juice, or sugar beet juice. In case the fermentation system is very specific, pure carbohydrates are used. These could be either glucose, sucrose, glycerol, or any other pure sugar. For the production of any proteins and nucleic acids, fixed nitrogen sources are required in the form of bulk protein. These include soy meal, peptone, tryptone, salts of ammonia or nitrate. For the production of phospholipids in cellular membranes and for the production of nucleic acids, phosphorus is required, which is added to the nutrient medium. To supply micronutrients and vitamins, addition of yeast extract has been found to be effective, which is required for the growth of the microorganism. Inorganic nutrients, iron, zinc, copper, manganese, molybdenum, and cobalt are supplied by the unrefined carbon and nitrogen sources. The fermentation processes in which gas is produced in large amounts, addition of an antifoam agent into the system has been found to be effective. Mineral buffering salts, such as carbonates and phosphates, can be used for the stabilization of pH to optimum conditions. An optimal fermentation medium is vital for any successful process.

## 7.7 Production of Biomass

For processes such as single cell protein, Baker's yeast, *Lactobacillus*, and *E. coli*, production of large biomass is required. For maximum biomass production, it is important to optimize the physical parameters (temperature and light) and the chemical constituents in the culture media.

## 7.8 Various Types of Fermentation Processes Used in Industrial Biotechnology

Monciardini (2014) discussed the production of many metabolites by fermentation. The fermentation systems, their characteristics, and techniques are explained by Marcel (2010).

**Batch Fermentation**

Any product formation by this method has a known quantity of nutrients in the culture media. The medium is usually taken in conical flasks, which is kept on an orbital shaker at a speed of 80–120 rpm. There are two types of fermentation. In any batch fermentation the media is initially sterilized. This is then inoculated with desired microbial culture, which is usually 2–5% of the total volume. Usually, the primary metabolites are formed during the logarithmic growth phase, which tend to reduce in the stationary growth phase as had been observed during the ethanol formation by *Saccharomyces cerevisiae*. However, the secondary metabolites are produced in the stationary growth phase as shown for penicillin production by *Penicillium chrysogenum*.

When the nutrients in the medium get exhausted, the microbial growth stops. Sometimes, the toxic compounds start getting formed at this stage. In such processes, the substrate and the nutrients are all added at the time of microbial inoculation. The mixture is then incubated for optimized time to produce maximum end product. The United States produces around 84% of alcohol by using batch fermentation. This is mainly because the procedure is easy to handle and is cost-effective.

However, cleaning, sanitizing, and filling the fermentor is often quite challenging. The batch fermentation process can reduce the time of the process, energy, and results in high product yields. This does not happen in a continuous fermentation process. Many parameters such as the source of culture, size of the substrate particles, amount of buffer to be taken, and the gas venting, all play a major role in the productivity of the metabolite. In comparison to batch and continuous fermentation, a fed-batch fermentation process has been found to be more effective as it helps the microenvironment of microbes. During the fed-batch fermentation systems, small amounts of ingredients can be supplemented at specific optimized times. For liquid media sterilization, batch sterilization has been found to be effective. Steam is circulated into the coils/double jacket of the vessel. It can also be injected into the medium. Another method is to heat the vessel electrically. In batch culture it is easier to obtain the different phases of microbial growth, signifying the lag phase, log phase, stationary phase, and decline phase. As it is a closed system, it reduces the chances of contamination. However, in a continuous method, the product gets the heat treatment in a continuous flow. In batch autoclaves, for many sterilization systems to happen simultaneously, stacks of trays are used. Batch fermentations are being utilized in wastewater treatment in pharmaceutical industry and in small-scale production of metabolites such as beverage products. For all these products the substrates can be added initially and later the product can be harvested. This method has been found cost-effective for the formation of biomass (Baker's yeasts) and primary metabolites as lactic acid, citric acid, acetic acid, ethanol, and alcoholic beverages (wine and beer). Distilled alcoholic products such as brandy, whisky, and rum are also formed using batch fermentation process. Many sweeteners such as aspartate and amino acids are produced using batch fermentation processes.

(a) **Substrate Surface Fermentation (SSF)**

Substrates used in this surface fermentation are solid. The specific microorganism for the process is inoculated on this substrate medium. The culture utilizes the nutrients present in the medium for the growth. The solid substrates commonly used are rice bran, wheat bran, or any grain. These substrates do not need any processing. For optimization, major parameter is water activity requirement. As these substrates do not require much water, the problems of contamination do not occur. SSF needs very less power. In olden times, this technique had been used very commonly to check the growth of the soil for composting and silage production. Mushrooms have been produced by using wood as a substrate. Products such as miso, tempeh, and soy sauce have been produced using SSF. The method is commonly used for the degradation of toxic chemicals, degradation of agro-industrial wastes, biotransformation of agricultural wastes into high-valued products, biopulping, and many other secondary metabolites that include antibiotics, alkaloids, plant growth hormones, enzymes, organic acids, biopesticides, biosurfactants, biofuel, and aroma compounds to name a few.

SSF is a very cost-effective method that makes it very lucrative to be used by the industrialists. The procedure is being used by many countries around the world for the production of metabolites such as ontjom (Indonesia), shaohsing wine and kaoliang liquor (China), and sake (Japan).

In SSF the microorganisms are grown on moist solid particles. The water droplets are in between the particles and are therefore discontinuous. These get filled with the gas phase. However, in this process of fermentation, usually, the mixing of nutrients is not adequate and therefore the diffusion of nutrients gets limited. There is accumulation of heat and lack of perfect control systems. This makes the method good only for low-valued products as they do not need much monitoring and control. For fungi, the hyphae get exposed directly to air and therefore get desiccated. Even with these limitations, SSF has many benefits as compared to the submerged fermentation. Due to the low cost of this technology, researchers have been utilizing it for the formation of many products. These include enzymes such as amylases, proteases, lipases, pectinases, tannases, cellulases, and rennet; pigments, aromas, and flavor compounds; organic acids, growth hormones, antibiotics, bioherbicides, and many others.

By this method the products can be obtained in high volumes, the energy requirements are less, aeration method is easy and oxygen can be supplied as per the requirement for the microbial biomass, harvesting of biomass is very simple, and controlling the temperature is also very easy as probe is not required. Nutrients are supplied in such a way to produce maximum yield of the product. There could be gradient difference of nutrient particles into the substrate, causing shear damage to the microorganism. SSF has been found more effective than SLF (substrate liquid fermentation) as they are able to combat adverse conditions. The use of filamentous fungi by SSF has been found very beneficial as they can penetrate into the solid substrates, making it conducive for growth.

SSF/SLF fermentors are of six types, which are given as follows:

- Tray
- Packed bed
- Rotary drum
- Swing
- Stirred vessel
- Air solid fluidized bed

Tray type is simple as the substrate is spread in a tray horizontally, which is kept in a temperature-regulated room.

(b) **Submerged Fermentation by Substrate Liquid Fermentation (SLF)**

Submerged fermentor uses liquid substrate and saves space. The microorganism is inoculated into a liquid nutrient media for the growth. The fermentation vessel can be designed to a volume required for each batch. The vessel is sterilized before putting the media and after adding it. This avoids contamination. As the control probes can be added to the reactor, the optimization for maximum product yield becomes easier. In case of extracellular product, the separation is easier and microbial cells can be removed by centrifugation. However, in case of intracellular products, the cells have to be broken (details are given in Chap. 13 of this book).

The most commonly used fermentor is the Aerated Stirred Tank Batch Fermentor.

### 7.8.1 Advantages and Disadvantages of Batch Fermentation

**Advantages of batch fermentation**

- Many products can be produced.
- Contamination is low as nutrients are added only initially.
- Mutation of cells is low as process time is less.
- The process is cost-efficient.
- Yields of the product are high.

**Disadvantages of batch fermentation**

- The microorganism gets exposed to the environment conditions.
- As it is a batch fermentation process, each time after completion of the reaction there is a need to restart. This is time-consuming as cleaning and refilling has to be done.
- Around the end of the reaction, many toxic compounds get formed, which may reduce the cell growth.
- Fermentation vessel needs sterilization in every batch.
- Labor intensive process.
- Cost of process operation increases due to subculturing again and again.

### 7.8.2 Fed-Batch Fermentation

In this process of fermentation, the nutrients are added to the fermentation medium at a period, which is fixed after optimization of all parameters. The nutrient inflow into the vessel can be adjusted and can be added at a controlled rate. This is an excellent method to produce compounds by organisms that grow very slowly and are dependent on a specific nutrient composition of carbon and nitrogen. It has advantages as the optimization of parameters are easy and the product yield is high. However, it is not possible to maintain a specific growth rate as that makes the process expensive.

Fed-batch fermentation processes are of three types, which is given as follows:

- Surface fermentation
- Submerged fermentation
- Solid substrate/submerged state fermentation

  In these methods, different types of substrate feeding are done. The nutrient or the precursor can be either added at the start or at any other time of the process or during a specific growth phase of the culture as and when it is required as per the optimization conditions. The addition can be done by supplementing it into the medium. In this type of fermentation, it is very important to monitor the end product and remove it from time to time to prevent any type of feedback inhibition. Monitoring should be done to check the product so that at no time the substrate concentration becomes high, which may cause catabolic repression. Understanding of the product metabolic pathway is very vital before using these fermentors. During the scaling up, optimization of many parameters is required. These are the source of the nutrient, metabolite compartmentalization, rate of agitation, amount of aeration, the capacity of the fermentor, and the amount of broth to be added. Understanding of the component chemistry and the selection of right sterilization method are important. This helps in the prevention of any decomposition of the components. When all the conditions are monitored in a precise manner, it improves the product quality. The feeding can be made successfully using various probes. For slow metabolism, glycerol is used as a carbon source. The biomass production depends on the media composition.

It has been found that the choice of carbon/nitrogen sources, inorganic salts, and the trace elements influence the accumulation of the metabolites. Therefore, to obtain maximum metabolite yields, the correct selection of the sources and their optimization are very important. Using statistical techniques is very beneficial as many parameters can be optimized at a time and are very useful to study their interactions. The fractional factorial designs known as Plackett–Burman, Box–Behnken, and Robust Taguchi can be applied by using Minitab or Sigma Plot programs. These methods have been found very useful for the optimization of parameters. The fed-batch fermentation can be monitored manually/automatically. This method has been found good for the production of bioactive metabolites such as enzymes, antibiotics, vitamins, alkaloids, and phenols from bacteria, fungi, and algal cells. Lipids, oils, biodiesel, biogas, and bioethanol are the common products produced by fed-batch fermentation. Work done by Devasya (2017) showed the economical production of biofuel both by batch and fed-batch fermentation. Biodiesel production from a microalgae (blue-green algae, *Nannochloropsis gaditana*) has been found successful. Production of docosahexaenoic acid (DHA), a polyunsaturated fatty acid that is used in food and pharmaceuticals, can be produced using a marine alga *Crypthecodinium cohnii* by fed-batch fermentation method.

Using fed-batch fermentation, the heterotrophic marine algal cells have been able to produce docosahexaenoic acid, which is widely used by the pharmaceutical industry. For maximum yield of the product, the culture viscosity and stirring rates are controlled using probes.

### 7.8.3 Continuous Fermentation

In this method the growth nutrients are sterilized previously. They are added continuously to the fermentation vessel, and an equal amount of converted end product with microorganisms is harvested simultaneously during the process. This avoids feedback inhibition. The microbial growth depends on the selection of the substrate, which helps the organism to grow exponentially. The fresh medium is added in a continuous manner. The chemostat and turbidostat are good examples for this method of fermentation. In a chemostat, the nutrient medium is fed into the culture vessel equivalent to the amount of the media removed. Both are done simultaneously. In a turbidostat, a photocell is fitted, which regulates the flow of nutrient medium and maintains a constant cell density or turbidity. The nutrients are present in excess amounts and the cell density is monitored by a photocell device. The optical sensing device (the photo cell) measures the absorbance of the culture density (turbidity) in the growth vessel. Change in turbidity either reduces or increases the passage of light through the culture that controls the flow of nutrients in the turbidostat system and the wastes are removed. For measuring the turbidity, the growth of the cells is kept at a constant level. The chemical environment is also kept the same. Parameters such as nutrient depletion, accumulation of toxic substances, and excess cells in the vessel initiate the stationary phase. The method has many advantages, which is listed as follows:

I. Maintenance of microbial growth for a long time.
II. It is simple to study metabolic process of the organisms.
III. Steady state of microbial growth can be obtained easily.
IV. Product yields are high.

There are many disadvantages as well:

- Problem at any step can close the entire process.
- Chances of contamination are high.
- Viscose fungi cannot be used in these processes.
- The properties of the culture strain can change with time.
- There can be cell aggregation that does not allow the steady state of the microbial growth.

### 7.8.4 Semicontinuous Fermentation

This is a process that comes in between the batch and continuous fermentation. After preparation of the growth media, it is inoculated with a specific microorganism inoculum with a specific cell count. In this method the fresh medium is added at regular intervals. However, no media is removed from the used fermentor vessel. In a semicontinuous method, the cultures can be kept in the logarithmic phase for a longer time. The vessel in which the reaction takes place is called as a semicontinuous reactor or a semibatch reactor (SBR). There are many products that cannot be produced by continuous fermentation, as the operating parameters need to be controlled. The method is beneficial for the products that need continuous bubbling of a gaseous reactant in the medium being used. The method is found to be good as it does not have any side reaction and the reaction is highly exothermic. The method gives higher yield of the product/s and is cost-effective as the reaction rate is high and gets completed in shorter time. Also, the purification cost is much less as compared to other fermentation techniques. Vaporization is utilized to remove the reactants from the reactor. This is called as reactive distillation. The most common ones are acetylation and esterification reactions. The method has been widely used in industrial bioprocessing. The process produces less wastes as compared to other processes and makes it environmentally more efficient. According to Link and Borchert (2015), many of the fed-batch culture products are now being produced by semicontinuous fermentation method. Flocculation is used for enhancing the sedimentation of the cells. The metal ions to be used for flocculation differs from cell to cell. For example, for *P. pasttoris* cells, $Fe^{3+}$ and $Ca^{2+}$ have been found good for sedimentation and to separate the cells into two fractions being active and inactive (Denggang et al. 2021). Recently, much work has been done using semicontinuous method for onion fermentation. Scientists have found this method very effective for the production of vinegar from onions (Lee et al. 2017).

The finished metabolite produced by this method have been found to have higher shelf-life, which also have better taste and flavor. The nutritional value of foods and beverages has also been found to be better. It also reduces the toxicity and improves the digestibility in products.

### 7.8.5 Production of Extracellular Metabolites

While growing, the microorganisms produce different primary and secondary metabolites into the medium. As these metabolites are in the medium and not inside the cells, extraction of these compounds is a challenging task. Fouillaud and Dufossé (2022) discussed the microbial secondary metabolism and the formation of metabolites. Many analytical tools have been developed for their extraction, which involve different types of sample preparation. The metabolites with low molecular weight can be studied by using extracellular metabolomics, which mainly involves the culture media. This is termed as exometabolome. It includes the metabolites outside the microbial cell and components of the microbial culture media present in the supernatant. It is also called as metabolic footprinting. The most common methods to separate them are centrifugation and filtration. Using this method, the scientists have been successful in the study of heterologous protein production to study the biomarkers in the fermentation process, to understand the formation of aroma in wine, and the metabolic interactions among different yeast strains. Care should be taken to avoid any cross-contamination between intra- and extracellular metabolites. The secretion and yields of these metabolites from the medium are influenced by the physical factors of the environment and chemical constituents of the media. Therefore, on optimization of the medium parameters, the product yields can be maximized. This method has the ability to find out the formation of any xenobiotic compounds, which can be harmful to the humans and plants. In such cases, metabolic engineering techniques can be helpful. Footprinting along with Omics have been found to be helpful in the identification of unknown genes and pathways.

The best example is the production of biofuel from plant lignocellulose, which is a waste in agriculture. The production of ethanol from *Fusarium* sp. by the fermentation of cellulose (Panagiotou et al. 2005) is another good example. The technique is also useful in functional genomics as it can identify the different strains of the same microorganism. Many studies have shown that the phenotypic characters among the different strains of a microorganism can identify the production of a particular metabolite.

This method helps in the identification and analysis of the extracellular quorum sensing compounds being produced. This has been worked out for *Pseudomonas aeruginosa*. The organism causes a chronic lung infection. This happens due to the production of a biofilm from the microorganism. Due to this the antibiotics are not able to go into the lungs and act against the pathogenic microorganism as it develops resistance to antibiotics. With the technique of microbial, scientists have been able to develop medicines that can prevent the formation of QS (quorum sensing) compounds. Shobharani and Renu Agrawal (2010) worked on the interception of quorum sensing signal molecule by furanone compounds to enhance the shelf life of fermented milk.

With recent improvements in analytical tools, it has been possible to know the intracellular metabolites. The study is innovative as it helps in the understanding of the metabolism taking place inside the microorganism. However, the study is not very precise and needs further work in this line. With recent advances, dynamic metabolic footprinting has been very informative. This can analyze the production and secretion of a particular metabolite into the medium at a specific time of the culture growth phase. This method also allows to understand the specific physical as well as the chemical parameters both at the same time and at a particular moment in the medium. It produces data at different time points. Therefore, it is very informative and helps in phenotyping. It also analyzes the contaminants in the media at different time intervals. This helps to remove/avoid such contamination by interventions, which can ultimately improve the product quality and save the complete batch fermentation. In this regard, a study by Sue et al. (2011) has been done for the detection of any contamination in microalgae fermentation broth. If the nutrient media contains high amount of salts, sugars, proteins, and lipids, then it might interfere in the formation/quantity of extracellular metabolites. It can also hamper the instrument of analysis. Therefore, it is suggested to separate the metabolite using liquid–liquid technique or by using a solid matrix phase as it can hold the metabolite in it. The solvent used in the method can be evaporated and then the product sample can be concentrated. Liquid–liquid separation method is not very popular as it is very time-consuming and most of the times the separation is incomplete, which results in low recovery of the product. Among solid-phase extraction methods, solid-phase extraction (SPE) and solid-phase microextraction (SPME) are widely used, which give better recovery.

Separation methods have been dealt separately in this book (see Chap. 12). These days a kit known as "Metabo Tools" can be obtained from the market, which can give useful results by integrating extracellular metabolomic data obtained from microorganisms, plants, and animals (Aurich et al. 2016). These tools have the ability to predict the pathways taking place inside the microbial cells, which is based on the extracellular metabolites. However, a total result of the microbial metabolism can be obtained only if both intra- and extracellular metabolomic datasets are combined together. Primary metabolites are produced during the growth phase of the organism.

Citric acid is produced by the strains of *Aspergillus niger* as part of the citric acid cycle, glutamate by *Micrococcus* species, and the species of *Corynebacterium* produce lysine, threonine, tryptophan, and other amino acids. When the products are produced during the stationary phase, they are called as secondary metabolites. These include products such as penicillin, cyclosporin A, gibberellin, and lovastatin.

### 7.8.6 Production of Intracellular Compounds

When the product is produced inside the cells (intracellular), the cells need to be ruptured (lysed) after the fermentation process is over. For the maximum yield of the product, all the various parameters should be optimized. After the product is formed, the other cellular proteins are separated from the lysate. Enzymes such as catalase, amylase, protease, pectinase, cellulase, hemicellulase, lipase, lactase, and streptokinase are commonly produced by the microbial cells that are intracellular. The common intracellular products include insulin, hepatitis B vaccine, interferon, granulocyte colony stimulating factor, and streptokinase that are produced using recombinant proteins.

A recent technique known as metabolomics is very helpful in the detection, identification, and quantification of the small organic molecules produced by living cells (Zhang et al. 2013). It helps to analyze the intracellular and extracellular microbial metabolites. The intracellular metabolites are present inside a mechanical barrier, the cell membrane. This covers the complete cell. After the identification of any metabolite, it is important to quantify the product. Also, it is vital to know the economic feasibility of any microbial product if the production has to be taken on a large scale. Initially, the product is extracted from the cells by using different solvents. These could be either organic, inorganic nonaqueous, or a mixture of both. The solvents make the cell wall permeable or porous. This helps in the efficient recovery of the product.

For extraction, the complete breakdown of the cell wall should not be done as that usually extracts all the different molecules of various sizes, which may also carry the undesirable ones. The efficacy of the method/solvent can be judged by comparing the different methods. The purity of the metabolite can be evaluated by spiking the product sample with the standard one.

For any intracellular metabolite, initially the microorganism is grown in an optimized nutrient media after which the microbial cells can be separated. After extraction of intracellular metabolites, an analysis is made for the intracellular and extracellular metabolites using appropriate analytical tools. Many factors are involved for the maximum production of the metabolite and for the maximum conversion of the substrate to the product. The substrate concentration and the cofactor requirement need to be optimized for the maximum conversion to the desired product. Sometimes, negative feedback can take place. This can happen if, in case, any metabolite in the system acts as an inhibitor to the enzymes, which are required for the formation of the desired metabolite product. Many a times, activators are required that determine the rate of conversion. Most of the products need specific enzymes and cofactors at a particular optimized level to produce maximum product. The cell metabolism involves interactions between the enzymes and the compounds. The turnover of the metabolites in primary metabolism is rapid, whereas in the secondary metabolism due to the lower growth rate, the breakdown of cellular component is slow. For intracellular products the cell wall has to be removed. For this purpose, the study of the cell wall constituents is very important. The cell wall that protects the cellular components is generally made up of lipoproteins as well as polysaccharides. However, the constituents of the cell wall are different in yeasts, bacterial, and fungal cells. Yeasts contain low lipid and high proteins in the cell wall as compared to the bacteria. In case of filamentous fungi, the cell wall consists of chitin and glucans. In micro algae, the cell wall mainly consists of resistant biopolymers. Sometimes, they also have silicon deposits in their cell walls. As the cell wall constituents vary in different microorganisms, the techniques to remove the cell wall for the extraction of any intracellular metabolite also varies. In case of protozoa, the cell wall consists of cellulose or a combination of proteins and polysaccharides. The study of protozoan cell wall is important as they reside in the human cells as parasites-causing diseases.

When the extraction of the metabolite is done using a solvent, the metabolite gets diluted due to the high volume of the extracellular medium. This makes the detection and quantification very difficult.

To measure the intracellular metabolites accurately, quenching solutions are used. However, many a times they have been known to interfere. It has been found that when perchloric acid solution is used, it disrupts the cell envelope and releases the intracellular metabolites into the extracellular medium in such a way that it has both intra- and extracellular compounds. This makes the further studies very difficult.

Saez and Lagunas (1976) extracted the living cells from the cultured media by filtration and then immersing the biomass in liquid nitrogen. As the temperature of the liquid nitrogen is usually less, the metabolism of cells goes down. However, the extraction procedure gets more time, which in turn improves the efficiency of the metabolite extraction. This method has not been found good for the extraction of ATP, NADH, pyruvate, and glutamate, which have fast turnover rates.

A method (De Koning et al. 1992) in which the sample is taken into 60% (v/v) methanol solution, at (−) 40 °C has been found effective. In this method, the cell biomass is separated from the culture medium by centrifugation (3000 rpm, 20 min). An integrated approach is used to develop a rapid sampling strategy for the quantitative analysis (in vivo). The kinetic behavior is based on the measured concentrations of intracellular metabolites in *Saccharomyces cerevisiae*. Emphasis is laid on small sample sizes during sampling and analysis (Lange et al. 2001). In another study, the researchers have found that the biomass gets frozen in liquid nitrogen and forms ice crystals. Also, the cell membrane gets damaged. This allows the intracellular metabolites to go into the medium. Another method has been worked out (Christoph et al. 2004) for the quenching of bacterial cells in which the filtration is done under vacuum. The biomass is then washed with saline. However, the method is time-consuming and is not used commonly.

Quenching methods for different microorganisms are as follows:

**Bacterial Cells**

The best method for quenching of bacterial cells is by using perchloric acid solution, cold-methanol solution, fast filtration combined with saline solution, glycerol–saline solution, and liquid nitrogen. However, the organic solvents can damage the cell wall matrix. Due to this the intracellular metabolites get leaked outside the cell.

In the cold shock treatment, there are very fast temperature changes, which make the intracellular metabolites to get leaked out. The work of Leder (1972) found a method to reduce the cold shock by a simultaneous hyperosmotic transition. This dries the periphery of the cell and reacts with the membrane lipids. It helps in keeping up the cell membrane intact and avoids leakage. Bacame-Valenzuela et al. (2020) optimized the production of a redox metabolite (pyocyanin) by *Pseudomonas aeruginosa* NEJ01R using a maize by-product. Nawaz et al. (2021) written an overview on the industrial and medical applications of bio-pigments synthesized by marine bacteria.

**Yeast Cells**

Villas-Boas and Bruheim (2007) studied the efficiencies of cold glycerol–saline solution, cold methanol–water solution, glycerol–water solution, glycerol–saline solution, and glycerol–mannitol solutions for the quenching of yeast cells. Cold glycerol–saline solution has been found to be the best for the recovery of intracellular metabolites.

**Filamentous Fungi**

Because of the diverse structure of the filamentous fungi, the quenching becomes very challenging. They are very viscous and due to the filamentous nature, they are heterogeneous. The use of liquid nitrogen has been found to stop the cell metabolism. However, the separation of intracellular and extracellular metabolites from the growing liquid broth becomes difficult. Methanol (60%, v/v) buffered with 200 mM triethanolamine at low temperature (−45 °C) has been used to quench the cells of *Aspergillus niger* (Ruijter and Visser 1996) and to prevent the leakage of intracellular metabolites into the medium. The biomass can be separated using vacuum filtration. Chuppa-Tostain et al. (2020) evaluated the filamentous fungi and yeasts for the biodegradation of sugarcane distillery wastewater. Kozlovskii et al. (2013) worked on *Penicillium* sp. and have shown them as producers of physiologically active compounds.

**Protozoa**

The cell structure of protozoa is very diverse. A protozoan, namely *Leishmania donovani*, which was in promastigote life phase, could be separated successfully from the culture broth by keeping the complete culture broth into an ice-bath (dry ice and ethanol) with a temperature of zero degree centigrade (0 °C) (De Souza et al. 2006). The cell biomass could be separated from the culture medium by using a solution of cold phosphate-buffered saline (0 °C). It was also observed that the leakage of intracellular metabolites into the culture medium was negligible and this method could be adopted to separate the cell biomass with maximum yield of the desired intracellular product.

**Microalgae**

The cell envelope of microalgae is very diverse. A study has been done using the cells of *Chlamydomonas reinhardtii* (Bolling and Fiehn 2005). The cell suspension of *C. reinhardtii* was sprayed with a solution of 32.5% methanol in water along with tris–acetate phosphate (TAP) macro salts. This method of quenching the microalgae cells has been found to be very effective and there is no leakage of intracellular metabolite into the extracellular media.

**Sampling from Fermentors/Bioreactors**

Figure 7.1 shows a typical bioreactor with different probe controls. Ezemba and Ezemba (2022) discussed different fermentation methods and reviewed well the different types of fermentors, design, and the uses of fermentors. They have also discussed on the importance of optimization of different parameters while taking up any fermentation process. There are three methods to take the sample of microbial culture from any fermentor or bioreactor. The techniques can be manual, semiautomatic, or completely automatic. While using manual technique, syringes are used to

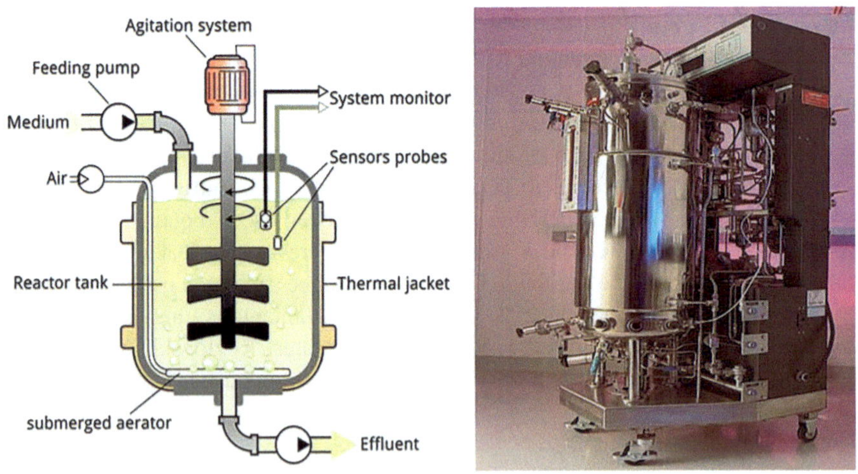


**Fig. 7.1** A typical bioreactor with different probe controls. (Source: Wikipedia)

take out the microbial culture sample. However, this method is very time-consuming. In order to avoid contamination and to save time, the microbial samples are removed by semi- or fully automatic sampling devices. Harrison and Maitra (1968) worked on many semiautomatic devices. Theobald et al. (1993) designed a special valve that can harvest the microbial cells aseptically at a very fast speed (needs only 0.5 s per sample). Larsson and Törnkvist (1996) invented an electronically controlled valve with an electronic timer. According to this method, the system is set up in such a way that the valve opens only for a specific time to release the biological sample, which is connected to the quenching solution through a tube. *Escherichia coli* and *Saccharomyces cerevisiae* cultures have been tried by this method. However, as human intervention is required to insert and collect the testing tubes, it makes the method very time-consuming. Later, a fully automated device was developed by Schaefer et al. (1999). According to this method, a high pressure is developed inside the vessel that opens the magnetic valve. This allows a continuous sampling of the culture broth (at 3.3 mL/s) into the sample tubes that contain cold quenching solution. However, each of the method has advantages and disadvantages. Therefore, it is important to recheck and optimize the method for the desired product.

**Stopped-Flow Systems**

The stimulus-response or continuous perturbation are good methods to study cell metabolism (Mashego et al. 2006). The concentration of metabolites is measured at varying time intervals by disturbing the cell metabolic state. This helps to understand the reaction rates and pathways. The only drawback comes when multiple stimuli have to be studied. A definite time is required for the cell culture to get mixed in the perturbing solution. The time depends on some factors such as the speed of the stirrer being used, the amount of culture being used, and the viscosity of the liquid. As the reactors are very big is size, it also needs perturbing agent in large amount. This increases the cost and therefore the method is not used on a regular basis. This has brought forward the sampling systems known as stopped flow. According to this the perturbations can be made in a secondary flask, or outlet, which is outside the bioreactor. A known volume of the cell culture is pushed into the secondary flask along with the stimulus solution. As the secondary flask is small, the time taken for mixing gets reduced many folds and the whole pulse-response process can be conducted outside the bioreactor. This helps in maintaining the steady state and allows the sampling and quenching of microbial cells at a high frequency and in a reproducible manner.

De Koning et al. (1992) were successful in analyzing the changes in glycolytic metabolites by a stopped-flow system. However, it did not allow the analysis of a continuous growth culture. Buziol et al. (2002) developed a new bioreactor coupled with rapid stopped-flow sampling technique to measure the metabolite dynamics on a subsecond time scale. By using this bioreactor, individual sampling has been made possible and the time taken for sample harvesting is also very less (100 ms). However, it does not allow the exchange of oxygen during the sample processing, which usually causes a lot of oxygen limitation.

Visser et al. (2002) invented a system for continuous-pulse experiments by using abioscope. The method utilizes an oxygen-permeable silicon tubing, which helps in an oxygen exchange. All the different methods have been well reviewed by Pinu et al. (2017).

### 7.8.7 Agricultural Feed

Fungi help to break down agro-industrial waste products, which can be used as an animal feed. The industrial fermentations are mostly batch fermentations that start by using a pure starter culture. The preculture, which should be in the logarithmic phase, is then inoculated into the fermentor with about 5–10% of the working volume of the production fermentor.

### 7.8.8 Different Types of Fermentors

- **Airlift fermentor/tower reactor** It is used for the gas–liquid or gas–liquid–solid contact devices. It divides the fluid volume into two zones to improve the circulation, oxygen transfer, and equalize forces in the reactor. Only one zone is sparged with gas. It provides a controlled liquid flow in a recycling system using pumps. The sterilization is easy, which has low energy requirements and is cost-effective.
- **Bubble column fermentor** It consists of a cylindrical vessel equipped with a gas sparger that pushes gas bubbles into a liquid phase or a liquid–solid suspension. To the base of the column, either air or gas is introduced with the help of perforated pipes or plates, or metal micro porous sparger. The rheological properties of the fluid and the gas flow rate have a significant influence on the mixing of $O_2$ and other performance factors. To improve the mass transfer, there are horizontal perforated plates, vertical baffles, and corrugated sheet packings, which are placed inside the vessel. These reactors have low maintenance and are cost-effective. These are usually used for biological wastewater treatment, chemical, petrochemical, and biochemical industries.
- **Fluidized-bed fermentor** It is made up of packed beds with smaller particles, which prevents clogging, high liquid pressure drops, channeling, and bed compaction associated with packed bed reactors. Catalyst is laid at the bottom of the reactor and the reactants are pumped into the reactor through a distributor pump to make the bed fluidized. The cells are immobilized small particles, which move with the fluid. Hence, they increase the mass transfer, oxygen transfer, and nutrition to the cells.
- **Packed or fixed-bed fermentor** This has a bed of solid particles having biocatalyst on or within the matrix of solids. It can be run in the submerged mode or in the trickle flow mode. The air is introduced through a sieve that supports the substrate. The conversion rate is high, and the operation is easy. It has high contact between the reactant and catalyst and can endure high temperatures and

pressures with cost efficiency. Therefore, this method is efficient and is cost-effective.

## 7.9 Fermentor Designs for Major Microbial Products in Large-Scale Industries

(a) **Wine**

Wine is a product formed by the alcoholic fermentation of fresh grape juice. At present, France is the biggest producer of wine, which produces more than 50 million hectoliters. It is also the biggest exporter of wine in the world. The climatic conditions of the country make it very conducive to obtain good grape crops.

Wine fermentation from grapes is very common around the globe (Table 7.1). There are two types of wine: white and red wine. White wine does not contain the peel and the seeds. As soon as the juice (must) is obtained from the grapes it is treated with $SO_2$. This helps in inactivating the enzymes in order to stop the growth of the microorganisms. Enzyme inactivation can also be done by taking the juice at high temperature for a short time. However, each reaction defines the quality of the grapes. Boulton et al. (1996) discussed on the principles of wine making.

### 7.9.1 Optimization of Parameters for Wine Production

(a) **Brix, Temperature, and pH**

In case the clarification of the juice is not good, the taste of the wine becomes grassy. The Brix plays an important role in the quality of wine. In white wine the Brix is fixed to 22.5 °B. The ratio of acid and sugar is necessary to be checked and regulated continuously for a better product. Sugar concentration can be regulated by osmosis and cryoextraction techniques. Acidity needs to be optimized as per the product. Studies have shown that for white wines a temperature between 18 and 24 °C has been found good. In order to obtain fruity flavor in wine (which consists of esters and acetates of high alcohol) fermentation should be done at a lower temperature.

**Table 7.1** Major wine-producing countries around the world

| Wine-producing country (2023) | Approximate amount (thousand liters) |
|---|---|
| France | 46 |
| Italy | 44 |
| Spain | 31 |
| The United States | 25 |

Red wines are produced from red/black grapes. In this case, the "must" is obtained along with skin and seeds. The time of maceration is very important and needs optimization. In case maceration time is increased, it has been found that the concentration of phenolic compounds increases, which improves the quality of wine and shelf storage. Red wines are prepared at 24–27 °C. At lower temperatures the color of these wines is very light. It is the anthocyanin pigment along with some cofactors that stimulate the fermentation. The quantity of anthocyanin in the grapes and the particulate matter in the "must" are critical factors in the formation of red wine. In a heading down system, the "must" is put into a tank in such a way that the grape skin goes below the surface of the juice. However, due to the build-up of $CO_2$, the complete extraction gets prevented. For this purpose, a chimney is incorporated to remove the gas. With recent developments, newer and modified methods are being adopted in which the liquid is automatically recycled through the cap in "Autopiegeage" tanks, which is an improvement over the manual method.

Carbonic maceration is another improved method in which the flavor of red wines is much better in quality. In this method the grapes are subjected to $CO_2$ environment at high temperatures (20–35 °C) for a period of 8–10 days.

This is followed by maceration using enzymes. In this process the produced carbon dioxide gas remains in the tank itself due to the pressure resulting in extracting the desired color and flavor. Maceration is followed by pressing. Due to the presence of sugars in the press run juice, there is formation of acetic acid. Therefore, it is advisable to ferment both the juice fractions separately. These can be mixed as per the need of the consumer. It is a common practice to use free run juice for the production of the light wine known as Nouveau style wine and wine from press run juice as Beaujolais wine, which is made without the skin and fermented at 15–20 °C. It has high concentration of volatile compounds such as benzaldehyde, ethyl salicylate, vinylbenzene, and ethyl-9-decanoate. These wines cannot be stored.

There is a process called as "Thermovinification" in which wine is produced automatically. In this process the grapes are heated at 60–80 °C for a period of 20–30 min with maximum anthocyanin extraction. Many studies have proved that extraction yield depends on the temperature used, time of treatment, and the contact time between the grape skin and the juice. During this process the extraction is maximum from the peels/skins of the grapes. Therefore, these wines usually have an unpleasant green stemmy notes (off flavor) and are also bitter. Therefore, this method is not adopted for the formation of fine wines.

In another method the grapes are heated to 85 °C for a period of 5 min only and are then taken under vacuum (60 mbar) for flash expansion. This results in converting water inside the cells into vapor phase, which produces a differential pressure across the cell wall. This pressure ultimately ruptures the cell wall. The extraction of anthocyanin and total phenol yields are very high using this method.

The quality of wine depends on the size, shape, design, and construction material of the fermentor. Basically, the vessel should not be made of a toxic material and should not be porous. There are two types of vessels named as vats and tanks.

**Vats** These are open-topped vessels. These are common for red wine production as it needs an access to the seeds and skins of the grapes during fermentation. Fermentors (200 hL or more) of high capacity are cost-effective and are used on a commercial basis. The fermentor can be either cylindrical, V-shaped, or square in shape.

**Tanks** The vessel tops are sealed in tanks. Recently, an automatic fermentation tank for red wine has been patented in the United States. This is long-lasting, impervious, and is corrosion proof. The heat transfer is very fast in tanks with aromatic neutrality. The tanks can be sized or shaped as per the requirement by the user. The tanks are usually made up of stainless steel, with a jacket in which a coolant is circulated. This coolant helps in controlling the temperature. The tank is periodically pumped with the fermenting juice using an external cooling coil or the cooling coils are inserted into the fermenting juice. To make the vessel cost-effective, fiber glass is used in place of stainless steel. However, it does not have the heat conducting properties. Cement is another cost-effective material. However, it has low heat conductivity and mobility. Moreover, the cement vessel needs an inner coating with wax or tiles. With this lining, the calcium uptake gets stopped during fermentation. Wooden vessels have limited life, and the worst thing is that they cannot be sterilized. The vessel is not inert and allows air to get inside. This causes oxidation. When wines are prepared in barrels, the concentration of higher alcohols is more but the quantities of fatty acid esters are very low. This affects the overall fruity flavor of the wine. When wine production is done in oak barrels, the phenolic compounds from the oak wood gets extracted out, which gets into the product and hampers the taste of wine.

Fermentors have also been made using highest food grade specialized plastics, which is made up of multilayered polyurethane. Fermentabag™ has been designed for red wine fermentation. Carbonic maceration is preferred and the vessel can be filled between 90% and 97% of the full capacity. From these bags, samples can be drawn during the fermentation process. The wines prepared by this method can be stored for a period of 4–6 weeks after fermentation.

Recently, immobilized microbial cells are being used for wine fermentation. However, gas diffusion is a problem in these gases. When $CO_2$ gas gets accumulated and the cells are in the void volume, there is an increase in the back pressure. Sometimes, the gel beads get ruptured during the process. Large quantities of $SO_2$ (80–100 mg/L) is passed in order to avoid any contamination.

For continuous wine production, a horizontal multistage bioreactor is used, which has replaceable immobilized plates (bioplates). These bioplates contain immobilized (using calcium alginate as the carrier) live microbial cells that are taken on a membrane and, in turn, the membrane is attached to a glass plate. The continuous fermentation has been found to have higher amino acid content as compared to the wine produced by batch fermentation. As the cells get entrapped in the membrane, the (carbon dioxide) gas released during fermentation gets diffused from the plates. This prevents the disruption of the cells. Continuous wine making with higher yields has been demonstrated successfully using a glass tower bioreactor. The use of enzyme β-glucosidase during wine making has shown to enhance the flavor in wine. The stability of the enzymes can be studied by using an ultrafiltration membrane in the reactor. The enzyme can be separated and reused.

For the fermentation of wine, a culture of *Saccharomyces cerevisiae* is used in the pure form with an inoculation rate of 3–5% (v/v). This should have a cell count of $1 \times 10^6$ cells/mL when taken in the grape juice. When the alcohol content increases, the uptake of ammonium ions and many of the amino acids goes down. It has been found that when the ethanol content becomes 13–15%, the process of fermentation stops. Hydrogen sulfide is produced when the grape juice has low content of free amino nitrogen. In such cases, if diammonium phosphate (200 ppm) is added in the beginning and later, 100 ppm during the course of fermentation, then $H_2S$ formation gets reduced to a large extent.

(b) **Polyhydroxyalkanoates**

Plastics have become an essential material to be used in various applications. The use in many disposable items has been causing much pollution and climate change. The degradation of plastics takes a long time, which is more than 100 years. This has caused a great concern toward the health of human beings and animals. Degradation in landfills is possible, but even that takes a long time. Incineration is another method, but it is very expensive and produces many toxic chemicals. Looking into the problems of these petroleum-based polymers, scientists have come up with polymers that are of biological origin. These are the derivatives of starch such as polylactic acid, cellulosic polymers, and polyhydroxyalkanoates (PHA). These can also be produced from renewable resources (Nath et al. 2008). The usage of these material depends on the development of effective technologies, which are also economical. Nath et al. (2008) worked extensively on the production of PHAs using microbes. It is the cost of carbon source that makes the process expensive. Therefore, the use of waste materials as a carbon source is being tried. Out of many wastes tried, the use of solid agro-industrial residues by using the method of solid state fermentation (SSF) has been found to be effective.

PHA are the polymers of hydroxyalkanoic acids. These get accumulated inside the cells of prokaryotic microorganisms (eubacteria and archaea). They are in the form of small granules. According to the studies (Steinbüchel 2001; Choi and Lee 1999), it has been discovered that these are basically the reserve material of carbon and energy. It happens when the cells have high amounts of carbon, but the concentration of either nitrogen or phosphorus are limiting (Koller et al. 2008; Anderson and Dawes 1990). At that time, the granules of PHA get accumulated inside the cells. Initially, it was Lemoigne (1926) who identified the formation of PHA as poly(3-hydroxybutyrate) [P(3HB)] from a bacterium, *Bacillus megaterium*.

Many years later, PHA was discovered (Wallen and Rohwedder 1974) from the sewage sludge, which consisted of monomer units of 3-hydroxyvalerate (3 HV) and 3-hydroxyhexanoate (3HHx). With the continuous research studies, presently there are around 90 microbial genera that have the ability to produce and accumulate such polyester compounds in their cells (Zinn et al. 2001).

When these molecules have 3-hydroxy acids, they are optically active. They are stereospecific with R configuration (Sudesh et al. 2000). According to Kim and Lenz (2001), the molecular mass of PHAs varies between 100 and 1000 kDa.

The molecular mass of these compounds depends on the microbial species, the cultural conditions, the pH, and the type of carbon source used (Anderson et al. 1992). The classification is done based on the length of the monomer. A chain of four to five monomers is called as a short chain of PHA, whereas a chain of more than six monomers is called as a midchain of PHA. There is a hybrid polymer that contains both short chain and medium chain monomer units. Studies have shown (Fukui and Doi 1997) the formation of poly(3-hydroxybutyrate-*co*-3-hydroxyhexanoate), which are formed in the cells of *Aeromonas caviae*. According to Crank et al. (2004), properties of the polymer gets vastly affected with the hydrophobicity, melting point, glass transition temperature, degree of crystallinity, and the length of the monomers. Although P(3HB) has good thermoplastic properties, it has poor oxygen permeability and is brittle (Sudesh et al. 2000). The added advantage is that it is biocompatible as it is present in the blood of eukaryotes (Wiggam et al. 1997; Reusch 2000).

It is being used for wounds (sutures, skin substitutes), in various medical applications (heart valves and vascular grafts), orthopedic treatment (scaffolds for cartilage, screws, bone grafting), in controlled drug delivery, and in various stents (Chen et al. 2005; Zinn et al. 2001). The bottles from this material can be easily degraded when they are incubated for a period of 8 weeks in aerobic sewage sludge (Jendrossek 2001). On commercial scale (1950s), it was the North American company (Grace W.R. Co.) that ventured into the production of PHA. Imperial Chemical Industries (ICI, UK; 1970) started the production with trade name Biopol®. The process was patented and purchased by various companies such as Zeneca, Monsanto, and Metabolix Inc. (USA). Metabolix utilizes corn sugar for the production of 50,000 tons of PHA/year with a trade name Mirel™ (Kram 2008). Biomer (Germany) produced PHA for medical products by Chemistry Innovation (Castilho et al. 2009). Japan-based Mitsubishi Gas Chemicals produced P(3HB) from methanol. Many other plants have come after this company was established. The global market is increasing every day. In 2023 it was valued for US$ 7.9 billion and is projected to be around US$ 20.9 billion by 2028. It is proposed to grow almost 21.3% in the forecast time duration. As these are biodegradable polymers and are bio-based, they come under GRAS category. Ultimately, these bioplastics get decomposed into carbon dioxide, water, and humus.

Presently, cost is a matter of concern. A lot of research is underway to reduce the production costs by using cost-effective substrates and to optimize the fermentation strategies. Even molecular techniques such as recombinant DNA technology and metabolic engineering are being tried for better yields. The use of whey reduces the production cost to a large extent. However, the use of agro-industrial wastes needs to be worked out in a detailed manner.

In a study (Yilmaz and Beyatli 2005), the scientists used molasses from sugarcane and beet root as carbon source. Beet molasses was used at a concentration of 1% (w/v). *Bacillus cereus* was used as an inoculant. This produced 0.1 g/L PHA. However, when molasses was used at 3%, the production of P(3HB) increased by 70%; however, the yields of polymer decreased to almost half. Using a recombinant strain of *E. coli* in fed-batch fermentation technique has shown higher conversion rates (Table 7.2).

For the production of PHAs, the most commonly used carbon source is usually starch.

An effective and low-cost method for the liquefaction and saccharification of starch has been worked out (Kim 2000). The investigation has been done on the cultures of *Azotobacter chroococcum* for the production of P(3HB) directly from soluble starch. Using a stirred bioreactor under fed-batch cultivation with oxygen-limiting condition has shown to enhance the production of P(3HB) (Table 7.3).

The work done with *C. necator* produced good yields of P(3HB) (51.1 g/L) polymer, which has economic significance.

Cellulose hydrolysates and xylose are other carbon sources that have been worked out by the scientists for economical production of PHA (Silva et al. 2004). When bagasse hydrolysate was used as the carbon source, the PHA yields increased to a large extent. The yields were higher with *Burkholderia sacchari* (62%) and *Burkholderia cepacia* (53%) (Table 7.4).

Whey-based carbon sources have also been tried for the production of PHA on large scales (Ahn et al. 2000). These have been found to be effective substrates for giving good yields. They are a waste/by-product that is left after the cheese is prepared. Whey usually remains back in large quantities in cheese manufacturing industries (80–90%, v/v) (Table 7.5).

A comparison of PHA production by *Pseudomonas hydrogenovora* from pure sugars (glucose and galactose) and from hydrolyzed whey permeate has shown that this inexpensive surplus product that is also advantageous (Kollar et al. 2008).

### (c) Citric Acid Production

Citric acid is produced in the tricarboxylic acid (TCA) cycle. This has many uses. It is widely used in foods and pharmaceuticals. It has high antioxidant properties. It had a market of US$ 3.2 billion in 2023 and is expected to increase by 5.1% (Market report world.com, 2020). It is commonly produced by bacteria, fungi, and yeasts. *Bacillus licheniformis*, *Arthrobacter paraffinens*, *Corynebacterium* sp.,

**Table 7.2** PHA / PHA with copolymer production using molasses and sucrose with different microorganisms

| Strain | PHA (g/kg) | Reference |
|---|---|---|
| *Bhurkolderia* sp. | 80-100 | Nonato et al. (2001) |
| Recombinant *E. coli* | 80 | Liu et al. (1998) |
| *Azotobacter* | 66 | Page et al. (1992) |

**Table 7.3** PHA production using starch as carbon source

| Strain | PHA (g/kg) | Reference |
|---|---|---|
| *C. necator* | 51.1 | Koutinas et al. (2007) |
| *Azotobacter* | 46 | Kim (2000) |

**Table 7.4** PHA production using cellulosic material as carbon source

| Strain | PHA produced (g/kg) | Reference |
|---|---|---|
| *Burkholderia sacchari* | 34.8 | Silva et al. (2004) |
| *Burkholderia cepacia* | 32.5 | Silva et al. (2004) |

*Bacillus subtilis*, *Brevibacterium flavum* are the common bacteria which produce citric acid. Among fungi, *Aspergillus niger*, *A. aculeatus*, *A. awamori*, *A. carbonarius*, *A. wentii*, *A. foetidus*, and *Penicillium janthinellum* can produce citric acid. Many yeasts genera such as *Saccahromicopsis lipolytica*, *Candida tropicalis*, *C. Oleophila*, *C. Guilliermondii*, *C. Parapsilosis*, *C. Citroformans*, *Hansenula anamola*, *Yarrowia lipolytica*, *Torulopsis*, *Hansenula*, *Debaromyces*, *Torula*, *Pichia*, *Kloekera*, and *Zygosaccharomyces* can produce citric acid.

It can be produced by surface and submerged fermentation. The microorganism, raw materials, and substrate used are very important. According to Mattey et al. (1999), a pretreatment is needed to remove any trace metals present. Zinc and iron can be precipitated by using potassium ferrocyanide and enzymes. Many agri-wastes can be used to produce the product (Soccol et al. 2006). Surface fermentation involves labor but consumes less energy. The substrate is solid and it is moistened (70%), optimum pH between 4.5 and 6.0, and temperature between 28 and 30 °C. The process can be completed in 4–5 days (Drysdale and McKay 1995). The product can be removed from the fermentation media by solvent extraction and purification methods. Immobilized cells of the microorganisms have shown to use low sugar. A high nitrogen and low phosphate medium (molasses) is good.

(d) **Amino Acids Production**

Initially, L-glutamic acid was produced from bacterium *Corynebacterium glutamicum*. Amino acids are the building blocks and therefore are in high demand. Many amino acids are produced using auxotrophic mutants. Permeability is an important factor in amino acid formation. The optimization of the fermentation parameters should be done before starting the process. Glucose and sucrose are good as carbon sources. Nitrogen ammonium salts or ammonia gas have been found useful. Biotin is supplemented into the medium for glutamic acid production.

**Table 7.5** PHA production using whey as carbon source

| Strain | PHA (g/kg) | Reference |
|---|---|---|
| Recombinant *E. coli* (*Azotobacter* sp. genes) | 51.1 | Nikel et al. (2006) |
| Recombinant *E. coli* (*A. latus* genes) | 96.2 | Ahn et al. (2000) |
| *Pseudomonas hydrogenovora* | 1.27 | Kollar et al. (2008) |

## 7.9.2 Common Carbon Sources in Medium

**Oils, Fatty Acids, and Glycerol in Culture Media as Carbon Source**

The study by Fukui and Doi (1998) is very interesting in which they used olive oil, corn oil, and palm oil for the production of PHAs using a wild-type strain of *C. necator* H16. They produced high polymer content (80%), but the concentration (2.9–3.4 g/L) was found to go low after 72 h. In another study (Kahar et al. 2004), two strains of *C. necator* strains were used for the production PHA by using soybean oil in a 10-L fed-batch bioreactor. The wild-type strain of *C. necator*, produced P(3HB) at a high concentration of 85–95 g/L with a conversion rate of 72–76%. However, the recombinant strain produced P(3HB) along with 3HHx and the yields reached a high concentration of 91–102 g/L. The studies, however, proved that the process can be cost-effective using these oils and the cost of PHA can come down to US$ 0.4/kg, which makes an industrial process very viable. Fed-batch process gives better results. Cromwick et al. (1996) tried using low-cost triglycerides as carbon source for PHA production. The yields were found to be low and not cost-effective. There are studies (Cavalheiro et al. 2009; Koller et al. 2005) using glycerol-rich by-products for the production of PHA. However, they are not cost-effective.

**Use of Gases as Substrates**

Using $CO_2$ as a gaseous carbon source has been worked out by many workers. Gas is obtained from the exhaust of factories or thermal power stations. $CO_2$ is separated using a pressure swing adsorption system. In a study done by Taga et al. (1997) using *C. necator* has resulted in good conversion rate for PHA in which a mixture of gases of $CO_2$, $O_2$, and $H_2$ were used in a fed-batch process. In order to avoid explosion, the concentration of gases should be optimized. Listewnik et al. (2007) worked on the production of P(3HB) from natural gas. They designed a process that could produce 500 tons of PHA/year with a cost of €6.35/kg. This makes the production of PHA using petroleum base as substrate very attractive.

**Use of Organic Matter from Wastes/Wastewaters as a Carbon Source**

A centrifuged fermented organic waste has been tried as the sole carbon source (Ganzeveld et al. 1999) for the production of P(3HB-*co*-3 HV). The polymer yield obtained was 1.13 g/L in a duration of 45 h of fermentation.

In another study, Rusendi and Sheppard (1995) used the wastes of potato obtained from the potato chips factory as a carbon source for the production of P(3HB). Initially, they were hydrolyzed with the help of enzymes (α-amylase and amyloglucosidase). This was then diluted and some mineral salts were added. Ultimately, the culture medium was adjusted in such a way that it contains about 13.6 g/L of glucose. For the process, a cyclone bioreactor was used. It resulted with a yield of 5 g/L and a polymer content of 76.9%.

Scientists have proposed the use of open mixed microbial cultures, which could help in using mixed substrates. This might also eliminate the step of sterilization, making the process cost-effective. The research work has concentrated on the submerged fermentation (SMF) processes for PHA production.

Recently, solid state fermentation method has been studied for the production of PHA. SSF is basically a method in which the microorganisms are grown on the moist solid particles of substrate. The interparticle spaces are filled with a continuous gas phase. The water used in these spaces should be to a minimum but should be enough for the microbial growth to take place (Moo-Young et al. 1983). In this method, cheap and inexpensive substrates can be used, which also includes various agricultural wastes as cassava bagasse, rice bran, and remains of the vegetable oil extraction. This approach is helpful toward a better environment.

Oliveira et al. (2004) used SSF technique for PHA production using cultures of *C. necator* and *A. latus* on soy cake and with sugarcane molasses and glucose at different concentrations. With 2.5% (m/m) molasses, a 2.3-fold higher polymer content was produced. Oliveira et al. (2007) characterized P(3HB) samples produced by SSF in non-supplemented but in a medium supplemented with molasses soy cake. They characterized using different techniques such as X-ray diffraction, differential scanning calorimeter, nuclear magnetic resonance, infra-red spectroscopy, gel permeation chromatography, and polarizing optical microscopy.

With all these researches, the yields of PHA have not been accepted for commercial production. Therefore, it is important to enhance the yields of PHA by altering the genes that are involved in the PHA production process and the modification is done at the molecular level. Recently, researchers have used CRISPR/CAS9 technology to modify the genes. Chimeric enzymes have been prepared that are more stable. Work has been done to enhance the expression of proteins that can give higher levels of PHA. Borrero-de Acuña et al. (2023) did the rational engineering of natural polyhydroxyalkanoates-producing microorganisms for their improved synthesis and recovery. CRISPR-based genome editing, which is derived from the RNA-guided immune systems, is present in many bacteria. With this the organism can be manipulated at genetic level. The method is very efficient and does not damage the organism. Using this technique, scientists have worked on microorganisms such as *Streptococcus pneumoniae*, *E. coli*, *Streptomyces*, *Lactobacillus reuteri*, *Clostridium*, *Bacillus subtilis*, and *Corynebacterium glutamicum* to produce maximum yields of PHA. An electroporation-based CRISPR/Cas9 technique has been developed in *R. eutropha* (Xiong et al. 2018), resulting in higher yields of PHA. Using the cells of *H. bluephagenesis* TD01 (Qin et al. 2018), the engineered strain produced P(3HB-*co*-3 HV) more efficiently and PHA production was increased by 16-folds.

For the recovery of the products, many different methods are being used. Some examples are solvent-based, chemical, enzymatic, and mechanical methods, which are cost-effective. The use of green solvents under pressure or the use of ionic liquids/supercritical solvents/hypotonic cell disintegration have been found to be helpful in the recovery of PHA (Koller 2020). The process can be made cost-effective if the conditions are optimized in the right manner. Extraction can be done using any of the solutions of sodium hypochlorite, dimethyl sulfoxide (DMSO), or using organic solvents. Ultrasound method for extracting PHA in *Pseudomonas putida* has been worked out (Arikawa et al. 2017; Ishak et al. 2016).

Super critical fluids (SCF) are fluids with low viscosity, which can be a gas, high density such as a liquid, and almost nonexistent surface tension. Such fluids help in cell disruption of bacteria and yeast and they do not damage the cell. Extraction by using supercritical $CO_2$ ($scCO_2$) is a very good method. $scCO_2$ penetrates and expands within the cell, which disrupts the cell wall. No heat is generated and the process is also cost-effective (Khosravi-Daravi et al. 2004).

### 7.9.3 Fermentor Operation

It is important that a fermentor should be completely sealed during processing. The fermentor should be designed in such a way that it can be cleaned easily and should have good heat exchange performance. In order to avoid corrosion and leaching of toxic metals into the medium, the fermentation tanks/bioreactors are made up of stainless steel. Industrial fermentations use both batch and continuous culture methods.

The steps to follow before starting a fermentor are as follows:

Initially, cleaning of the electrodes (pH and DO) is done and then they are calibrated. The calibrated electrodes are inserted into their respective ports and are tightened with the respective screw. The motor, all the electrodes (pH electrode, DO electrode, defoaming electrode), and the ports are wrapped in tin foil. An air filter should be installed from the place of air inlet. The connection pipe and the sampling pipe should be clamped. The exhaust pipeline needs to be controlled but should be opened with a screw clip. It should be wrapped with eight layers of gauze. After the fermentor is sterilized, it is cooled and all the circuits of the motor are connected after which the electrodes can be installed. The important electrodes in any fermentor are the pH electrode, DO electrode, defoaming electrode, and the temperature electrode. Then water is circulated through the water pipeline. The air is then allowed to enter through the air filter into the tank. The tank is raised to the required temperature and all the parameters are set for the reaction. The pressure of the tank is adjusted to less than 0.02 MPa for inoculation, which is increased later. Then the fermentation process can be started. In this method, sterilization of the fermentor and the growth medium before inoculation are extremely important. Usually, the outside jacket is heated and the steam is injected into the reactor. The gas exit valve needs to be closed, which helps to create a pressure in order to heat the system above 100 °C. In an alternative way, the medium can be sterilized outside the fermentor. This can be supplied with the help of sterile lines. A sparger is used for the generation of air bubbles. These air bubbles help in the diffusion of oxygen throughout the reactor volume in aerobic fermentations. Sometimes, agitation of the medium is done with the help of a stirrer, which fastens the process. Once the reaction time is completed, the fermentation is stopped by switching off the main power. This is followed by taking out the fermentation broth. The tank and the electrodes are then cleaned with distilled water.

During the fermentation, as the cell density increases, the level of nutrients declines in batch culture. However, in continuous culture method, optimal conditions for fermentation are maintained for a longer time by programming cycles of nutrient injection and other changes in the fermentation conditions. Mostly, the fermentation products are secondary metabolites, formed in the stationary phase of the organism. Results using continuous culture can be obtained by programming successive cycles of nutrient injection, followed by starvation and secondary metabolite synthesis as in the production of ethanol and single cell protein (mycoprotein).

### 7.9.4 Submerged Fermentations in a Fermentor

In this type of a fermentor the microorganisms are grown in a liquid medium. Submerged fermentation is a process involving the development of microorganisms in a liquid broth. This liquid broth contains nutrients and produces different types of secondary metabolites. For the production of any desired secondary product/metabolite, a specific microorganism has to be taken for fermentation. The rich nutrient broth (sterilized) is put into the sterilized tank into which microbes are inoculated. The cell number is fixed after optimization of parameters. The submerged fermentation process needs a lot of oxygen. During the reaction, the microorganisms interact with the nutrients in the broth. The substrate gets broken down into bioactive metabolites/compounds. In case of extracellular metabolites, they are secreted into the fermentation broth. As the process is aerobic, the $O_2$ requirement increases as the reaction progresses. Therefore, some variables like temperature are maintained at a constant value during the process, but $O_2$ concentration is controlled at a changing level for maximum product yield. The environmental variables and the state variables are usually controlled by using online measuring equipment, which is linked to a computer. With recent developments in computer science, the computer controls with appropriate algorithms, help in implementing the feeding pattern, and provides accurate control of the inlet and outlet flows. Routine sampling of the culture is also possible by computerization. Aerobic batch culture needs more $O_2$ as the culture grows. All variables can be controlled online that are linked to a computer. This provides accurate controls to the inlet and outlet flows. Noninvasive integrated dual optical chemical sensors provide simultaneous measurement of %DOT and pH. The oxygen sensor is based on the selective sensor luminescence. However, for pH sensor, the luminescence is recorded through a dual-signal process that uses an added sensor (integrated reference sensor). The partial pressure of $CO_2$ in the culture can be monitored by opticochemical noninvasive $CO_2$ sensors with a fiber optic $CO_2$ transmitter. The construction of noninvasive probes have been very helpful. To recover the biological products from the fermentation broth involves multistep.

The cell metabolism depends on the pathway enzyme(s) and the regulatory protein(s). This improves the yield of an industrial fermentation product. By changing the medium composition or process conditions, the metabolism of cells can be adapted as per the requirement.

Metabolite production using the submerged fermentation is usually done by fed-batch fermentation and continuous fermentation.

Most of the bio-industries use fed-batch fermentation. The growth rate of the microbial culture is maintained by supplementation of the nutrients as and when required during the fermentation process. This helps in maintaining the growth rate of the culture and reduces the risk of overflowing of the metabolites.

However, in continuous system, the nutrients are supplemented continuously to the fermentation reaction tank. These are added at the same rate as that of the nutrient getting converted to the product. All variables (pH, time, temperature, $O_2$, and $CO_2$) are kept at optimized conditions for metabolite production. Therefore, it gives a steady rate of metabolite production. The metabolite production by this manner is cost-effective with high yields.

## 7.10 Bioreactor

**Bioreactor Definition**
A bioreactor is a type of fermentation vessel used for the production of various chemicals and biological reactions. It is a closed container with adequate arrangement for aeration, agitation, temperature, and pH control. It drains or removes the waste biomass of cultured eukaryotic cells along with their products. A bioreactor should have an agitator to mix the cells with the medium, aeration in case of aerobic processes, regulation valves for temperature, pH, pressure, nutrient feeding, and an outlet to remove cells/medium.

**Bioreactor Design** The design of a bioreactor is based on the organism, conditions for maximum product yield, and its scale of production, which should be economically feasible. A bioreactor is used mainly for the eukaryotic cells (mammalian, insect cells, etc.). There are various types of bioreactors, namely continuous stirred tank bioreactor, airlift bioreactor, and fluidized-bed bioreactor.

The optimization of various parameters needs to be done before taking up the production.

(a) Organism phases of growth, fermentation medium, and oxygen
(b) Production of biomass

(c) Production of extracellular metabolites
(d) Primary and secondary metabolites, production of intracellular components, and transformation of the substrate.

It has been observed that in case of scaling up of any laboratory process, all parameters should be optimized. The results of a lab scale process and when the same is taken for a large-scale operation usually have a lot of variation. Therefore, if the laboratory experiments have to be scaled up, then optimization of parameters should be taken up on a large scale.

A bioreactor has the ability to regulate an agitator system, an oxygen delivery system, a foam control system, temperature and pH, sampling ports, cleaning and sterilization system, and lines for charging and to empty out the reactor. The bioreactor is selected depending on the type of fermentation to be undertaken. In case fermentation is done using either yeast or bacteria strains (which involves the conversion of starch to sugar and then into lactic acid), then it does not require heat. Usually, the fermented products prepared by fermentation include chemicals such as acetic acid, citric acid, and ethanol. For any industrial process it is advantageous to have less fermentation time. This depends on the concentration and the growth phase of the organism, number of cells used, cellular components, enzymes, temperature, and pH in aerobic fermentation. There are many useful enzymes produced by fermentation processes using different types of fermentors/bioreactors, which are of great commercial use. These mainly include lipase, invertase, and rennet, which are produced using fermentation processes with genetically modified microbes. Single-cell proteins, Baker's yeast, and starter cultures for lactic acid are used in cheese making, which are produced by fermentation and utilize various organisms such as bacteria, yeasts, molds, algae, animal cells, and plant cells.

### 7.10.1 Parts of the Bioreactor and their Functions

The material used should be noncorrosive, with no toxic substances entering the fermentation media, and the reactor should be able to endure steam sterilization process and also high pressure. It should be resistant to pH changes. The sizes of the bioreactor can vary depending on the application. Bioreactors are available in all sizes from few $mm^3$ to 500 $m^3$ in which the biological-based materials are converted into desirable products.

The exterior is fitted with a cooling jacket that seals the vessel and provides water for cooling, thermostatically controlled baths or internal coils that produce heat, and silicone jackets that remove excess heat. Aeration system is important for oxygen availability, which consists of two separate aeration devices (sparger and impeller) to ensure proper aeration in the bioreactor. The stirring helps in mixing the cells through the liquid culture medium.

The sealing assembly is used for the sealing of the stirrer shaft to offer proper agitation. This can be in the form of packed gland seal, mechanical seal, or magnetic drives. The baffles are made up of metal strips that are incorporated into the bioreactor to prevent aeration.

**Impellers** provide uniform suspension of plant/eucaryotic cells in different nutrient medium and are made up of impeller blades attached to a motor on the lid that reduces the size of air bubbles and distributes them uniformly into the fermentation media. These are of two types: disc turbines or variable pitch open turbine. The sterile air comes into the bioreactor vessel with a sparger. The sparger pipes contain small holes of about 5–10 mm, through which pressurized air is released. Three types of spargers are used, which are described as porous, nozzle or combined sparger–agitator. The feed ports are tubes made up of silicone and they help in adding the nutrients and acid/alkali to the bioreactor. The level of foam in the vessel is minimized to avoid contamination. This is done by using foam sensing and a control unit placed on the top with an inlet into the bioreactor. To control the movement of liquid in the vessel, a valve is used in the bioreactor which could be any one of the following types: globe, butterfly, ball, or diaphragm type along with a safety valve. Controls for temperature, oxygen concentration, pH, cell mass, essential nutrient levels, and product concentration are provided. For process monitoring and data collection, the bioreactors are usually coupled with modern automated and semiautomated computers and databases. The parameters can be monitored and controlled with the help of sensors.

### Parts of a Bioreactor

**Reaction Vessel** This is usually in a cylindrical shape that is closed at the top and at the bottom. Many different pipes and valves are connected to the vessel. Due to the sensors attached to the vessel, there are many advantages as it helps to work under the optimized conditions that can be controlled. These bioreactors are either made up of glass or stainless steel material. Both these materials do not have any toxicity and are corrosion proof. The glass vessel is mainly used in small-scale industries, whereas stainless steel vessel is used in large-scale industries as it can resist pressure and corrosion.

### Heating and Cooling Apparatus

On the exterior side of the bioreactor vessel a, cooling jacket is fitted. This provides cool water. To provide heat the bioreactor is provided with thermostatically controlled baths or with internal coils. However, to remove excess heat, it is provided with silicone jacket. For the removal of the heat generated, it is important to have a cooling jacket.

### Aeration System

For proper aeration and for oxygen availability to the culture, a good aeration system is required. This has two devices: sparger and impeller. The stirrer helps in mixing the gas bubbles with the liquid culture medium and the eukaryotic cells.

**Sealing Assembly**
The sealing of the stirrer shaft is done using a sealing assembly that provides proper agitation. These are categorized as packed gland seal, mechanical seal, and magnetic drives.

**Baffles**
The baffles are either vertically arranged vanes or circular or elongated metal plates. These are kept inside the vessel that controls the radial swirl inside the bioreactor. It converts the rotational flow to axial mixing. It helps in improving the aeration in the bioreactor.

**Impeller**
These help in making a homogeneous suspension of cells in different nutrient media. They are provided with blades that are attached to a motor on the lid. These blades reduce the size of the air bubbles so that they can be distributed uniformly in the fermentation media. Different types of impellers are being used. These are disc turbines and variable pitch open turbine type.

**Sparger**
It is through a sparger that the sterile air is introduced into the bioreactor vessel. This provides proper aeration inside the vessel. It is provided with holes (5–10 mm). Through these holes, air is pushed into the bioreactor under pressure. These are of three types: porous sparger, nozzle sparger, and combined sparger–agitator.

**Feed Ports**
For adding nutrients, acid or alkali to the bioreactor these ports are used, which are usually made up of silicone.

**Foam-Control in Bioreactor**
Foam formation should be at the minimum as it can help in reducing the contamination. The foam control has two units: foam sensing and a control unit. This total unit is placed on the top of the bioreactor that is provided with an inlet to the bioreactor.

**Valves in Bioreactor**
A bioreactor is mainly provided with five types of valves. These help in controlling the movement of liquid in the vessel. They can be globe valve, butterfly valve, ball valve, and a diaphragm valve. In addition to this, there is also a safety valve that controls the air under pressure to operate.

**Controlling Devices for Environmental Factors**
Probes/sensors help in controlling the temperature, oxygen concentration, pH, cell mass, essential nutrient levels, and product concentration.

**Use of Computers in Bioreactors**
The bioreactors are connected to the computer for collecting data and for real-time process details, which can be monitored immediately.

For industries the various bioreactors used are: Continuous stirred tank that is made up of a cylindrical vessel with a central shaft controlled by a motor that supports one or more agitators (impellers). The sparger, in combination with impellers (agitators), allows the improved gas distribution throughout the vessel. A stirred tank bioreactor can be operated continuously, is economical, and is easy to clean.

**Photobioreactors** There are around one million species of phototrophic microbes around the globe. Around 50% of the total photosynthesis in the world is done by these microorganisms. The most common ones are of the genus *spirulina*, from which metabolites are produced using bioreactors. The algae have a sequestration rate of 1.83 kg of $CO_2$ per kg of dry biomass. Therefore, it is very beneficial in the improvement of the environment. They are also used in the treatment of affluent and can be used as biofertilizers. The yields can be enhanced by optimization of the conditions for which mathematical models have proved very beneficial. For optimization, all physical and chemical factors play a major role and closed photobioreactors (PBRs) have been found to be efficient. Being a closed and illuminated vessel, the microbial culture does not come in direct contact with the environment. This helps in the prevention of any contamination. $CO_2$ and water loss are also to a minimum. The intensity of light should be kept constant. Usually, sunlight is the primary source; however, due to the changes in the intensity of sun light, artificial light is provided by tungsten lamps, fluorescent lamps, and light emitting diodes (LEDs). Tungsten lamps provide illumination of 35–100 lm/W, but lose intensity over time. However, LED light has been found to be more efficient. A major problem is the fouling of the bioreactor due to the formation of biofilms on the transmitting surfaces. This reduces the productivity and makes the process very expensive. It should be seen that the high rates of mass transfer are obtained without damaging the cells. For this purpose, the reactor is provided with foam for which the conditions and the concentration should be optimized. Other factors that need optimization are to reduce the shear stress. There are different types of photobioreactors. These are stirred tank, bubble column, flat panel, horizontal tubular, bag PBR, membrane PBR, pyramid, and nature inspired.

In stirred tank the agitation is done with the impellers, which could be axial, radial, or mixed. The impellers can be of different shapes and sizes. Tangential agitation is reduced with the help of baffles. For large-scale production, such tanks are not advisable due to the low surface-to-volume ratio. A large amount of electricity is required to move the impellers, making the process expensive. Fed-batch bioreactors have been found to be more cost-effective.

In flat panel PBRs, there are two parallel and transparent flat panels, which are placed on a frame in such a way that it forms a rectangular channel. This increases the illumination area per volume of the PBR, ultimately increasing the efficiency.

To increase the growth of phototrophic microorganisms, the inside of the PBR can be colored with dyes or a color coating. Immobilization of cells on a grid helps in increasing the surface area of the reactor and increases the efficiency of the process. In batch process, the microbial growth becomes limited as the nutrients get depleted and many a times there is product inhibition. Therefore, when production is done in a continuous mode, it helps in higher yields of the product and enhances the process efficiency.

**Tubular PBRs** can be either vertical, horizontal, near-horizontal, or spiral/helical. They are made up of glass or plastic. The aeration is done through a pump. They are used very commonly for the outdoor large-scale cultivation. In vertical designs there is an advantage of light getting penetrated directly. The heat and mass transfer rates are high, making the process very efficient. The only constraint is the reflection of light during the mid-day.

**Horizontal tubular PBRs** These help in allowing high biomass densities. For maximum light penetration, acrylic pipes are used, which have an internal diameter of around 60 mm. *Chlorella* sp. is grown in Germany at large scales using horizontal tubular PBRs. The only problem is of oxygen accumulation due to photo inhibition. This has been modified to near horizontal PBR in which it is inclined to 45°. This helps the air bubble to rise faster and the gas transfer coefficients are better.

**Bag type PBR** These are usually made up of plastic bags that are transparent. Usually, they are made of polyethylene. However, it has a limitation, as the bag needs to be replaced again and again.

**Pyramid PBR** These are being used for algae cultivation. They are pyramid in shape and are made up of acrylic. The fluid is mixed using an airlift system.

**Nature-inspired PBRs** These help in increasing the surface-to-volume ratio. These bioreactors can be of various shapes found in nature. Under optimized flow, this type of PBR is quite efficient.

Scientists have also developed hybrid photobioreactors. This combines the usefulness of various reactors, ultimately increasing the efficiency of the process. It is proposed that if technologies as light filters, illumination type, aeration, and mixing methods are combined and optimized, it can bring newer, more efficient strategies for the large-scale cultivation of biomass and the formation of products in large quantities that will be cost-effective.

**Membrane bioreactors (MBRs)** These bioreactors are based on the membrane technology that employs filtering using a biological mechanism. A mechanical border is made with the help of the membrane between the liquid and the solid. Poostchi et al. (2012) studied the filtration performance of mesh filter bioreactor for the synthetic municipal wastewater treatment. They studied the specific aeration demand on the suspended effluent solids. The filtrate quality has been found to decrease with high aeration due to the dynamic membrane. The total filtration resistance was found to go down with increased aeration.

The membrane helps in removing the unwanted particles. However, it causes membrane fouling, which is very expensive. The membrane bioreactor (MBR) has been found very effective for municipal and industrial wastewater treatment. However, the application of MBRs causes membrane fouling. This reduces the efficiency of the membrane. Due to these reasons, the maintenance and operating costs become very high. Iorhemen et al. (2016) worked on these aspects of biofoulants, organic foulants, and inorganic foulants to reduce the problems. They have also studied the various factors that are responsible for membrane fouling. It was observed that the fouling could be controlled with the addition of coagulants and adsorbents, combining the aerobic granules with MBRs, introducing the granular materials with air in the MBR tank and also by quorum quenching.

However, the optimization of coagulant concentration is important for each process. Also, the addition of aerobic granules with MBRs (targeting biofoulants, and organic foulants) has proved better filtration and a reduction in fouling rate to a large extent. Quorum quenching has been found to be another good method to control fouling.

Many refineries in the Middle East are using MBR to obtain treated water for irrigation. A company in Oman produces 78,000 $m^3$/day of water by MBR, which is used for irrigation. The method has been found to be very cost-effective. Veolia Water Solutions and Technologies, Brazil, has an MBR process at its Henrique Large refinery to treat waste water. Petro™ system is designed to treat hydrocarbons from the wastewater coming from petroleum and petrochemical companies using MBR.

The use of tubular ceramic membranes has been tried for the removal of suspended solids from liquor. COD can be reduced in waste water by using MBR. However, an RO step needs to be followed after the filtration (Table 7.6).

There are many products produced by fermentations. These include cheese, fermented dairy products, lactic acid fermentations of vegetables, fermented sausages, fermented bakery products, wine and brandy, beer fermentation, alcoholic fermented beverages. There are many oriental fermented foods such as soy sauce, miso, tempeh, ontjom, hamanatto, sufu, natto, idli, angkak, and fermented fish products.

**Table 7.6** Type of bioreactors used for specific products

| Type of bioreactor | Product name |
|---|---|
| Stirred tank | Antibiotics, enzymes |
| Bubble | Algae |
| Photobioreactor | Quality of water |
| Membrane bioreactor | Waste water |

## 7.11 Major Fermentation Processes

### 7.11.1 Production of Baker's Yeast

The nutrient source is usually molasses, which is the by-product of the sugar refining. Large tanks are used for fermentation and after the yeast gets filled in the tank, it is harvested by centrifugation, giving an off-white liquid known as yeast cream. Yeast strains used in the production of ale and bread are from a single species, *Saccharomyces cerevisiae*. Lager beer strains have been prepared using *Saccharomyces uvarum* (*S. carlsbergensis*) in Central Europe and in the United States. As this strain is not tolerant to high osmotic pressure in dough it cannot be used in baking. However, *S. cerevisiae* is a distiller's yeast and it cannot be used for bread baking. Also, it has separation problems and therefore the bakeries do not use it. In dough the yeast grows very slowly, which needs a long fermentation time. The number of yeast cells used per gram of dough are the deciding factor for beer or wine fermentation (Table 7.7).

For the production of yeast, higher aeration has been found to be good for enhanced productivity. Aerobic process produces more yeast with little alcohol and then there is no need to recover any ethanol. The active, dry yeast is extruded in the form of fine strands from the cake. For drying the yeast, tunnel driers on endless steel mash belts, rotary louver driers, or airlift driers are used. Compressed yeast has a solid content of about 30%, whereas active dry yeast contains 90–95% yeast solids.

**Yeast Strains** Baker's yeasts are strains of *S. cerevisiae*. They are propagated by pure culture methods in the laboratories of yeast producers. Suitable strains are also available at known public culture collection centers. In the brewing, distilleries, and wine industries, the yeast strains are considered as proprietary item but not in the bakery industry. However, a producer can obtain protection for a proprietary strain by applying appropriate patents. Although as per the present literature, patent protection has not been obtained for all the commercially available yeast strains. In the cases, where the strain is improved by hybridization, it becomes difficult to evaluate a number of hybrids for yield, growth rate, stability on storage of the compressed yeast, ability to withstand drying process, and other qualities. However, an improved osmotolerant yeast strain has been obtained by the hybridization of *S. cerevisiae* with *Saccharomyces rosei*.

**Table 7.7** Yeast fermentations (cfu/mL) in different products

| Parameter during processing | Bread | Beer | Wine | Whiskey |
|---|---|---|---|---|
| Initial yeast (cfu/mL/g) | 270 | <10 | <10 | <10 |
| Fermentation time (days) | <5 h | <10 | <10 | <3 |

CFU/mL is colony forming units in 1 mL

**Raw Materials and Process Requirement for the Production of Baker's Yeast**

**Carbon and Energy Sources** The most widely used carbon sources are cane and beet molasses with a fermentable sugar concentration between 50% and 55%, and a Brix of about 80°. Molasses being a by-product of the sugar industry has many variations in the composition. Yeast manufacturers have no control over the composition of the molasses. The pH of the molasses is usually in the range of 6.5–8.5. All yeast strains ferment and assimilate sucrose very rapidly in the form of invert sugars (glucose plus fructose) due to the high invertase activity. The fructose moiety of the trisaccharide (raffinose) has the ability to get fermented by the Baker's yeast. The residual melibiose can be fermented only by some strains. Other compounds present in molasses (acetic acid, lactic acid, succinic acid, tartaric acid, and glycerol) get assimilated in the presence of mono- and diglycerides. Some amino acids may also serve both as carbon as well as nitrogen sources.

**Nitrogen Sources** The nitrogenous compounds in cane or beet molasses varies in concentration. Ammonia nitrogen and nitrogen from many of the amino acids get assimilated as glutamic and α-aminobutyric acid. These acids serve as sources of assimilable nitrogen. However, aspartic acid, alanine, glycine, and lysine serve as partial sources. Usually, the required nitrogen is supplied by the supplementation of ammonium salts, liquid ammonia, or urea. The nitrogen present in nitrates does not get assimilated.

**Vitamins** Usually the Baker's yeast needs biotin to grow. Compressed yeast contains about 0.75–2.5 ppm of biotin. Cane molasses can provide about 0.5–0.8 ppm but beet molasses has very low concentration (0.01–0.02 ppm) of biotin. Therefore, a minimum of 20% of cane molasses needs to be blended with beet molasses for the preparation of the feed wort, or otherwise the feed needs to be supplemented with synthetic biotin. When urea is used, as a source of nitrogen, it has been observed that in such cases, higher amounts of biotin are needed. L(+)-aspartic acid can partially replace biotin; however, L(+)-aspartic acid plus oleic acid can completely replace biotin in the growth media for the production of Baker's yeast that are present in molasses in sufficient quantities. Supplementation of thiamine is required for the optimum growth of yeast cells.

**Minerals** Addition of phosphates are required for the proper growth of yeast cells. The composition required for yeast growth usually involves 2.5–3.5% of $P_2O_5$ along with 7–9.5% nitrogen (dwt). Commonly, the phosphorus is added in the form of phosphoric acid, alkali phosphate salts, or ammonium phosphate. Molasses provide potassium, calcium, sodium, and sulfur. However, magnesium salts have to be supplemented from outside. Yeast ash contains 0.4–0.5% sodium as $NaO_2$ and 0.2–0.25% sulfates as $SO_3$. When sodium chloride is added to the yeast cream for filtration, the sodium concentration becomes high. However, for better growth of yeast, it is necessary to supply the trace elements such as Fe, Zn, Cu, Mn, and Mo.

**Fermentation Activators and Inhibitors** Flour milling waste and sludge from aerobic digesters have been found to enhance the growth and are called as activators for yeast growth. Indolyl acetic acid is also a growth stimulant for yeast but this acid cannot be utilized. $SO_2$ inhibits the growth of yeast, but up to 800 ppm comes under the tolerable limit.

**Other Carbon and Energy Sources** Any sugar such as sucrose, maltose, glucose, fructose, and mannose can be fermented by yeasts, but lactose cannot be fermented by yeasts. Galactose is fermented very slowly. These sugars are usually obtained from raw materials such as sugarcane juice or molasses, grape juice concentrates, date juice, wood hydrolysates, starch hydrolysates, or waste sulfite liquor. The liquor from paper pulp industry contains a mixture of hexoses and pentoses at very low concentrations. *S. cerevisiae* assimilates only the hexoses, and consequently very large volumes of liquor have to be passed through the fermentors. The use of molasses is not advisable as the recovery of sucrose from beet or cane juice is much better. There is a new process in which a major portion of sucrose can be removed from molasses. This results in a substrate with very low concentration of fermentable sugars and high concentration of inorganic and organic (non-sugar) compounds. The fermentation industries that use molasses produce effluents with very high BOD. Therefore, the use of alternate carbohydrate sources is being sought out for the production of Baker's yeast. Usually, the starchy materials from cereal grains like corn (maize) are used. However, these starchy grains need conversion to fermentable sugars, which utilize either the bacterial or fungal amylases.

**Principles of Growth in Fermentor for Yeasts**

(i) **Growth rate** In anaerobic batch fermentations, initially the microbial growth is slow, which is followed by a period of exponential growth. This is usually termed as stormy in the beer and wine industry. This is followed by the latent phase in which the growth declines or stops. In the latent phase, the fermentation takes place at a reduced rate till the substrate is used completely. In aerobic continuous fermentations, there is an exponential growth throughout the length of the fermentation. For such fermentations the increase in yeast cell mass during the exponential growth can be expressed, as the generation time or the specific growth rate constant. Generation time is the time required for doubling of the yeast population. There is a specific growth rate during the exponential phase, and for continuous fermentations its value is identical with that of the dilution rate. The total time for such fermentation is generally between 8 and 20 h. During the fermentation, the generation time increases from 3 to 5 h and finally to 7 h. This leads to an eight-fold multiplication of yeast cells in a 15 h fermentation period.

(ii) **Oxygen requirement and aeration** Under anaerobic conditions the maximum of theoretical yield coefficient is 0.075, which is equal to a yield of 7.5 kg of yeast solids from 100 kg of fermentable sugar. Under aerobic conditions of fermentation, the maximum yield coefficient is around 0.54. Other important conditions for maximal yields are the growth rate. This should not exceed a value of 0.2, and the amount of substrate present at any one time must not exceed a given limiting value. When the growth rates are very high, the formation of carbon dioxide also becomes more and the RQ increases. This helps in the formation of ethanol from the fermentable sugar. Therefore, it is vital to keep the growth rates below 0.20. For yeast growth, the oxygen needed is around 1 g of $O_2$ per g of yeast solids. This can be supplied either in the form of pure oxygen gas or in the form of hydrogen peroxide. However, in practice, it is always supplied by blowing air in the liquid fermentor contents. The capacity of an aeration system to transfer oxygen from air into the liquid phase is expressed in terms of its volumetric oxygen transfer coefficient, $K_La$ ($h^{-1}$), where $K_L$ is the oxygen transfer coefficient (m/h) and $a$ is the interfacial area between air bubbles and liquid per unit volume of liquid ($m^2/m^3$ or 1/m). The oxygen transfer rate (OTR) is the amount of oxygen that can be transferred into 1 L of the culture in 1 h. OTR plays an important role during fermentation. The oxygen transfer rate also plays a vital role in the selection, design, and scaling-up of a bioprocess. The rate of oxygen transfer in a fermentation system is usually expressed in terms of millimoles $O_2$/L/h. The standard oxygen transfer rate (SOTR) is the rate at which the oxygen is transferred in tap water at 20 °C and zero dissolved oxygen. An oxygen transfer rate of 140 mM/L/h is required to produce 4.5 g of yeast solids/L/h. The efficiency of such systems depends on the oxygen utilization. If the dispersion of the incoming air is good, there are smaller bubbles, which provide a larger bubble surface area by mechanical agitation at the point of air inlet to the fermentor. In such systems, 40–50% of the available $O_2$ can get transferred to the liquid phase and can be used by the growing yeast. The oxygen transfer rates can be analyzed using sulfite oxidation method. The method measures the rate of oxidation of sulfite to sulfate in a solution in the presence of a metal catalyst. The method is very useful to determine the effect of variables (agitator speed, addition of surfactants, etc.) and oxygen transfer in a fermentor. The rate of oxygen transfer depends on the catalyst used (Co or Cu) and other extraneous factors. When the surface-active agents are added, they affect the rate of oxygen transfer. This affects the oxygen transfer coefficient ($K_L$), which may differ from that on the bubble surface area ($a$). Therefore, the sulfite oxidation measurements are not very accurate. It can be obtained only by varying the aeration rate during the actual fermentations and by determining the effect of these variations on the yield of the yeast. Oxygen electrodes in the fermentor can determine the level of dissolved oxygen during fermentation. Calibration can be done by aerating the fermentor medium in the absence of yeast and assuming an oxygen concentration of 7.7 ppm at 30 °C. During the fermentation process these values are usually below 0.4 ppm (or about 5% that of the saturation level). If the fermentor is aerated with gas mixtures containing a very high percentage of oxygen, then there can be disturbance in the yeast metabolism as ethanol is formed and the ability of the yeast to ferment and to respire glucose becomes less. It has been thought that oxygen transfer from the gas to the fermentor liquid is the limiting step in oxygen transfer rates and in the rate of oxygen uptake by the yeast cells. Einsele et al. (1978) worked with continuous, aerobic fermentations of Baker's yeast. Mixing times have been determined based on the response of intracellular NADH to a glucose step change during the continuous fermentation.

(iii) **Concentration of fermentable sugars** Under the anaerobic conditions the rate of glucose fermentation toward the production of ethanol and $CO_2$ becomes faster, as compared to the aerobic conditions. This is called as the Pasteur effect. The level of enzymes, the electron transport system, and the citric acid cycle all become slow in reactions. The respiratory coefficient measures the amount of glucose repression in respiration. Maximum yields can be obtained by limiting the growth rate to 0.2 or less and by incremental feeding that gives a RQ (respiratory quotient) of approximately 1 and ethanol formation is almost negligible.

(iv) **pH and temperature** Baker's yeast can tolerate pH levels between 3.6 and 6 with optimum levels between 4.5 and 5.0. At lower pH the contamination of bacteria is very low. Usually, the fermentation processes are initiated at a low pH and then the pH is raised slowly, with the addition of ammonia or alkali. A pH of 4.1 has been found to be optimum for yeast production from *S. cerevisiae*.

(v) **Yield and development of heat** Yield is expressed as the yield coefficient, Ys (0.5 for yeast), i.e., grams of yeast solids per gram of substrate used (percentage of the molasses). Some amount of substrate is used for the maintenance of the cells. The solid content of the compressed yeast is 30% in the United States, whereas it is 27–28% in Europe.

(vi) **Osmotic pressure** The high osmotic pressure inhibits the yeast growth. *S. cerevisiae* is not an osmophilic yeast. The growth conditions affect the osmotolerance of the yeast. It has been found that the Baker's yeast grown with incremental feeding is more osmotolerant and the addition of salt brings down the specific growth coefficient of the Baker's yeast.

**Yeast Concentration in the Fermentor** The growth rate in yeast decreases rapidly when the concentration of yeast is more than 3–4%. This may be due to the exhaustion of growth factors or the nutrients. Also that the nutrients are not able to reach the surface of the yeast cells as there is an increase in the viscosity. If mixing is not proper, it increases the concentration of inhibitors. An increase in the concentration of $CO_2$ in the fermentor results in lower growth rate of yeast. At a concentration of 10% yeast solids, the cell volume should be 25% of the liquid fermentor volume. In case the solid concentration is high (22–23%), then the mass cannot be pumped.

**Periodicity and Budding** The supply of molasses is reduced just before the end of the fermentation period. This helps in the maturation of the yeast. A study on *Schizosaccharomyces pombe* has been found that it can be grown in a synchronous way. The rate of glycolysis increases in a linear fashion, which doubles at each division. However, this does not happen in *S. cerevisiae*.

### Aerobic Growth of Baker's Yeast and Optimization for Fermentation

(a) **Preparation of the substrate** Molasses is used with 80° Brix, which is diluted to a Brix of 40° and the pH is adjusted to about 5.0 with acids (sulfuric acid) and fermentable sugar of 50–55%. In fed-batch fermentations, it is diluted and clarified to remove insoluble solids. Desludger centrifuge is used to remove the insoluble solids. In case of beet molasses, clarification can be done by filtration. However, it is difficult to filter cane molasses. After clarification it is sterilized at a high temperature for a short time and is then cooled. Heating of molasses or the diluted molasses wort should not be done for a long time to prevent darkening and destruction of the sugars.

(b) **Fermentation tanks** All tanks and pipes are made up of stainless steel. Usually, it is sized for a volume of 200 $m^3$. Also, the ratio of height to diameter is very important. Oxygen transfer is better when the height of the liquid column is more. In case the liquid height is 3–5 m, then compressors should be used to supply the air. This helps to overcome the hydrostatic pressure in the liquid column. For squat fermentors, air blowers or self-priming aerators are used.

(c) **Aeration** Addition of oxygen is difficult and is a limiting factor. However, high oxygen transfer rates are possible by using specially designed aeration equipment. In the United States, air is fed into the fermentor through perforated tubes at the bottom of the tank. This usually consists of a horizontal tube with 24 side tubes provided with 30,000 holes of 1.5 mm diameter. Holes of very small diameter are required for better oxygen transfer, which can create air bubbles of small diameter. Distribution of air bubbles is assisted with the help of internal agitators. The agitators reduce the requirement of air volume but have a higher total energy requirement. Air requirements are often stated in terms of volume of air per fermentor volume per min (VVM). Fermentors with simple air spargers have less oxygen transfer rates with a higher usable fermentor volume than systems in which air bubbles are finely dispersed and have a long dwelling time.

(d) **Cooling** Internal cooling coils or external heat exchangers are used for cooling. Fermentation reaction for the production of Baker's yeast is usually carried out at 28–30 °C. It has been observed that an energy level of 3.5 kcal is generated per gram of yeast solids produced (aerobically), which requires cooling.

(e) **Defoaming** In highly aerated Baker's yeast fermentation process, a lot of foam is formed. Therefore, anti-foaming agents such as silicones, fatty acid derivatives, or other edible surface-active materials are added during the process. This reduces the effects of oxygen transfer by reducing the oxygen transfer coefficient and increases the total interfacial area between the air bubbles and the fermentor liquid.

(f) **Feed rates** With a constant feed rate in fed-batch fermentation, exponential growth does not take place. With a large yeast population, even at lower growth rates the productivity (grams of yeast solids produced/liter fermentor volume/hour) can be more as compared to a small population with a high growth rate. The time taken for a Baker's yeast fermentation is around 10–13 h (four-fold multiplication) and 16–20 h for an eight-fold multiplication. Initial yeast cell concentration can be increased if the fermentation process has to be completed in shorter time. Ethanol formation should be very less (<0.1%), which indirectly increases the yield of the Baker's yeast containing 30% solids.

(g) **Harvesting of the yeast cells** This is done by using a centrifuge that is vertical and nozzle type and it develops a $G$ of 4000–5000. The yeast cells can be separated from the fermentor liquor, which has a water content of 62 g per 100 g of cells and a density of 1.133 $g/cm^3$. The yeast concentration gets multiplied up to three times after the first centrifugation. With additional action, a concentration of 18–20% yeast solids can be obtained and the resultant is called as yeast cream. This is filtered with plate and frame filter presses that contain rotary continuous vacuum filters. It can be stored for many days in the refrigerator maintaining the same yeast quality. The filter surface of the rotary filter must be coated with an edible material such as potato starch or sago palm starch, which has a sufficiently large granule size to permit efficient filtration. Such filters produce a press cake of about 27–28% solids. Higher solid contents can be achieved by this equipment if the cream is salted just before filtration and the moisture content in the cells depend on the osmotic effect of the salt treatment. Salt can be removed with water sprays if required to be removed. This can be done directly on the filter. The mass of yeast cells obtained is called as the press cake. Small amounts of emulsifiers (mono- or diglycerides, sorbitan esters, or lecithin) and/or cutting oils are mixed with the yeast press cake, resulting in a lighter color to the yeast cake. These are extruded through nozzles in the form of thick strands with a rectangular cross section and can be cut into appropriate lengths of known shape for packaged Baker's yeast, which usually weighs about 1 lb or 500 g for commercial purposes. The 500 g (1 lb) cakes are wrapped in wax paper and are packaged with a weight of 22.5 kg (50 lb) in one case. Admixture of hydrophobic and hydrophilic additives can improve the free-flowing characteristics of the compressed yeast sold to the Baker's. Fast cooling for 24–48 h is required after packaging and therefore refrigeration is needed, which has a circulation of cold air. The bags should be arranged in a way to circulate the air freely. It is stored in the refrigerator until used.

(h) **Stability of compressed yeast** Temperatures of 5–8 °C have been found good for shelf storage of compressed yeast. Low concentrations of nitrogen and a low percentage of budding cells (<5–10%) favor storage stability. A liquid yeast cream has a shelf life of 10 days at 4–6 °C. However, it remains good only for one day if kept at 20 °C. High trehalose content indicates full maturation of the cells, which occurs at reduced growth rate toward the end of the fermentation. With higher trehalose content, the storage stability of the yeast increases. If the nitrogen concentration goes above 8% (based on solids), it reduces the shelf life. Live and dead cells can be detected using fluorochromes and primuline.

(i) **Contaminants** Many microbial contaminants such as lactic acid bacteria (*Lactobacillus* or *Leuconostoc*), coliforms, and *E. coli* can be found. Many molds grow easily on yeast cakes that have been stored for 3–4 weeks or longer even in the refrigerator. Species such as *Saccharomyces paradoxus*, *Candida utilis*, *Torulopsis minor*, *Candida krusei*, *Candida mycoderma*, and *Candida krusei* are present. Also, *C. mycoderma*, *C. tropicalis*, *Trichosporon cutaneum*, *Torulopsis*, *Candida*, and *Rhodotorula mucilaginosa* have been found as contaminants.

(j) **Automatic control of the fermentation** To control the pH, the process unit should be attached with a sensor, a recorder, and a device that activates a solenoid valve. The addition of molasses feed also needs a controlling system. It has been found that less feeding of molasses can lead to a loss of productivity, and overfeeding of molasses can lead to the production of ethanol and a loss of yield. This can be done using molasses wort. For continuous monitoring of ethanol in the effluent, the unit should be provided with a sensor. The level of dissolved oxygen in the feed can be monitored using a sensitive oxygen electrode. The oxygen level helps in monitoring the amount of feed to be supplemented. In case the oxygen level falls, then less amount of molasses should be added. The respiratory quotient (RQ) is the rate of carbon dioxide evolution divided by the rate of oxygen uptake, which is used for regulating the molasses feed. For respiration of a carbohydrate, the $Q_{CO2}$ should be equal to the $Q_{O2}$, i.e., the RQ should be 1.

(k) **Continuous fermentation** In this method, a minimum of two vessels are required in a series. However, contamination is a big problem in a continuous fermentation process.

(l) **Effects of ultrasound** When ultrasound is used, an increase in the fermentation rate of bread doughs has been observed. This helps in the reduction of fermentation time from 25% to 30%.

(m) **Active dry yeast** Baker's yeast is a vegetative microorganism that can be dried without much loss in the cell viability. Active dry yeast (ADY) has replaced the compressed yeast as it has higher storage stability.

(n) **Drying methods** The compressed yeast is dried by crumbling or extruding it in the form of small strands, which are then spread out on a paper in a dry room. However, it has been observed that the dried yeast has lower baking activity.

**Continuous belt tunnel driers** These are widely used for the commercial production of ADY. In this method, the strands of extruded press cake are deposited on the wire mesh screen of a belt that carries the yeast through several drying chambers. An airflow is pushed through the yeast bed in an upward and downward direction in an alternate manner. Drying takes about 2–4 h with an air inlet temperatures of 28–42 °C. Airlift (fluidized bed) driers are used in the commercial production of ADY in a continuous or batch fermentation process. Generally, the airlift driers can be operated with shorter drying periods. The drying of yeast press cake is done using a vibrating fluidized-bed method.

**Roto-louver type driers** These are used for the production of ADY in the form of small, round pellets. Drying times are generally between 10 and 20 h in a drum that carries the yeast press cake on the inside and rotates at about 4 rpm. Hot air is blown through louvers on the surface of the drum, which passes through the yeast bed. Vacuum-drying can also be carried out where liquid yeast cream is spread on wide steel belts. Spray-drying is not advisable and freeze-drying brings losses in viability. Traditionally, ADY is made without the use of any additives. Sucrose esters and sorbitan esters are used as additives only if the yeast is dry and needs rehydration. Leaching of solids from the yeast can be entirely prevented by vapor rehydration.

When the yeast is dried by tunnel type drier and packaging is done under an inert atmosphere/vacuum, it loses approximately 1% of its activity per month. However, the annual loss does not exceed 10%. Airlift dried ADY usually has very small cylindrical particles with rounded ends and a diameter of less than 1 mm. It has a moisture content of 4–6% and is less stable in the presence of air. Therefore, it is advisable to have a protective packaging, which should also be made mandatory. Yeast is packaged in aluminum foil pouches laminated to a heat sealable plastic film known as Pliofilm or Saran. The inert gas commonly used in packaging is nitrogen. However, bigger foil bags are usually sealed under vacuum. Tin cans containing 10–15 kg ADY are usually flushed with nitrogen before sealing so that ADY can absorb the carbon dioxide gas.

(o) **Yeast concentration in the fermentor** There are many factors that influence the concentration of yeast in the fermentor. Basically, it is the moisture, temperature, and the nutrient availability in the culture media. For this purpose, the physical and chemical parameters of the culture media should be optimized before taking up the process. This helps to obtain maximum yields of the yeast. The inorganic constituents in the media, the copper concentration, and the level of vitamin E, all these parameters play a major role toward the growth of yeast in the fermentation process. It is the level of glucose that decides the production of yeast. However, at a particular level, it reaches a saturation gradient and stops carbon dioxide production (Hewitson and Hill 2018). Guadalupe-Daqui et al. (2023) worked out the fermentation kinetics, fermentation metabolic products, and yeast stress indicators in brewing fermentations under partial vacuum along with the atmospheric pressure controls. It was found that the partial vacuum reduced the solubility of carbon dioxide in the media and decreased the time needed for carbon dioxide saturation.

(p) **Periodicity and budding** A new cell gets formed as a small bud on the parent body. Later, this gets detached from the parent, which is a replica of the parent cell; the nucleus of the parent yeast also divides into two, and one of the nuclei shifts into the bud, developing into a new organism. Budding in yeast cells depends on the nutrients, physical and chemical parameters, metal activators, and minerals provided. Under optimum conditions the periodicity of budding is usually 90 min.

(q) **Chemical composition of ADY** The regular ADY has a moisture level of 7.5–8.5%. Airlift dried yeasts and protected ADYs usually have lower moisture levels (4–6%). The nitrogen level is around 7% (based on solids) except for some highly active yeasts with nitrogen levels up to 9.5%. The concentration of phosphorus is expressed as $P_2O_5$, which is generally one-third of the nitrogen.

### 7.11.2 Production of Sourdough

Sourdough starter is a wild yeast and the product is made from flour and water. The main difference of sourdough starter is the time taken to leave the bread as the wild yeast in sourdough starter takes many hours, sometimes even days, to raise the bread.

### 7.11.3 Soda Crackers

A mixture of flour with yeast and a small amount of water is the base for soda cracker. This mixture is fermented for 18 h to form a sponge. The remainder of the flour is then added to the sponge together with additional water, salt, and an amount of sodium bicarbonate, which is sufficient to neutralize the acids in the flour and the ones obtained after the fermentation. Around 5 h is required to produce a relatively stiff dough. The dough is rolled into sheets, docked, and baked for 3–6 min, depending upon the temperature employed in the oven.

### 7.11.4 Production of Alcohol Fuel

Apart from fuel alcohol, oil from the grains can easily be converted into a fuel. Recently, biodiesel is becoming very important as an alternative fuel for diesel engines. It is produced from renewable biological sources, which is a mixture of mono-alkyl esters. It can be produced from sources such as vegetable oils, animal fats, waste cooking oil, and algae oil.

### 7.11.5 Fuel from Corn Oil

It is the main source of fuel alcohol production in the United States. Corn is composed of endosperm (82–84%), germ (9–10%), pericarp (5–6%), and tip cap (1%). Although corn germ is a by-product of milling industries, it is used to produce corn germ oil. This is used directly as salad and cooking oil or as an ingredient to prepare margarine, spread, and other foods. Normally, the corn kernel contains about 3–6% of the oil, of which 85% is present in the germ, followed by aleurone (12%) and endosperm (5%). Total lipid content of germ varies between 18% and 41%. The corn germ oil is produced by the separation of germ from the whole corn kernel by dry milling or wet milling process. The wet milled germ produces higher yield of oil, and therefore it is the main procedure utilized for the commercial production of corn oil in the United States.

Dry milling uses less energy and removes the germ and pericarp from the endosperm, and therefore products such as flour, corn meal, and flaking grits are produced by this method. However, the oil content is low (20–25%) due to the incomplete separation of germ. In wet milling, corn is steeped in water containing sulfur dioxide ($SO_2$; 0.02% for 24–48 h) at 48–52 °C. This swells and softens the kernel and solubilizes the proteins from the germ for better separation of corn components. After steeping, drained corn is coarsely milled to separate germ from other proportions, and the separated germ can be further processed to extract oil. Alternatively, methanol is a simple alcohol that can be mixed with gasoline or petrol, water, and most of the organic compounds. It can be produced from plants, fruits, and animal wastes through anaerobic metabolism by many bacteria or coal. Synthetically, a mixture of $CO_2$, CO, and $H_2$ can react with each other to form methanol and water as a by-product. Ethanol is another fuel produced from sugar-cane molasses that produces carbon dioxide and water and is an eco-friendly fuel for transportation vehicles. Yeasts help in converting carbohydrates to ethanol. Industrial methanol or ethanol are blended to produce gasohol. Usually, a 10% ethanol is blended with 90% gasoline, which is known as low-level ethanol blends (E10). This is used in conventional vehicles. However, when 85% ethanol is blended with 15% gasoline, it is known as high-level blend, as E85. This is used in motorized vehicles like flexible fuel vehicles (Pasotti et al. 2015). Methods for genetic optimization of biocatalysts for biofuel production from dairy waste has been worked out. Pasotti et al. (2015) discussed many different methods for the genetic optimization of biocatalysts in the production of biofuels from dairy waste through synthetic biology. High-titer lactic acid production by *Pediococcus acidilactici* PA204 from corn stover has been worked out by Zhang et al. (2020) using fed-batch simultaneous saccharification and fermentation.

### FAQs

1. How can antifoam be removed from the broth?
2. Why are the levels of C and N sources generally higher in fermentors than in shakers?
3. How can the fermentor volume be calculated?
4. How can the yeasts cells be unflocculated?
5. What type of antifoam agents can be used for oil-based fermentation process in bioreactor?
6. What are the different limitations of PHA produced by microbial sources?
7. How much time does PHA take to decompose? Is it beneficial to produce PHA by microbial sources?
8. Which are the conditions for the maximum growth of yeast by microbes?
9. What are the factors that affect the growth of yeasts?
10. What is the pH range for yeasts to be viable?

## References


Ahn WS, Park SJ, Lee SY (2000) Production of poly (3-hydroxybutyrate) by fed-batch culture of recombinant *Escherichia coli* with a highly concentrated whey solution. Appl Environ Microbiol 66(8):3624–3627

Anderson AJ, Dawes EA (1990) Occurrence, metabolism, metabolic role, and industrial uses of bacterial polyhydroxyalkanoates. Microbiol Rev 54(4):450–472. https://doi.org/10.1128/mr.54.4.450-472.1990

Anderson AJ, Williams DR, Taidi B, Dawes EA, Ewing DF (1992) Studies on copolyester synthesis by *Rhodococcus ruber* and factors influencing the molecular mass of polyhydroxybutyrate accumulated by *Methylobacterium extorquens* and *Alcaligenes eutrophus*. FEMS Microbiol Rev 9(2–4):93–101

Arikawa H, Sato S, Fujiki T, Matsumoto K (2017) Simple and rapid method for isolation and quantitation of Polyhydroxyalkanoate by SDS-sonication treatment. J Biosci Bioeng 20:1–5

Aurich MK, Fleming RM, Thiele I (2016) MetaboTools: a comprehensive toolbox for analysis of genome-scale metabolic models. Front Physiol 3(7):327. https://doi.org/10.3389/fphys.2016.00327

Bacame-Valenzuela FJ, Pérez-Garcia JA, Figueroa-Magallón ML, Espejel-Ayala F, Ortiz-Frade LA, Reyes-Vidal Y (2020) Optimised production of a redox metabolite (pyocyanin) by *pseudomonas aeruginosa* nej01r using a maize by-product. Microorganisms 8:1559

Bansod TA (2021) Fermentation technology. World J Pharma Res 10(3):260–299. ISSN 2277–7105

Bolling C, Fiehn O (2005) Metabolite profiling of *Chlamydomonas reinhardtii* under nutrient deprivation. Plant Physiol 139:1995–2005. https://doi.org/10.1104/pp.105.071589

Borrero-de Acuña JM, Poblete-Castro I (2023) Rational engineering of natural polyhydroxyalkanoates producing microorganisms for improved synthesis and recovery. Microbiol Biotechnol 16(2):262–285. https://doi.org/10.1111/1751-7915.14109

Boulton RB, Singleton VL, Bisson IF, Kunkee RE (1996) Principles and practices of wine making. Chapman and hall, New York

Buziol S, Bashir I, Baumeister A et al (2002) New bioreactor-coupled rapid stopped-flow sampling technique for measurements of metabolite dynamics on a subsecond time scale. Biotechnol Bioeng 80:632–636. https://doi.org/10.1002/bit.10427

Castilho LR, Mitchell DA, Freire DM (2009) Production of polyhydroxyalkanoates (PHAs) from waste materials and by-products by submerged and solid-state fermentation. Bioresour Technol 100(23):5996–6009

Cavalheiro JM, de Almeida MCM, Grandfils C, Da Fonseca MMR (2009) Poly (3-hydroxybutyrate) production by *Cupriavidus necator* using waste glycerol. Process Biochem 44(5):509–515

Chen GQ, Wu Q (2005) The application of polyhydroxyalkanoates as tissue engineering materials. Biomaterials 26(33):6565–6578

Choi J, Lee SY (1999) Factors affecting the economics of polyhydroxyalkanoate production by bacterial fermentation. Appl Microbiol Biotechnol 51(1):13–21

Christoph W, Jens OK, Patrick K, Tina B, Elmar H (2004) Impact of the cold shock phenomenon on quantification of intracellular metabolites in bacteria. Anal Biochem 327(1):135–139. ISSN0003-2697. https://doi.org/10.1016/j.ab.2004.01.002

Chuppa-Tostain G, Tan M, Adelard L, Shum-Cheong-Sing A, François JM, Caro Y, Petit T (2020) Evaluation of filamentous fungi and yeasts for the biodegradation of sugarcane distillery wastewater. Microorganisms 8:1588

Costa E, Teixidó N, Usall J, Atarés E, Viñas I (2002) The effect of nitrogen and carbon sources on growth of the biocontrol agent *Pantoea agglomerans* strain CPA-2. Lett Appl Microbiol 35(2):117–120

Crank M, Patel M, Marscheider-Weidemann F, Schleich J, Hüsing B, Angerer G (2004) Techno-economic feasibility of large-scale production of bio-based polymers in Europe (PRO-BIP). Final report

Cromwick AM, Foglia T, Lenz RW (1996) Document details - The microbial production of poly(hydroxyalkanoates) from tallow. Appl Microbiol Biotechnol 46(5–6):464–469

De Souza DP, Saunders EC, McConville MJ, Likic VA (2006) Progressive peak clustering in GC-MS metabolomic experiments applied to leishmania parasites. Bioinformatics 22:1391–1396. https://doi.org/10.1093/bioinformatics/btl085

Denggang W, Wenjie L, Xinying Z, Shuli L, Ying L (2021) Green process: improved semi-continuous fermentation of *Pichia pastoris* based on the principle of vitality cell separation. Front Bioeng Biotechnol 9. https://doi.org/10.3389/fbioe.2021.777774

Devasya RP (2017) Batch and fed batch cultivation and harvesting of nano *Chloropsis gaditana* for environmental applications. Electronic thesis and Dissertation repository. 4670. The university of western Ontario supervisor Dr Amarjeet Bassi, the University of Western Ontario graduate program in chemical and biochemical engineering. A thesis submitted in partial fulfilment of the requirements for the degree in doctor of philosophy

Drysdale CR, McKay AM (1995) Citric acid production by Aspergillus Niger in surface culture on inulin. Lett Appl Microbiol 4:252–254. https://doi.org/10.1111/j.1472-765x.1995.tb00440.x. PMID: 7766122

Einsele A, Ristroph DL, Humphrey A (1978) Mixing times and glucose uptake measured with a fluorometer. Biotechnol Bioeng 20:1487–1492

Ezemba C, Ezemba A (2022) Fermentation, types of fermentors, design and uses of fermentors and optimization of fermentation process. In: Insights in industrial microbiology. Book1. LAP Lambert academic publishing, Republic of Moldova Europe, pp 79–119. ISBN: 9786202056496

Fouillaud M, Dufossé L (2022) Microbial secondary metabolism and biotechnology. Microorganisms 10(1):123

Frost GM, Moss DA (1987) Production of enzymes by fermentation. In: Rehm HJ, Reed G, Kennedy JF (eds) Biotechnol, vol 7a. VCH, Weinheim, pp 65–211

Fukui T, Doi Y (1997) Cloning and analysis of the poly (3-hydroxybutyrate-co-3-hydroxyhexanoate) biosynthesis genes of *Aeromonas caviae*. J Bacteriol 179(15):4821–4830

Fukui T, Doi Y (1998) Efficient production of polyhydroxyalkanoates from plant oils by *Alcaligenes eutrophus* and its recombinant strain. Appl Microbiol Biotechnol 49:333–336

Ganzeveld KJ, van Hagen A, van Agteren MH, de Koning W, Uiterkamp AJ (1999) Upgrading of organic waste: production of the copolymer poly-3-hydroxybutyrate-co-valerate by *Ralstonia eutrophus* with organic waste as sole carbon source. J Clean Prod 7(6):413–419

Ghasem DN (2007) Chapter 5-Growth kinetics. In: Ghasem DN (ed) Biochemical engineering and biotechnology. Elsevier, pp 81–141

Giulia G (2017) The facts about steam chemical indicators. Milestones 3:06–39

Guadalupe-Daqui M, Goodrich-Schneider RM, Sarnoski PJ, Carriglio JC, Sims CA, Pearson BJ, MacIntosh AJ (2023) The effect of $CO_2$ concentration on yeast fermentation: rates, metabolic products, and yeast stress indicators. J Ind Microbiol Biotechnol 50(1):kuad001. https://doi.org/10.1093/jimb/kuad001

Harrison DE, Maitra P (1968) Role of ATP in control of energy metabolism in growing bacteria. J Gen Microbiol 53:S7–S8

Hewitson J, Hill C (2018) How does sugar affect yeast growth? https://www.saps.org.uk/saps-associates/browse-q-and-a/169-q-a-a-how-does-sugar-affect-yeast-growth. Accessed 10/1/18

Iorhemen OT, Hamza RA, Tay JH (2016) Membrane bioreactor (MBR) technology for wastewater treatment and reclamation: membrane fouling. Membranes (Basel) 6(2):33. https://doi.org/10.3390/membranes6020033

Ishak AK, Annuar MSM, Heidelberg T, Gumel AM (2016) Ultrasound-assisted rapid extraction of bacterial intracellular medium-chain-length poly(3-hydroxyalkanoates) (mcl- PHAs) in medium mixture of solvent/marginal non-solvent. Arab J Sci Eng 41:33–44

Jay J (2000) Fermentation and fermented dairy products. In: Modern food microbiology, pp 113–130. https://doi.org/10.1007/978-1-4615-4427-2_7

Jendrossek D (2001) Microbial degradation of polyesters. Biopolyesters:293–325

Kahar P, Tsuge T, Taguchi K, Doi Y (2004) High yield production of polyhydroxyalkanoates from soybean oil by *Ralstonia eutropha* and its recombinant strain. Polym Degrad Stab 83(1):79–86

Kennedy M, Krouse D (1999) Strategies for improving fermentation medium performance: a review. J Ind Microbiol Biotech 23:456–475

Khosravi DK, Vasheghani FE, Shojaosadati SD, Yamini Y (2004) Effect of process variables on supercritical fluid disruption of *Ralstonia eutropha* cells for poly-(R)-hydroxybutyrate recovery. Biotechnol Prog 20:1757–1765

Kim BS (2000) Document details-production of poly(3-hydroxybutyrate) from inexpensive substrates. Enzyme Microb Technol 27(10):774–777

Kim YB, Lenz RW (2001) Polyesters from microorganisms. Biopolyesters:51–79

Koller M (2020) Established and advanced approaches for recovery of microbial polyhydroxyalkanoate (PHA) biopolyesters from surrounding microbial biomass. Euro Biotech J 4(3):113–126

Koller M, Bona R, Braunegg G, Hermann C et al (2005) Production of polyhydroxyalkanoates from agricultural waste and surplus materials. Biomacromolecules 6(2):561–565

Koller M, Bona R, Chiellini E, Fernandes EG, Horvat P, Kutschera C, Braunegg G (2008) Polyhydroxyalkanoate production from whey by *Pseudomonas hydrogenovora*. Bioresour Technol 99(11):4854–4863

Koning D, Van W, Dam K (1992) A method for the determination of changes of glycolytic metabolites in yeast on a subsecond time scale using extraction at neutral pH. Anal Biochem 204:118–123

Koutinas AA, Xu Y, Wang R, Webb C (2007) Polyhydroxybutyrate production from a novel feedstock derived from a wheat-based biorefinery. Enzyme Microb Technol 40(5):1035–1044

Kozlovskii AG, Zhelifonova VP, Antipova TV (2013) Fungi of the genus *Penicillium* as producers of physiologically active compounds (review). Appl Biochem Microbiol 49(1):1–10

Kram JW (2008) Metabolix grows plastic (producing) plants. Biomass Magazine October

Lange HC, Eman MZ, Visser G, Diana D, Frank JC, Johannes M, Joost H (2001) Improved rapid sampling for in vivo kinetics of intracellular metabolites in *Saccharomyces cerevisiae*. Biotechnol Bioeng 75:406–415. https://doi.org/10.1002/bit.10048

Larsson G, Tornkvist M (1996) Rapid sampling, cell inactivation and evaluation of low extracellular glucose concentrations during fed-batch cultivation. J Biotechnol 49:69–82. https://doi.org/10.1016/0168-1656(96)01534-9

Leder IG (1972) Interrelated effects of cold shock and osmotic pressure on the permeability of the *Escherichia coli* membrane to permease accumulated substrates. J Bacteriol 111(1):211–219. https://doi.org/10.1128/jb.111.1.211-219.1972

Lee S, Lee JA, Park GG, Jang JK, Park YS (2017) Semi continuous fermentation of onion vinegar and its functional properties. Molecules 22(8):1313. https://doi.org/10.3390/molecules22081313

Lemoigne M (1926) Produit de déshydratation et de polymérisation de l'acide β-oxybutyrique. Bull Soc Chim Biol 8:770–782

Link PM, Borchert LH (2015) Coastline Reports 24. In: Link PM, Borchert LH (eds) Blickpunkt Küste: Aktuelle Forschungsansätze in der Meeres- und Küstengeographie. ISSN 0928-2734. ISBN 978-3-939206-17-0

Listewnik HF, Wendlandt KD, Jechorek M, Mirschel G (2007) Process design for the microbial synthesis of poly-β-hydroxybutyrate (PHB) from natural gas. Eng Life Sci 7(3):278–282

Liu F, Li W, Ridgway D, Gu T, Shen Z (1998) Production of poly-β-hydroxybutyrate on molasses by recombinant *Escherichia coli*. Biotechnol Lett 20:345–348

Marcel GA (2010) Characteristics and techniques of fermentation systems. In: Food fermentation and biotechnology. Chap.7.1. Asia Tech Pub, pp 183–227

Mashego MR, van Gulik WM, Vinke JL, Visser D, Heijnen JJ (2006) In vivo kinetics with rapid perturbation experiments in *Saccharomyces cerevisiae* using a second-generation BioScope. Metab Eng 8(4):370–383. https://doi.org/10.1016/j.ymben.2006.02.002

Mason A, Kenneth H, Peter R, Johnson GB, Singer SR (2013) Foundations of life: chemistry, cells, and genetics: selected materials from biology, 10th edn. McGraw-Hill Education, Boston. ISBN 978-0077775803. OCLC 846845827

Mattey MICHAEL, Kristiansen BJØRN (1999) A brief introduction to citric acid biotechnology. In: Citric acid biotechnology. Taylor & Francis, pp 1–10

Monciardini P, Iorio M, Maffioli S, Sosio M, Donadio S (2014) Discovering new bioactive molecules from microbial sources. Microb Biotechnol 7:209–220

Moo-Young M, Moreira AR, Tengerdy RP (1983) Principles of solid-substrate fermentation. In: The filamentous fungi, vol 4. Edward Arnold, London

Nath A, Dixit M, Bandiya A, Chavda S, Desai AJ (2008) Enhanced PHB production and scale up studies using cheese whey in fed batch culture of *Methylobacterium* sp. ZP24. Bioresour Technol 99:5749–5755

Nawaz A, Chaudhary R, Shah Z, Dufossé L, Fouillaud M, Mukhtar H, ul Haq I (2021) An overview on industrial and medical applications of biopigments synthesized by marine bacteria. Microorganisms 9:11

Nikel PI, De Almeida A, Melillo EC, Galvagno MA, Pettinari MJ (2006) New recombinant *Escherichia coli* strain tailored for the production of poly (3-hydroxybutyrate) from agroindustrial by-products. Appl Environ Microbiol 72(6):3949–3954

Nonato RV, Mantelatto P, Rossell C (2001) Integrated production of biodegradable plastic, sugar and ethanol. Appl Microbiol Biotechnol 57:1–5

Oliveira FC, Freire DMG, Castilho LR (2004) Production of poly(3- hydroxybutyrate) by solid-state fermentation with *Ralstonia eutropha*. Biotechnol Lett 26:1851–1855

Oliveira FC, Dias ML, Castilho LR, Freire DMG (2007) Characterization of poly(3-hydroxybutyrate) produced by *Cupriavidus necator* in solid-state fermentation. Bioresour Technol 98:633–638

Page WJ, Manchak J, Rudy B (1992) Formation of poly(hydroxybutyrate-co-hydroxyvalerate) by *Azotobacter vinelandii* UWD. Appl Environ Microbiol 58:2866–2873

Panagiotou G, Christakopoulos P, Olsson L (2005) Simultaneous saccharification and fermentation of cellulose by *Fusarium oxysporum* F3—growth characteristics and metabolite profiling. Enzyme Microb Technol 36(5–6):693–699. ISSN 0141-0229, https://doi.org/10.1016/j.enzmictec.2004.12.029. https://www.sciencedirect.com/science/article/pii/S0141022905000037

Pasotti L, Susanna Z, Michela C, Nicolo P, Ilaria M et al (2015) Methods for genetic optimization of biocatalysts for biofuel production from dairy waste through synthetic biology. In: Engineering Med Biol Soc (EMBC), 37th Ann Internl Conf IEEE:953–956

Pinu FR, Villas-Boas SG, Aggio R (2017) Analysis of intracellular metabolites from microorganisms: quenching and extraction protocols. Metabolites 7(4):53. https://doi.org/10.3390/metabo7040053

Poostchi AAM, Mohammad R, Fariba SM (2012) Low cost monofilament mesh filter used in membrane bioreactor process: filtration characteristics and resistance analysis. Desalination 286:429–435. https://doi.org/10.1016/j.desal.2011.12.002

Qin Q, Chen L, Yiqing Z, Tian Y, Jin Y, Yingying G, Guo QC (2018) CRISPR/Cas9 editing genome of extremophile *Halomonas* spp. Metab Eng 47:219–229. ISSN 1096-7176. https://doi.org/10.1016/j.ymben.2018.03.018

Reusch RN (2000) Transmembrane ion transport by polyphosphate/poly-(R)-3-hydroxybutyrate complexes. Biochemistry 65(3):280–295

Ruijter GJG, Visser J (1996) Determination of intermediary metabolites in *Aspergillus Niger*. J Microbiol Methods 25:295–302. https://doi.org/10.1016/0167-7012(95)00104-2

Rusendi D, Sheppard JD (1995) Hydrolysis of potato processing waste for the production of poly-β-hydroxybutyrate. Bioresour Technol 54(2):191–196

Saez MJ, Lagunas R (1976) Determination of intermediary metabolites in yeast. Critical examination of the effect of sampling conditions and recommendations for obtaining true levels. Mol Cell Biochem 13:73–78

Schaefer U, Boos W, Takors R, Weuster-Botz D (1999) Automated sampling device for monitoring intracellular metabolite dynamics. Anal Biochem 270:88–96. https://doi.org/10.1006/abio.1999.4048

Shobharani P, Agrawal R (2010) Interception of quorum sensing signal molecule by furanone to enhance shelf life of fermented milk. Food Control 21(1):61–69

Shobharani P, Ramesh B, Agrawal R (2006) Strain improvement by mutagenesis and optimum conditions for culture parameters by response surface methodology for lactose tolerance in a novel native culture isolate *Leuconostoc mesenteroides* sub sp. Res J Biotechnol 1(2):5–11

Silva LF, Taciro MK, Michelin RME, Carter JM, Pradella JGC, Gomez JGC (2004) Poly-3-hydroxybutyrate (P3HB) production by bacteria from xylose, glucose and sugarcane bagasse hydrolysate. J Ind Microbiol Biotechnol 31(6):245–254

Soccol CR et al (2006) Citric acid production. Food Technol Biotechnol 44(2):141–149

Steinbüchel A (2001) Perspectives for biotechnological production and utilization of biopolymers: metabolic engineering of polyhydroxyalkanoate biosynthesis pathways as a successful example. Macromol Biosci 1:1–24

Sudesh K, Abe H, Doi Y (2000) Synthesis, structure and properties of polyhydroxyalkanoates: biological polyesters. Prog Polymer Sci 25(10):1503–1555. ISSN 0079-6700. https://doi.org/10.1016/S0079-6700(00)00035-6

Sue T, Obolonkin V, Griffiths H, Villas-Bôas SG (2011) An exometabolomics approach to monitoring microbial contamination in microalgal fermentation processes by using metabolic footprint analysis. Appl Environ Microbiol 77(21):7605–7610. https://doi.org/10.1128/AEM.00469-11

Taga N, Tanaka K, Ishizaki A (1997) Effects of rheological change by addition of carboxymethylcellulose in culture media of an air-lift fermentor on poly-d-3-hydroxybutyric acid productivity in autotrophic culture of hydrogen-oxidizing bacterium, *Alcaligenes eutrophus*. Biotechnol Bioeng 53(5):529–533

Theobald U, Mailinger W, Reuss M, Rizzi M (1993) In vivo analysis of glucose-induced fast changes in yeast adenine nucleotide pool applying a rapid sampling technique. Anal Biochem 214:31–37. https://doi.org/10.1006/abio.1993.1452

Villas-Bôas SG, Bruheim P (2007) Cold glycerol-saline: the promising quenching solution for accurate intracellular metabolite analysis of microbial cells. Anal Biochem 370(1):87–97. https://doi.org/10.1016/j.ab.2007.06.028

Visser D, van Zuylen GA, van Dam JC, Oudshoorn A, Eman MR, Ras C, van Gulik WM, Frank J, van Dedem GWK, Heijnen JJ (2002) Rapid sampling for analysis of in vivo kinetics using the bioscope: a system for continuous-pulse experiments. Biotechnol Bioeng 79:674–681. https://doi.org/10.1002/bit.10328

Volesky B, Luong HJT (1985) CRC Crit Rev Biotechnol 2:119–146

Wallen LL, Rohwedder WK (1974) Poly- beta-hydroxyalkanoate from activated sludge. Environ Sci Technol 8(6):576–579

Wiggam MI, O'Kane MJ, Harper R, Atkinson AB, Hadden DR, Trimble ER, Bell PM (1997) Treatment of diabetic ketoacidosis using normalization of blood 3-hydroxybutyrate concentration as the endpoint of emergency management: a randomized controlled study. Diabetes Care 20(9):1347–1352

Xiong B, Li Z, Liu L et al (2018) Genome editing of *Ralstonia eutropha* using an electroporation-based CRISPR-Cas9 technique. Biotechnol Biofuels 11:172. https://doi.org/10.1186/s13068-018-1170-4

Yİlmaz MİRAÇ, Beyatli Y (2005) Poly-β-hydroxybutyrate (PHB) production by a *Bacillus cereus* M5 strain in sugar beet molasses. Sugar Industry 130:109–112

Zhang A, Sun H, Xu H, Qiu S, Wang X (2013) Cell metabolomics. OMICS 17(10):495–501. https://doi.org/10.1089/omi.2012.0090

Zhang Z, Li Y, Zhang J, Peng N, Liang Y, Zhao S (2020) High-titer lactic acid production by *Pediococcus acidilactici* PA204 from corn stover through fed-batch simultaneous saccharification and fermentation. Microorganisms 8:1491

Zinn M, Witholt B, Egli T (2001) Occurrence, synthesis and medical application of bacterial polyhydroxyalkanoate. Adv Drug Deliv Rev 53(1):5–21

# Industrially Important Secondary Metabolites

8

**Abstract**

The most important products that have relevance to human health are produced by microorganisms during fermentation. These are formed in the idiophase of growth and are known as idiolites. The ability to produce secondary metabolites is usually lost on mutation. These compounds have a low molecular weight. However, they have diverse biological activity. With recent biogenetic tools, it has been possible to modify the biosynthetic pathway in order to enhance the yields of secondary metabolites. This chapter deals with the major secondary metabolites that can be produced from microorganisms using fermentation technology processes. The chapter also discusses the various challenges in microbial processes and provides with inducers that can be used to enhance the yield of the secondary metabolites.

**Keywords**

Industrial · Secondary metabolites · Microorganisms · Biotechnology · Screening · Extraction · Bioactive properties

## 8.1 Introduction

The secondary metabolites (SM) are very beneficial to the health. Traditionally, it has been a practice to use the fermented food products. These products have metabolites produced by the fermentation of bacteria/fungi/yeasts. The production of these metabolites are regulated by the media nutrients, rate of microbial growth, regulation of feedback inhibition, and inactivation and induction of cellular enzymes. For SM production, initially, an improved strain of the isolated microorganism is inoculated into a growth medium, which is then transferred from the flask to a small fermentor or a seed culture. In the growth phase, it is inoculated into a


R. Agrawal, *Textbook of Industrial Microbiology*,
https://doi.org/10.1007/978-981-97-9582-6_8

fermentor tank with production medium. The parameters such as the medium composition, pH, temperature, and agitation and aeration rate need to be optimized in order to produce maximum yield of the product. The metabolism is very strain specific and the biologically active molecules have different structures and mode of action. The common secondary metabolites include pesticides, pigments, toxins, pheromones, enzyme inhibitors, immunomodulating agents, receptor antagonists and agonists, antitumor, anticancer agents, and cholesterol-regulating agents. Other properties of these secondary metabolites are antimicrobial, antiviral, antioxidant, vasorelaxant/contractant, diuretic, or laxative. Around 40% of the microbial secondary metabolites have been approved by the FDA. Till recently, more than 10,000 antibiotics have been discovered, but most of them are either toxic or inactive. There are many unusual chemical structures of many of these secondary metabolites. The potential of microorganisms to produce secondary metabolites is very high. Many of these metabolites help in improving human health and wellness (Ruiz et al. 2010). Actinobacteria and fungi have shown a lot of promise in this direction (O'Brien and Wright 2011). Usually, the cryptic gene clusters that are not active get activated and create chemical diversity (Xu et al. 2019). These secondary metabolites are also known as idiolites as they are produced during the idiophase, which come after the growth is completed. Most of the secondary metabolites are produced either at the end or near the stationary growth phase of the microorganism. These products have no function in growth and are produced by narrow taxonomic groups of organisms. The chemical structures are varied and are of different types. These include β-lactam rings, cyclic peptides, depsipeptides, polyacetylenes, and polyenes.

Agrawal et al. (1999) have shown the improvement in the biotransformation of alpha pinene to verbenol in *Aspergillus* sp. and *Penicillium* sp., which were induced by mutation using chemicals and ultraviolet irradiation.

## 8.2 Classification of Secondary Metabolites

(i) **Terpenes** These are a class of natural products composed of isoprene units. These include hydrocarbons and terpenoids, which are basically oxygenated hydrocarbons. The general molecular formula of terpenes are multiples of $(C_5H_8)_n$, where "*n*" is the number of linked isoprene units. Therefore, terpenes are also termed as isoprenoid compounds. These compounds are classified based on the number of isoprene units present in their structures. Divyashree et al. (2006) studied the biotransformation of chiral terpenes by *A. niger* and *Pseudomonas putida*.

(ii) **Phenolics** These are chemical compounds that have an aromatic ring structure and has a one or more hydroxyl groups. These are common secondary metabolites of plants, which can be simple molecules as phenolic acid to highly polymerized substances like tannins. The classes of phenolics have been characterized on the basis of their basic skeleton.

Agrawal et al. (2003) studied the biotransformation of ferulic acid to vanillin by locally isolated bacterial cultures.

(iii) **Alkaloids** The structure includes two nitrogen atoms, both of which are derived from amino acids. The nitrogen atom in the pyridine ring is derived from aspartate, whereas the nitrogen atom that in the pyrrolidine ring is derived from arginine or ornithine.

These are classified into two broad categories: nonheterocyclic and atypical alkaloids, as hordenine or *N*-methyltyramine, colchicine, and taxol. Heterocyclic alkaloids are quinine, caffeine, and nicotine.

Examples of alkaloids produced by plants are hyoscyamine, which is present in *Datura stramonium* and atropine, which is present in *Atropa belladonna*.

- Cocaine is present in *Erythroxylum coca* the Coca plant
- Scopolamine is present in the *Solanaceae* (night shade, plant family).
- Codeine and morphine are present in *Papaver somniferum*, the opium poppy.
- Vincristine and vinblastine are well-known mitotic inhibitors found in *Catharanthus roseus*, the rosy periwinkle.

Most of the alkaloids affect the central nervous system by binding to neurotransmitter receptors. Glucosinolates are secondary metabolites that include both sulfur and nitrogen atoms and are derived from glucose, an amino acid, and sulfate. These have been studied with microbial cells.

(iv) **Fungal secondary metabolites** The three main classes of fungal secondary metabolites are polyketides, nonribosomal peptides, terpenes, and compounds from shikimic acid. The most common fungal secondary metabolite is penicillin discovered by Fleming in 1928 (Dias et al. 2012). Fleming (1929) studied the antibacterial action of *Penicillium*, with special reference to their use in the isolation of *B. influenzae*. Other important fungal secondary metabolites are Lovastatin, a polyketide from *Pleurotus ostreatus* (oyster mushrooms) and *Aspergillus terreus*; Aflatoxin B1, a polyketide from *Aspergillus flavus*; and Ciclosporin. Sometimes, fungal secondary metabolites are dangerous to humans. *Claviceps purpurea*, a member of the ergot group of fungi, which grows on rye, can result into death when ingested. The build-up of poisonous alkaloids is found in *C. purpurea*, which leads to the symptoms such as seizures and spasms, diarrhea, paresthesias, itching, psychosis, or gangrene (Arakelyan 2018). Elander et al. (1986) studied the different genetic aspects of overproduction of antibiotics and other secondary metabolites, strain improvement, and process control strategies.

Vidya and Agrawal (2003) studied the production of verbenol, a high-valued food flavorant from improved culture strain of *A. niger* using protoplast fusion.

(v) **Nonribosomal peptides** These are commonly produced by bacteria and fungi. A prenylated nonribosomal peptide from *Claviceps purpurea*. *Streptomyces* strains produce Echinomycin 5,7,4-trihydroxy-3-3,5-trimethoxy flavone, which has shown to have activity against Methicillin-resistant *Staphylococcus aureus*. Lovastatin was the first FDA-approved secondary metabolite that has the ability to lower the level of cholesterol.

The major producers of nonribosomal peptides (NRP) are fungi and bacteria. The enzyme nonribosomal peptide synthetase, which is a multidomain, multimodular enzyme (NRPS), is responsible for the production of NRPs (Soltani 2016). These bioactive molecules possess biological activities and are also beneficial in pharmaceutical products. NRPs include siderophores, pigments, and toxins. Many of NRP products are available in the market. A microbial strain, namely *P. brocae* MA-231, has shown the production of sulfide diketopiperazine derivatives and penicibrocazines A–E. In a study by Martínez-Núñez and Lopez (2016), it was found that NRP products can be produced either from proteinogenic amino acids or nonproteinogenic amino acids. These NRP products have wide range of properties and include compounds such as siderophores, pigments (indigoidine), antibiotics (actinomycin, penicillin, cephalosporin, vancomycin), cytostatics (bleomycin), immunosuppressants (cyclosporines), and anticancer agents (Cragg et al. 2009). NRPSs are also known to produce harmful toxins such as HC-toxin and AM-toxin. These are phytotoxins and are known to destroy the crops (Xu et al. 2021). Penicillin formation from the fungus *Penicillium* sp. is an NRP. Fierro et al. (1995) mapped the genome of *P. chrysogenum*. It was then discovered that the gene clusters for penicillin biosynthesis were located on chromosome I. When two amino acids (**L**-α-aminoadipic acid and **L**-cysteine and **L**-valine) condense together, they form the tripeptide δ-**L**-α-aminoadipyl-**L**-cysteinyl-**D**-valine. When this gets, it forms isopenicillin N and then gets converted to penicillins (Martin et al. 2010). Ziemons et al. (2017) studied the production of penicillin in industrial strain of *Penicillium chrysogenum* P2niaD18 and found that the production is not dependent on the copy number of biosynthesis genes.

(vi) **Ribosomal peptides** These include Microcins, which has a bacteriocin as microcin V from *Escherichia coli*. Thiostrepton, a ribosomal peptide, is obtained from several strains of streptomycetes, e.g., *Streptomyces azureus*.
(vii) **Glucosides** Nojirimycin is an iminosugar obtained from a class of *Streptomyces* species.
(viii) **Alkaloids** Tetrodotoxin is a neurotoxin, which is produced from *Pseudoalteromonas* and other bacteria living in symbiosis with animals such as puffer fish.
(ix) **Polyketides** This class of secondary metabolite is very important. These compounds are produced by varied microorganisms such as bacteria, fungi, plants, and animals (Staunton and Weissman 2001). The importance of these metabolites lies in their biological and pharmacological activities. These properties make them very useful in medicines (Seca and Pinto 2019). The synthesis of these compounds takes place by using the substrate acyl-coenzyme A. This substrate uses the enzyme polyketide synthase (PKSs) to form polyketides.

Woo et al. (2010) studied the polyketide synthase genes, which are multi-domain proteins in *Penicillium marneffei*, are quite varied. They observed that altogether around 25 pks genes were involved. For maximum production of polyketide, the conditions during fermentation have been optimized (Woo et al. 2019). Pharmaceutically important bioactive polyketide compounds commonly used in many drugs (Ochratoxins A and B) have been isolated successfully by different scientific works (Frisvad et al. 2004) and citrinin by Devi et al. (2009).

(x) **Ergot alkaloids** The ergot alkaloids are basically mycotoxins. The formation of ergot alkaloids is initiated from prenylated tryptophan. This later gets processed into other alkaloids. The formation is common in many different fungi. These secondary metabolites contain nitrogen in their molecule. This comes from indole alkaloids. Frisvad et al. (2018) studied the mycotoxin and secondary metabolite potential of *Aspergillus niger*, *Aspergillus oryzae*, and *Trichoderma reesei*.

The formation of ergot alkaloids in *Penicillium* sp. and other ascomycetous fungi (*Claviceps*, *Epichloë*, and *Aspergillus* species) is very common. They are divided into three groups: clavines, ergoamides, and ergopeptides. All these three alkaloids are synthesized by *Penicillium* fungi (Kozlovskii et al. 2012). The first group consists of the derivatives of 6-*N*-methylergoline, which includes festuclavine, epicostaclavine, costaclavine, fumigaclavines, and isofumigaclavines A and B with a fully saturated D ring. The second group consists of ergolines with a double bond at the 8,9-position. This includes agroclavine, agroclavine-I, chanoclavine-I, chanoclavine-III, and isochanoclavine-I. Epoxyagroclavine-I can also be included in this group as an exception. The third group is made up of clavine alkaloids with modified rings C and D: rugulovasins A and B, aurantioclavine, and α-cyclopiazonic acid (CPA).

**Alkaloid Biosynthesis**

These are organic compounds that are produced naturally and have a minimum of one nitrogen atom and has an amine type of structure. Most of the alkaloids are basic in nature; however, some may be neutral or weak acid. With the advent of recent modern technologies, it has been possible to understand that during alkaloid synthesis, the secondary pathways are connected with the host central metabolism. This happens with the help of many enzymes. Alkaloids are classified based on the substance that starts the pathway of synthesis. These include amino acids, terpenoids, and purines. Many amino acids such as ornithine, lysine, phenylalanine, tyrosine, tryptophan, and histidine work as precursors in the formation of alkaloids. Earlier researchers (Liscombe et al. 2005; Takemura et al. 2010, 2012) have shown that newer molecules are formed with the modification of gene expression of transcriptional regulators. Biosynthesis of alkaloid takes place by Mannich reaction in which the C–C bond formation takes place. Alkaloids are used as anesthetics, cardioprotective agents, and for their anti-inflammatory properties. Most commonly used alkaloids are morphine, strychnine, quinine, ephedrine, and nicotine.

There are many environmental stresses and plant defense mechanisms that regulate secondary metabolism. Benzylisoquinoline alkaloid (BIA)-specific cytochrome P450 enzymes are important in BIA biosynthesis. Genes in phenylpropanoid biosynthesis do not form a gene cluster. However, when formed, it is due to the sequential gene duplications, genome rearrangement, and gene loss. PUFA (polyunsaturated fatty acids) affects the membrane fluidity and the role in signaling is also affected.

Basically, alkaloid biosynthesis involves accumulation of precursors such as amine and aldehyde when cations are formed followed by Mannich reaction.

BIA metabolism in alkaloids has a special position as it is very common. There are many enzymes involved during the metabolism. Studies have shown that the corresponding cDNAs have been isolated, which belong to *S*-adenosylmethionine-dependent *O*- and *N*-methyltransferases, CYP-dependent monooxygenases, FAD-linked oxidoreductases, acyl-CoA-dependent acetyltransferases, 2-oxoglutarate/$Fe^{2+}$-dependent dioxygenases, and three different subgroups of NADPH-dependent reductases (aldo-keto reductases, short-chain dehydrogenase/reductases, and Rossman-fold reductases). With the BIA (benzylisoquinoline) gene repository, it has become possible to synthesize them in microbes, where the yields are much higher and can be used on a commercial scale.

There are 70 different types of deoxyhexose sugars found in natural metabolites. However, as secondary metabolites, there are 23,000 terpenoids. Many of the secondary metabolites are produced by nonribosomal systems and have a mass that is less than 1500 Da. The peptides that are obtained from ribosomes have higher molecular weight (3000–4000 Da). González-Jiménez et al. (2020) studied the proteins that are involved during the fermentation with strains of *Saccharomyces cerevisiae* in sparkling wine. The most important precursors used in secondary metabolism are acetyl-coenzyme A (CoA) and propionyl-CoA. Some are also produced from the intermediates of the shikimic acid pathway, the tricarboxylic acid cycle, and from amino acids. Omic sciences (genomic, transcriptomic, and metabolomic) are very helpful in strain characterization and in the improvement of the strains. Sweechin et al. (2015) worked on the bioprocess optimization for biomass production of probiotics, yeast *Saccharomyces boulardii*, in semi-industrial scale. Deshmukh et al. (2018) wrote an excellent review on the production of anticancer compounds from marine fungi. Cocultivation of bacteria and fungi have resulted in the production of new secondary metabolites. Argüelles et al. (2020) have shown a synergistic effect with a novel bi-factorial strategy against *Candida albicans* for an antifungal action. Williams (2009) have studied the potential of marine bacteria as a source of new therapeutics. Hilgers et al. (2021) studied the heterologous production of β-caryophyllene and evaluated the activity against plant pathogenic fungi. Hwang et al. (2020) worked on the pentaminomycins C–E. These are cyclic pentapeptides and their role as autophagy inducers has been studied. The genus *Penicillium* has more than 350 species and they produce many novel bioactive compounds. An antifungal antibiotic "griseofulvin" has been obtained from the mycelium of *Penicillium griseofulvum* Dierckx. It was also isolated from the culture filtrate of *Penicillium janczewskii* Zal, which caused abnormal curling of the hyphae of *Botrytis alii* and was later identified as griseofulvin. Griseofulvin has been isolated from fungi *Penicillium* species (*P. urticae*, *P. raistrickii*, *P. raciborskii*, *P. kapuscinskii*, *P. albidium*, *P. melinii*, *P. brefeldianum*, and some mutant strains of *P. patullum*). This metabolite is also produced by other genera of fungi. These include *Aspergillus versicolor* and *Nematospora coryli*. Griseofulvin consists of a tricyclic-spiro system, which is made up of a chlorine-substituted coumaranone and an enone containing a cyclohexane ring adjacent to the asymmetric spirane center (Oxford et al. 1940). Isochroman toxins are produced from *P. steckii* and tanzawaic acids from *P. citrinum*. *Penicillium* spp. can also produce other secondary metabolites as ergot alkaloids, diketopiperazines, quinolines, quinazolines, polyketides, camazulene, and azetidine. The secondary metabolites have unique properties to kill pathogenic microorganisms, can combat cancer, enhance the immunity, and are able to inhibit unwanted enzymes in the system along with many other biological activities. Many of the microorganisms produce antibiotics, which are highly useful in preventing harmful diseases. The production of many antibiotics has been explained in Chap. 2 of this book. Each antibiotic works in a specific manner to kill the pathogenic cells. A broad spectra antibiotic namely erythromycin is produced from *Saccharopolyspora erythraea*. This antibiotic is used to prevent many pathogenic organisms and is helpful to prevent many diseases. *Bacillus subtilis* is useful as it produces bacitracin, a topical antibiotic, which can kill many pathogens in a very less time.

According to the literature, it is the nonribosomal peptide synthetase that helps in the formation of a peptide. It is used as an antibiotic. Various parameters have been optimized for the maximum production of pyocyanin from *Pseudomonas aeruginosa* (Bacame-Valenzuela et al. 2020). A review article written by Fouillaud and Dufossé (2022) covers many of the microbial secondary metabolites. With recent discoveries, newer biomolecules are being produced (Monciardini et al. 2014). Recently, there is a general awareness to use natural biomolecules as antipathogenic, which can kill the pathogens and cure the human kind from harmful diseases without any side effects. With this in mind, much work is being done toward the production of bacteriocins. Work has been done using the culture of *Lactococcus lactis* Gh1 (Jawan et al. 2020) on the optimization of the culture conditions and medium compositions for the production of bacteriocins. The blue-green algae are another good source to produce beneficial secondary metabolites (Jeong et al. 2020). The endophytic fungi produce many beneficial health-promoting secondary metabolites. In this regard, the work of Khiralla et al. (2020) using *Curvularia papendorfii* cells is worth mentioning. The work isolated a new compound, polyhydroxy acid. This compound has activity both against bacteria and viruses (Table 8.1).

**Table 8.1** Major secondary metabolites produced at large scale by microorganisms

| Microorganism | Secondary metabolite | Reference |
|---|---|---|
| *Streptomyces* sp. | Lynamicins | Paulus et al. (2017) |
| *Nocardia uniformis* | Nocardicin | Igarashi et al. (2000) |
| *Actinomadura* sp. | Xylanase | Prakash et al. (2013) |

Recently, scientists have studied the great potential of marine microbes for producing valuable new molecules as secondary metabolites. Lebeau et al. (2020) discovered an alternative extraction and characterization of nitrogen containing azaphilone red pigments and ergosterol derivatives from the marine-derived fungus, *Talaromyces* sp., which has a lot of industrial importance. Sponges are a good source for producing compounds with antiaging properties. Saïd-Hassane et al. (2020) studied the marine sponge *Scopalina hapalia*. Marine epiphytic bacteria have been able to survive the constant pressures caused by the hydrostatic pressure, low oxygenation, and salinity. There is also competition due to other aquatic organisms (protozoans and nematodes) and hence they produce bioactive compounds. In this regard, Salikin et al. (2020) studied the potential for novel anti-nematode drug development. Nawaz et al. (2021) worked on the production of pigments (prodigiosin, astaxanthin, violacein, zeaxanthin, lutein/lycopene) from marine bacteria and are produced on a large scale. *Streptomyces* sp. GG23 strain (actinobacteria) has been isolated from an insect gut (*Tenebrio molitor*), which can produce cyclic pentapeptides pentaminomycins C, D, and E. In a study (Shamikh et al. 2020), it has been shown that the antibiotic compounds such as chlorocardicin, neocopiamycin A, and chicamycin B could be produced by *Micromonospora* sp. UA17 when it was cocultured with two mycolic acid-containing actinomycetes (*Gordonia* sp. UA19 and *Nocardia* sp. UA 23). These compounds were not formed in their monocultures. Voitsekhovskaia et al. (2020) worked on the formation of baikalomycins A–C, which is a new aquayamycin-type angucycline. They isolated a culture strain, *Streptomyces* sp. IB201691-2A, from the lake, namely Baikal. Iron overload disease can be cured and uptake of antibiotics has been made possible with the help of siderophores, as serratiochelins. This could happen as these compounds have a high affinity for ferric iron. They are usually produced by a pathogen (*Serratia marcescens*). Schneider et al. (2020) studied the bioactivity of serratiochelin a, a siderophore that was isolated from a coculture of *Serratia* sp. and *Shewanella* sp.

The use of microbial consortium has shown to produce new bioactive molecules that can improve human health. Marine strain of *Streptomyces* sp. (IB201691-2A) can synthesize three new angucyclines (baikalomycins A–C), along with rabelomycin and 5-hydroxy-rabelomycin, which have anticancer and antiproliferative activities. The crude extract from a culture of *Curvularia papendorfii*, an endophytic fungus, which was isolated from *Vernonia amygdalina*, a medicinal plant from Sudan, produces an antiviral metabolite. This can act against the human coronavirus HCoV 229E and the feline coronavirus FCV F9. It has also been found to have antibacterial activity against *Staphylococcus* sp. and anticancer activity against the human breast carcinoma MCF7 cell line. The aquatic microalgae *Aurantiochytrium* sp. can produce a compound known as docosahexaenoic acid (DHA), which is a n-3 long-chain polyunsaturated fatty acid and helps in improving health. In a study, Liu et al. (2020) produced a UV mutant of the microalgae for the production of DHA, which was many folds higher (1.90-fold higher) than the wild strain. The expression of many genes (CoAT, AT, ER, DH, and MT genes) have been found to be responsible for the enhanced production of DHA. Walter et al. (2020) studied the fermentation process to produce *N*-methylanthranilate from the bioengineered strain of *Corynebacterium glutamicum*.

Biofilms, which are made up of exopolysaccharides and protect the pathogenic bacteria, need to be removed in any process (Agrawal 2022). In this regard, the use of natural antimicrobial compounds and use of lytic phages is common. Papaianni et al. (2020) studied the lytic activity of Xccφ1 (*Xanthomonas campestris* pv. *campestris*-specific phage), which inhibits the formation of biofilms by modulating the gene pathways.

Hamed et al. (2020) studied the use of endophytic fungi that inhibit the nosocomial infections in textiles. In this regard, the use of *Aspergillus calidoustus* strain M113 has shown antimicrobial activity. Cyanobacteria, as well as all the photosynthetic organisms, can produce secondary metabolites by using solar energy and carbon dioxide.

**Natural Products with Special Role in Modern Medicine**

The contribution of natural secondary metabolites to modern medicine is very important. Secondary metabolites are mainly produced by the plants and many marine invertebrates as a chemical defensive mechanism. The studies undertaken in producing these compounds are important to understand (Williams 2009; Harborne 1994). Many of these products are naturally produced by microorganisms.

Bacteria produce signaling mechanism, which is known as quorum-sensing. This can control the cell growth and biofilm formation in pathogen microorganisms. Acyl homoserine lactones (AHLs) are the compounds (*N*-3-oxohexanoyl-L-homoserine lactone) from *Vibrio fisheri*, as well as another compound, furanone boronate diester (Autoinducer-2; Al-2), that activates the genes of virulence, spore formation, and for biofilm formation. A lot of research is being carried out on quorum sensing and to discover novel antimicrobial agents for controlling the virulence of deadly drug-resistant bacteria such as *Pseudomonas aeruginosa*, *Escherichia coli*, and *Staphylococcus aureus*. Recent study has shown the potential of natural products or modified natural products as inhibitors of biofilms. The potential of natural products that possess potential to inhibit indole-containing compounds has been discussed earlier (Melander et al. 2020).

Microbial secondary metabolites are the compounds produced mainly by actinomycetes and fungi in the stationary growth phase (idiophase). Apart from antibiotics, there are many other secondary metabolites (SM) that are used in pharmaceuticals, cosmetics, food, agriculture, and farming. All the compounds have properties of advantage, which include anti-inflammatory, hypotensive, antitumor, anticholesterolemic, and other beneficial effects. Many insecticides, plant growth regulators, herbicides, and pesticides are produced by liquid submerged fermentation. However, some are produced by solid-state fermentation. Strains can be improved for higher production of products. Most SM are of economic importance and are mainly produced by actinomycetes, particularly of the genus *Streptomyces* and by fungi. Mostly, after the primary metabolism, the secondary metabolites are obtained by:

(a) Metabolites derived from shikimic acid (aromatic amino acids) as ergot alkaloids, the antibiotic scandicidin, and chloramphenicol.
(b) Metabolites derived from amino acids as lactam antibiotics: penicillin, cephalosporins, and cephamycins, as well as cyclic peptide antibiotics such as gramicidin or the immunosuppressive agent cyclosporine.
(c) Metabolites derived from acetyl-CoA (other Kreb's cycle intermediates).

These microbial SM are subdivided into polyketides and terpenes. Polyketides include the antibiotic erythromycin, the insecticidal, antiparasitic compound such as avermectin, and the antitumor agent such as doxorubicin. Terpenes include the noncitotoxic antitumor agent taxol. Using sugars, antibiotics such as streptomycin and kanamycin can be produced (Smith and Berry 1976). Recently, many microbial SM have been identified, which are being used to combat diseases. In earlier days, only synthetic drugs were used, which had many side effects. However, now, innovative microbial metabolites are being used even in agriculture. The herbicides, insecticides, pesticides, plant growth regulators, and antiparasitic drugs are being commonly used in agriculture. The processes for the production of antibiotics are regulated by carbon. Usually, lactose is utilized for this purpose. Glucose is usually fed at a slow speed to avoid catabolite regulation. Soybean meal is used to avoid nitrogen (ammonium) regulation. In order to enhance the yield of any specific metabolite, a precursor is supplemented into the medium. Method of solid state fermentation (SSF) is very important for the production of secondary metabolites (Robinson et al. 2001). This system helps in the preparation of many fermented foods (Hesseltine 1977a, b). Various types of solid phases are used, which make the SSF method different and beneficial (Barrios-González and Mejía 1996). The support for the nutrients and water reservoir is usually an inert material. Commonly, either sugarcane bagasse pith or polyurethane is utilized for this purpose. The major contributors of secondary metabolites (SM) are fungi and actinomycetes, for which SSF is used. There are many advantages of SSF as compared to SmF. In SSF, the energy requirements are low and oxygen is transferred directly to the microorganism. In SSF, the optimization of parameters is similar as that of SmF. In both types of fermentations, it is important to optimize the initial moisture content, particle size, and medium concentration. It has been observed that high-producing strains in SmF usually produce lower yields in SSF (Shankaranand et al. 1992). In Western countries, SSF method is more commonly used. Many bioreactors and fermentors have been designed for the production of secondary metabolites (Mitchell et al. 2000; Mitchell 2002; Vadivelu et al. 2001; Junter et al. 2002).

Smitha et al. (2003) worked on the enhanced production of verbenol, a high-valued food flavorant, by an intergeneric fusant strain of *Aspergillus niger* and *Penicillium digitatum*. Agrawal and Richard (2000a, b, c) studied the factors affecting the bioconversion of alpha pinene to verbenone, an anticancer compound, by using resting cells of *Aspergillus niger*. Optimization of parameters is vital. In this regard, work has been done by Agrawal and Richard (2000a, b, c) on the conditions to biotransform alpha pinene to verbenone by a soil isolate *Penilillium* sp.

Bioconversion of alphapinene to verbeonone was also studied on the resting cells of *Aspergillus niger* (Agrawal and Richard 2000a, b, c).

Major microbial products and their uses are given in Table 8.2.

**Microbial Enzymes**

Enzymes are catalysts that can increase the rate of reaction without altering themselves. They have the ability to degrade various waste products into simpler compounds. The enzyme production from microorganisms is advantageous due to the ease of cultivation and yields are higher. These qualities have made the microbial processes to be utilized at commercial levels and by many industries. There are different uses of these microbial enzymes (Table 8.2). The details are given in Chap. 11 of this book.

Nazhat Ul Ainn and Renu Agrawal (2003) studied the solubilization and stabilization of alpha pinene hydroxylase from *Aspergillus niger*. This enzyme is very important in the production of verbenol and verbenone, an anticancer compound. Venkateshwaran et al. (1999) have shown the potential of producing catalase enzyme by using *Saccharomyces cerevisiae* CFR-201. Catalase protects the living cells from oxidative damage and therefore it plays an important role. Ogunremi et al. (2020) worked on the production and characterization of volatile compounds and phytase from potentially probiotic yeasts isolated from traditional fermented cereal foods in Nigeria (Table 8.3).

Details on microbial enzymes are given in Chap. 11 of this book.

**Table 8.2** Major microbial products

| Microorganism | Compound produced | Uses | Reference |
|---|---|---|---|
| *Streptomyces* spp. | Tetracycline | Antibacterial, antifungal | Qi et al. (2019) |
| *Penicillium* spp. | Penicillin | Antimicrobial | Devi et al. (2012) |
| *Amycolatopsis* sp. | Vancomycin | Antibacterial | Song et al. (2021) |
| *Streptomyces* spp. | Rapamycin | Anti-inflammatory | Ganesh et al. (2023) |
| *Streptomyces* spp. | Cahuitamycin | Biofilm inhibitory | Park et al. (2016) |
| *Lactococcus lactis* | Nisin | Bacteriocin | Saraiva et al. (2020) |
| *Strobilurus* sp. | Strobilurin | Antimicrobial | Feng et al. (2020) |
| *Aspergillus*, *Penicillium* sp., *Monascus purpureus*, *Pleurotus* spp. | Lovastatin | Inhibits cholesterol formation | Sadowska et al. (2023) |

**Table 8.3** Microorganisms and the production of enzymes

| Microorganism involved | Enzyme produced | Uses | Reference |
|---|---|---|---|
| Species of *Aspergillus*, *Penicillium*, *Bacillus*, and *Trichoderma* | Cellulase | Textiles, paper, and detergents | Korsa et al. (2023) |
| Species of *Aspergillus*, *Bacillus*, and *Streptomyces* | Amylase | Beverages and baking paper | Vojnovic et al. (2024) |
| Species of *Aspergillus* and *Candida* | Lipase | Dairy, paper, detergent, cosmetics, products for health, and leather | Ayesha et al. (2023) |
| Species of *Aspergillus*, *Serratia*, and *Pseudomonas* | Protease | Beverages, cosmetics, textiles, and detergents | Naeem et al. (2022) |

**Production of Food Substances Using Microorganisms**

Food substances can be prepared using microorganisms. This process is known as fermentation in which a microorganism converts the energy source into other edible substances. The food substances are usually separated by using different techniques. These can be physical, which include centrifugation and/or filtration from the intact microorganism and/or fermentation medium. The separated food material is then purified. In case the final food product is a liquid, then it has to be concentrated (details are given in Chap. 13 of this book). For increasing the shelf life of the product, stabilizers or preservatives are supplemented into the product. In case the food product is required in a powder form, then it is dried by various drying methods. Any food product obtained from the microorganism should be safe and the final product should be devoid of any microorganisms as they are consumed by human population. Flavoring compound was produced by using *Mucor* (fungal)cells, converting nerol to dihydrocarvyl acetate (Renu Agrawal et al. 2000). Ogunremi et al. (2015) developed a cereal-based functional food using cereal-mix substrate fermented with probiotic strain—*Pichia kudriavzevii* OG32.

## 8.3 Major Challenges in the Production of Secondary Metabolites at Large Scale

There are many challenges when the secondary metabolites are produced at the industrial scale. The work done with yeast, *Saccharomyces cerevisiae*, has made it possible to produce wine on industrial scale. Many volatile compounds are produced from the proteins in the yeasts and therefore for best yields a thorough proteomic analysis is very important, which specifies the quality of wine. Genetic modification of specific strains has helped in increasing the yields of the metabolites. The organism can be grown in a cost-effective media, which can reduce the cost and make the process cost-effective. The biosafety is a big challenge, which should be worked out before producing the metabolite at a large scale. Safe production of amino acids has been established utilizing *Corynebacterium glutamicum* for the food and feed industries. In many processes the engineered strain is helpful in producing higher yields of the metabolite. Starting from strain isolation, characterization, and improvement of the metabolite production, the extraction step is vital in industrial production. Newer extraction processes have been developed that use high pressure along with using eco-friendly solvents. Optimization of various physical and chemical parameters for the extraction of molecules is very important. The release of pollutants into the environment should be minimized and BOD and COD should be checked periodically. In this regard, it is important to have periodic monitoring by the officers of the pollution board. The use of natural biopreservative agents (bacteriocins) is becoming a common practice. The bacteriocins, which come under GRAS category, are nontoxic, nonimmunogenic, and thermo-resistant, can kill the pathogenic bacteria, and can be produced at large scale with ease. The knowledge of computers and statistical tools are required by the managers to efficiently monitor the large-scale operations in keeping the metabolite yield to maximum and effluent to a minimum. The use of computational tools in the optimization of various physicochemical parameters (initial pH, temperature, agitation speed, and fermentation time) as well as the production of pigments along with high biomass content in submerged fermentation by a marine-derived strain of *Talaromyces albobiverticillius* has been well studied by Venkatachalam et al. (2020). For optimization of various parameters, a factorial method can be used, which reduces the number of experiments to a large extent. CCRD or BBED experimental designs have proved very helpful. These methods have helped in the understanding and in the optimization among the multifactorial interactions between diverse parameters. The best culture conditions for maximum organism growth and high yields of the metabolite can be selected in minimum time. In sugarcane distillery, the spent wash (DSW) creates a lot of pollution in industrial effluents. This needs attention. In a study (Chuppa-Tostain et al. 2020), the species from *Aspergillus* and *Trametes* genera have been found to be effective to depollute the DSW.

Zhang et al. (2020) worked on a safe strain of *Pediococcus acidilactici* (PA204) that can utilize xylose from lignocellulose. With newer scientific developments, the use of microorganisms for the production of secondary metabolites is becoming a promising preposition. This is helping the mankind in curing many diseases and improving the health of humans. Patridge et al. (2016) analyzed some of the FDA-approved drugs and their derivatives.

Fusion of protoplasts among two cells with different desirable qualities is a good preposition to obtain better qualities and yield. It is another way to enhance the product yield. In this technique of protoplast fusion, the genome of two cells combines. This ultimately increases the microbial genetic diversity and expression of genes. However, the method is not widely used as the time duration is limited for the process of recombination. Shitut et al. (2020) highlighted the growth and propagation of engineered bacteria that lack a cell wall, and the gene expression could be based on the two parental genomes. This approach has been quite successful for the production of higher yields of antibiotics.

## 8.4 Inducers of Secondary Metabolites

Gram-negative bacteria contain autoinducers, which include the butyrolactones (butanolides) of the actinomycetes, called as the Nacylhomoserine lactones (HSLs). The oligopeptides of Gram-positive bacteria and B-factor [3-(1-butylphosphoryl) adenosine] of rifamycin production is well known in *Amycolatopsis mediterrane*. These inducers help in the development, sporulation, light emission, virulence, production of antibiotics, pigments, conjugation, and competence for genetic transformation. Actinomycete fermentations have inducers such as A factor (2-*S*-isocapryloyl-3-*R*-hydroxymethyl-butyrolactone), which play an important role in the production of metabolite. A factor induces both morphological and chemical differentiation in *Streptomyces griseus* and *Streptomyces bikiniensis*, which help in the formation of aerial mycelia, conidia, streptomycin synthases, and streptomycin. In *S. griseus*, A factor is produced just prior to streptomycin production and disappears before the product reaches the maximum level.

Secondary metabolites play an important role in human lives by improving the health, nutrition, and ultimately the economics of the country. The major products are antibiotics, pigments, toxins, effectors of ecological competition and symbiosis, pheromones, enzyme inhibitors, immunomodulating agents, receptor antagonists, pesticides, antitumor agents and growth promoters of animals and plants like gibberellic acid, antitumor agents, alkaloids, and other drugs. There are many factors that influence the formation of these compounds. The major factors are the nutrients supplied during the process. The rate of culture growth, the enzyme action, and the feedback control can change the rate of product formation. It is the transfer RNA that regulates the reaction. The genetic products are usually formed in the stationary phase of culture growth. The secondary metabolism is usually coded with clustered genes on chromosomal DNA or on plasmid DNA. Petit et al. (2010) studied the milk composition, milk fatty acid profile, digestion, and the ruminal fermentation in dairy cows when they were fed with whole flaxseed and Ca salts of flaxseed oil. They found that the levels of milk increased drastically. Sayed et al. (2020) demonstrated the formation of microbial natural products as the potential inhibitors of SARS-CoV-2 main protease (Mpro). Many pathways are yet to be worked out and understood in a better way.

**Catabolism of Carbohydrates**

There are four known pathways for the catabolism of carbohydrates up to pyruvic acid in bacteria, actinomycetes, fungi, and yeasts. These are the Embden–Meyerhof–Parnas pathway, the pentose phosphate pathways, the Entner–Duodoroff pathway, and the phosphoketolase pathway.

(i) **The Embden–Meyerhof–Parnas (EMP) pathway** In this pathway the glucose ($C_6$) gets reduced into pyruvate ($C_3$). It works both in aerobic and anaerobic conditions. In aerobic conditions, it works in the tricarboxylic acid cycle that oxidizes pyruvate to $CO_2$ and $H_2O$. However, in anaerobic conditions, pyruvate is fermented to various fermentation products with industrial significance.

(ii) **The pentose phosphate pathway (PP)** This is also known as the hexose monophosphate pathway (HMP) or the phosphogluconate pathway. While the EMP pathway provides pyruvate, a $C_3$ compound as its end product, there is no end product in the PP pathway. It produces many phosphates that include triose ($C_3$) pentose ($C_5$), hexose ($C_6$), and heptose ($C_7$). The energy is generated in the form of NADPH for biosynthetic and other purposes and pentose phosphates for nucleotide synthesis.

(iii) **The Entner Duodoroff pathway (ED)** This pathway is restricted to a few bacteria, especially *Pseudomonas*, but it is also carried out by some fungi. Some organisms break down glucose by this pathway.

(iv) **The phosphoketolase pathway** In this pathway, the fermentation of glucose produces lactic acid, ethanol, and $CO_2$. Pentoses are also fermented to lactic acid and acetic acid. An example is *Leuconostoc mesenteroides*.

## FAQs

1. What kind of microbes produce most of the commercially available biologically active secondary metabolites?
2. Which is the most significant method in secondary metabolite production?
3. Do all microorganisms produce secondary metabolites?
4. What are the most important industrial applications of secondary metabolites?
5. What are biologically active secondary metabolites from the fungi?
6. What are toxins produced as secondary metabolites by fungi called?
7. Are secondary metabolites produced by living organisms?
8. How are secondary metabolites from bacteria extracted?
9. Why is it important to optimize the fermentation parameters?
10. How does CCRD or other statistical methods help in the optimization of parameters?

## References


Agrawal R (2022) Biofilms. New India Publishing Agency, p 154. ISBN: 9789395319140, 9395319143

Agrawal R, Richard J (2000a) Bioconversion of alphapinene to verbeonone by resting cells of *Aspergillus Niger*. Appl Microbiol Biotechnol 53:335

Agrawal R, Richard J (2000b) Factors affecting bioconversion of alpha pinene to verbenone by resting cells of *Aspergillus Niger*. Appl Microbiol Biotechnol 53:335–337

Agrawal R, Richard J (2000c) Optimization of conditions for the biotransformation of alpha pinene to verbenone by a soil isolate *Penilillium* sp. J Food Sci Technol 37(4):430–432

Agrawal R, Nazhat Ul A, Deepika RJ (1999) Improvement in the biotransformation of alpha pinene to verbenol in *Aspergillus* sp. and *Penicillium* sp. induced by mutation using chemicals and ultra violet irradiation. Biotechnol Bioeng 63:249–252

Agrawal R, Deepika D, Vimala S (2000) Conversion of nerol to dihydrocarvyl acetate by *Mucor* sp. Ind J Microbiol 40:49

Agrawal R, Seetharam YN, Kelamani CR, Jyothishwaran G (2003) Biotransformation of ferulic acid to vanillin by locally isolated bacterial cultures. Ind J Microbiol 2:610–612

Arakelyan H (2018) Foodborne diseases-2. Ergotism (St Anthony's fire)

Argüelles A, Sánchez FR, Guirao AJP, Belda C, Lozano JA, Solano F, Argüelles JC (2020) Novel bi-factorial strategy against *Candida albicans* viability using carnosic acid and propolis: synergistic antifungal action. Microorganisms 8:749. https://doi.org/10.3390/microorganisms8050749

Ayesha S, Fatima I, Muhammad I (2023) Characterization of detergent-compatible lipases from *Candida albicans* and *Acremonium sclerotigenum* under solid-state fermentation. ACS Omega 8(36):32740–32751

Bacame-Valenzuela FJ, Pérez-Garcia JA, Figueroa-Magallón ML, Espejel-Ayala F, Ortiz-Frade LA, Reyes-Vidal Y (2020) Optimised production of a redox metabolite (pyocyanin) by *Pseudomonas aeruginosa* NEJ01R using a maize by-product. Microorganisms 8:1559. https://doi.org/10.3390/microorganisms8101559

Barrios-González J, Mejía A (1996) Production of secondary metabolites by solid-state fermentation. Biotechnol Annu Rev 2:85–121

Chuppa-Tostain G, Tan M, Adelard L, Shum-Cheong-Sing A, François J-M, Caro Y, Petit T (2020) Evaluation of filamentous fungi and yeasts for the biodegradation of sugarcane distillery wastewater. Microorganisms 8:1588. https://doi.org/10.3390/microorganisms8101588

Cragg GM, Grothaus PG, Newman DJ (2009) Impact of natural products on developing new anticancer agents. Chem Rev 109(7):3012–3043

Deshmukh SK, Prakash V, Ranjan N (2018) Marine fungi: a source of potential anticancer compounds. Front Microbiol 8:2536. https://doi.org/10.3389/fmicb.2017.02536

Devi L, Devi G, Kuniparambil R, Reeta RS, Carlos RS, Ashok P et al (2009) Solid-state fermentation. Current trends and future prospects. Ferment Microbiol Biotechnol. ISBN: 9781439855799. eBookISBN: 9781439855812

Devi P, Rodrigues C, Naik CG, D'Souza L (2012) Isolation and characterization of antibacterial compound from a mangrove-endophytic fungus, Penicillium chrysogenum MTCC 5108. Ind J Microbiol 52(4):617–623. https://doi.org/10.1007/s12088-012-0277-8

Dias DA, Urban S, Roessner U (2012) A historical overview of natural products in drug discovery. Meta 2(2):303–336

Divyashree MS, Jijomon G, Agrawal R (2006) Biotransformation of chiral terpenes by *A. Niger* and *Pseudomonas putida*. J Food Sci Technol 43(1):73–76

Elander RP, Vournakis JN (1986) In: Vanék Z, Butterworths HZ (eds) Genetic aspects of overproduction of antibiotics and other secondary metabolites, in overproduction of microbial metabolites. Strain improvement and process control strategies, London, pp 63–80

Feng Y, Huang Y, Zhan H, Bhatt P, Chen S (2020) An overview of strobilurin fungicide degradation: current status and future perspective. Front Microbiol 12(11):389

Fierro F et al (1995) The penicillin gene cluster is amplified in tandem repeats linked by conserved hexanucleotide sequences. Proc Natl Acad Sci USA 92:6200–6204

Fleming A (1929) The antibacterial action of a *Penicillium*, with special reference to their use in the isolation of *B. influenzae*. Br J Exp Pathol 10:226–236

Fouillaud M, Dufossé L (2022) Microbial secondary metabolism and biotechnology. Microorganisms 10(1):123. https://doi.org/10.3390/microorganisms10010123

Frisvad JC, Frank JM, Houbraken JAMP, Kuijpers AFA, Samson RA (2004) New ochratoxin A producing species of *Aspergillus* section Circumdati. Stud Mycol 50:23–43

Frisvad JC, Møller LLH, Larsen TO, Kumar R, Arnau J (2018) Safety of the fungal work horses of industrial biotechnology: update on the mycotoxin and secondary metabolite potential of *Aspergillus Niger*, *Aspergillus oryzae*, and *Trichoderma reesei*. Appl Microbiol Biotechnol 102:9481–9515

Ganesh SK, Subathra Devi C (2023) Molecular and therapeutic insights of rapamycin: a multi-faceted drug from *Streptomyces hygroscopicus*. Mol Biol Rep 50(4):3815–3833

González-Jiménez MC, García-Martínez T, Mauricio JC, Sánchez-León I, Puig-Pujol A, Moreno J, Moreno-García J (2020) Comparative study of the proteins involved in the fermentation-derived compounds in two strains of *Saccharomyces cerevisiae* during sparkling wine second fermentation. Microorganisms 8:1209. https://doi.org/10.3390/microorganisms8081209

Hamed AA, Soldatou S, Qader MM, Arjunan S, Miranda KJ, Casolari F, Pavesi C, Diyaolu OA, Thissera B, Eshelli M et al (2020) Screening fungal endophytes derived from under-explored Egyptian marine habitats for antimicrobial and antioxidant properties in factionalised textiles. Microorganisms 8:1617. https://doi.org/10.3390/microorganisms8101617

Harborne JB (1994) The flavonoids: advances in research since 1986. Chapman and Hall, London

Hesseltine CW (1977a) Solid state fermentation. Process Biochem 12:24–27

Hesseltine CW (1977b) Solid state fermentation. Process Biochem 12:29–32

Hilgers F, Habash SS, Loeschcke A, Ackermann YS, Neumann S et al (2021) Heterologous production of β-caryophyllene and evaluation of its activity against plant pathogenic fungi. Microorganisms 9:168. https://doi.org/10.3390/microorganisms9010168

Hwang S, Le LTHL, Jo SI, Shin J, Lee MJ, Oh DC, Pentaminomycins CE (2020) Cyclic pentapeptides as autophagy inducers from a mealworm beetle gut bacterium. Microorganisms 8:1390. https://doi.org/10.3390/microorganisms8091390

Igarashi Y, Ogawa M, Sato Y, Saito N, Yoshida R, Kunoh H, Onaka H, Furumai T (2000) Fistupyrone, a novel inhibitor of the infection of Chinese cabbage by *Alternaria brassicicola*, from *Streptomyces* sp.TP-A0569. J Antibiot 53:1117–1122. https://doi.org/10.7164/antibiotics.53.1117

Jawan R, Abbasiliasi S, Tan JS, Mustafa S, Halim M, Ariff AB (2020) Influence of culture conditions and medium compositions on the production of bacteriocin like inhibitory substances by *Lactococcus lactis* Gh1. Microorganisms 8:1454. https://doi.org/10.3390/microorganisms8101454

Jeong Y, Cho SH, Lee H, Choi HK, Kim DM, Lee CG, Cho S, Cho BK (2020) Current status and future strategies to increase secondary metabolite production from cyanobacteria. Microorganisms 8:1849

Junter GR et al (2002) Resistance training and intra-abdominal adipose tissue in older men and women. Med Sci Sports Exerc 34(6):1023–1028

Khiralla A, Spina R, Varbanov M, Philippot S et al (2020) Evaluation of antiviral, antibacterial and antiproliferative activities of the endophytic fungus *Curvularia papendorfii*, and isolation of a new polyhydroxyacid. Microorganisms 8:1353. https://doi.org/10.3390/microorganisms8091353

Korsa G, Konwarh R, Masi C et al (2023) Microbial cellulase production and its potential application for textile industries. Ann Microbiol 73:13. https://doi.org/10.1186/s13213-023-01715-w

Kozlovskii AZ, Antipova V, Tatiana (2012) Fungi of the genus *Penicillium* as producers of physiologically active compounds (Review). Appl Biochem Microbiol 49. https://doi.org/10.1134/S0003683813010092

Lebeau J, Petit T, Fouillaud M, Dufossé L, Caro Y (2020) Alternative extraction and characterization of nitrogen containing azaphilone red pigments and ergosterol derivatives from the marine-derived fungal *Talaromyces* sp. 30570 strain with industrial relevance. Microorganisms 8:1920. https://doi.org/10.3390/microorganisms8121920

Liscombe DK, Macleod BP, Loukanina N, Nandi OI, Facchini PJ (2005) Evidence for the monophyletic evolution of benzylisoquinoline alkaloid biosynthesis in angiosperms. Phytochemistry 66:1374–1393. https://doi.org/10.1016/j.phytochem.2005.04.029

Liu L, Hu Z, Li S, Yang H, Li S, Lv C, Zaynab M, Cheng CHK, Chen H, Yang X (2020) Comparative transcriptomic analysis uncovers genes responsible for the DHA enhancement in the mutant *Aurantiochytrium* sp. Microorganisms 8:529. https://doi.org/10.3390/microorganisms8040529

Martin N, Guez MA, Sette LD, Da Silva R, Gomes E (2010) Pectinase production by a Brazilian thermophilic fungus *Thermomucor indicae-seudaticae* N31 in solid-state and submerged fermentation. Mikrobiologiia 79(3):321–328

Martínez-Núñez MA, López VELY (2016) Nonribosomal peptides synthetases and their applications in industry. Sustain Chem Process 4:13. https://doi.org/10.1186/s40508-016-0057-6

Melander RJ, Basak AK, Melander C (2020) Natural products as inspiration for the development of bacterial antibiofilm agents. Nat Prod Rep 37(11):1454–1477

Mitchell RK (2002) Entrepreneurship and stakeholder theory, Business ethics quarterly, The Ruffin Series, pp 175–196

Mitchell RK, Smith B, Seawright KW, Morse EA (2000) Cross–cultural cognitions and the venture creation decision. Acad Manag J 43(5):974–993

Monciardini P, Iorio M, Maffioli S, Sosio M, Donadio S (2014) Discovering new bioactive molecules from microbial sources. Microb Biotechnol 7:209–220. https://doi.org/10.1111/1751-7915.12123

Naeem M, Manzoor S, Abid M-U-H, Tareen MBK, Asad M, Mushtaq S, Ehsan N, Amna D, Xu B, Hazafa A (2022) Fungal proteases as emerging biocatalysts to meet the current challenges and recent developments in biomedical therapies: an updated review. J Fungi 8(2):109

Nawaz A, Chaudhary R, Shah Z, Dufossé L, Fouillaud M, Mukhtar H, ul Haq I (2021) An overview on industrial and medical applications of bio-pigments synthesized by marine bacteria. Microorganisms 9:11

Nazhat Ul A, Agrawal R (2003) Studies on solubilization and stabilization of alpha pinene hydroxylase from *Aspergillus Niger*. J Food Sci Technol 40(1):63–66

O'Brien J, Wright GD (2011) An ecological perspective of microbial secondary metabolism. Curr Opin Biotechnol 22(4):552–558. https://doi.org/10.1016/j.copbio.2011.03.010

Ogunremi O, Agrawal R, Sanni A (2015) Development of cereal-based functional food using cereal-mix substrate fermented with probiotic strain—Pichia kudriavzevii OG32. Food Sci Nutr 3. https://doi.org/10.1002/fsn3.239

Ogunremi O, Renu A, Sanni A (2020) Production and characterization of volatile compounds and phytase from potentially probiotic yeasts isolated from traditional fermented cereal foods in Nigeria. J Genetic Eng Biotechnol 18:16. https://doi.org/10.1186/s43141-020-00031-z

Oxford AE, Lampen JO, Peterson WH (1940) Growth factor and other nutritional requirements of the acetone butanol organism *Cl. acetobutylicum*. Biochemist 34:1588–1597

Papaianni M, Ricciardelli A, Fulgione A, d'Errico G, Zoina A, Lorito M, Woo SL, Vinale F, Capparelli R (2020) Antibiofilm activity of a *Trichoderma* metabolite against *Xanthomonas campestris* pv. *campestris a*lone and in association with a phage. Microorganisms 8:620. https://doi.org/10.3390/microorganisms8050620

Park SR, Tripathi A, Wu J, Schultz PJ, Yim I, McQuade TJ, Yu F et al (2016) Discovery of cahuitamycins as biofilm inhibitors derived from a convergent biosynthetic pathway. Nat Commun 7:10710

Patridge E, Gareiss P, Kinch MS, Hoyer D (2016) An analysis of FDA-approved drugs: natural products and their derivatives. Drug Discov Today 21:204–207. https://doi.org/10.1016/j.drudis.2015.01.009

Paulus CY, Rebets B, Tokovenko S, Nadmid LP, Terekhova M, Myronovskyi SB, Zotchev C, Rückert SB, Zahler S (2017) New natural products identified by combined genomics-metabolomics profiling of marine *Streptomyces* sp. MP131-18. Sci Rep 7:1–11. https://doi.org/10.1038/srep42382

Petit HV (2010) Milk composition, milk fatty acid profile, digestion, and ruminal fermentation in dairy cows fed whole flaxseed and Ca salts of flaxseed oil. J Dairy Sci 93:3146–3157

Prakash D, Nawani N, Prakash M, Bodas M, Mandal A, Khetmalas M, Kapadnis B (2013) Actinomycetes: a repertory of green catalysts with a potential revenue resource. Biomed Res Int 264020. https://doi.org/10.1155/2013/264020

Qi D, Zou L, Zhou D, Chen Y, Gao Z, Feng R, Zhang M, Li K, Xie J, Wang W (2019) Taxonomy and broad-spectrum antifungal activity of *Streptomyces* sp. SCA3-4 isolated from rhizosphere soil of *Opuntia stricta*. Front Microbiol 10:1390

Robinson T, Singh D, Nigam P (2001) Solid-state fermentation: a promising microbial technology for secondary metabolite production. Appl Microbiol Biotechnol 55(3):284–289

Ruiz JR, Moran M, Arenas J, Lucia A (2010) Strenuous endurance exercise improves life expectancy: it's in our genes. Br J Sports Med 45:159–161

Sadowska A, Osiński P, Roztocka A, Kaczmarz-Chojnacka K, Zapora E, Sawicka D, Car H (2023) Statins from fungi to pharmacy. Int J Mol Sci 25(1):466

Said HC, Fouillaud M, Le GG, Sklirou AD, Boyer JB, Trougakos IP, Jerabek M, Bignon J, de Voogd NJ, Ouazzani J et al (2020) Microorganisms associated with the marine sponge *Scopalina hapalia*: a reservoir of bioactive molecules to slow down the aging process. Microorganisms 8:1262. https://doi.org/10.3390/microorganisms8091262

Salikin NH, Nappi J, Majzoub ME, Egan S (2020) Combating parasitic nematode infections, newly discovered antinematode compounds from marine epiphytic bacteria. Microorganisms 8:1963. https://doi.org/10.3390/microorganisms8121963

Saraiva MAF, Birri DJ, Brede DA, Baracat-Pereira MC, de Queiroz MV, Nes IF, de Moraes CA (2020) Nisin Z production by wild strains of *Lactococcus lactis* isolated from Brazilian (Italian Type) fermented sausage. Int J Microbiol 14:9309628

Sayed AM, Alhadrami HA, El-Gendy AO, Shamikh YI, Belbahri L, Hassan HM, Abdelmohsen UR, Rateb ME (2020) Microbial natural products as potential inhibitors of SARS-CoV-2 main protease (Mpro). Microorganisms 8:970. https://doi.org/10.3390/microorganisms8070970

Schneider Y, Jenssen M, Isaksson J, Hansen KØ, Andersen JH, Hansen EH (2020) Bioactivity of serratiochelin a, a siderophore isolated from a co-culture of *Serratia* sp. and *Shewanella* sp. Microorganisms 8:1042. https://doi.org/10.3390/microorganisms8071042

Seca AML, Pinto DCGA (2019) Biological Potential and Medical Use of Secondary Metabolites. Medicines (Basel) 12;6(2):66. https://doi.org/10.3390/medicines6020066

Shamikh YI, El Shamy AA, Gaber Y, Abdelmohsen UR, Madkour HA et al (2020) Actinomycetes from the red sea sponge *Coscinoderma mathewsi*: isolation, diversity, and potential for bioactive compounds discovery. Microorganisms 8:783. https://doi.org/10.3390/microorganisms8050783

Shankaranand VS, Matur R, Lonsane B (1992) Idiosyncrasies of solid-state fermentation systems in the biosynthesis of metabolites by some bacterial and fungal cultures. Process Biochem 27:33–36

Shitut S, Bergman GÖ, Kros A, Rozen DE, Claessen D (2020) Use of permanent wall-deficient cells as a system for the discovery of new-to-nature metabolites. Microorganisms 8:1897. https://doi.org/10.3390/microorganisms8121897

Smith JE, Berry DR (1976) The filamentous fungi. Vol I: Industrial mycology. Wiley, London

Smitha Rao CV, Rati Rao E, Agrawal R (2003) Enhanced production of verbenol, a high valued food flavourant by an intergeneric fusant strain of *Aspergillus Niger* and *Penicillium digitatum*. Biotechnol Appl Biochem 37(2):145–147

Soltani J (2016) Chapter 22 - Secondary metabolite diversity of the genus *Aspergillus*: recent advances. In: Gupta VK (ed) New and future developments in microbial biotechnology and bioengineering. Elsevier, pp 275–292. ISBN 9780444635051. https://doi.org/10.1016/B978-0-444-63505-1.00035-X

Song Z, Xu T, Wang J, Hou Y, Liu C, Liu S, Wu S (2021) Secondary metabolites of the genus *Amycolatopsis*: structures, bioactivities and biosynthesis. Molecules 26(7):1884. https://doi.org/10.3390/molecules26071884

Staunton J, Weissman KJ (2001) Polyketide biosynthesis: a millennium review. Nat Prod Rep 18(4):380–416. https://doi.org/10.1039/a909079g

Sweechin TO, Nor M, Roslinda AE, Nagib L, Ong R, Solleh M et al (2015) Bioprocess optimization for biomass production of probiotics yeast *Saccharomyces boulardii* in semi-industrial scale. J Chem Pharma Res 7:122–132

Takemura K et al (2010) Powder x-ray diffraction study of Ne up to 240 GPa: conference series. J Phys Conf Ser 215:012017

Takemura T, Ikezawa N, Iwasa K, Sato F (2012) Molecular cloning and characterization of a cytochrome CYP in sanguinarine biosynthesis from *E. californica* cells. Phytochemistry 91:100–108. https://doi.org/10.1016/j.phytochem.2012.02.013

Vadivelu K, Trichur Suryanarayanan S (2001) Occurrence and distribution of endophytic fungi in a mangrove community. Mycological Res 105(11):1388–1391. ISSN 0953-7562

Venkatachalam M, Shum-Chéong-Sing A, Dufossé L, Fouillaud M (2020) Statistical optimization of the physico-chemical parameters for pigment production in submerged fermentation of *Talaromyces albobiverticillius* 30548. Microorganisms 8:711. https://doi.org/10.3390/microorganisms8050711

Venkateshwaran G, Somashekar D, Prakash MH, Agrawal R, Basappa SC (1999) Production and utilization of catalase by using *Saccharomyces cerevisiae* CFR-201. Process Biochem 34:187–191

Vidya CM, Agrawal R (2003) Production of verbenol, a high valued food flavourant from improved culture strain of *A. Niger* using protoplast fusion. Appl Microbiol Biotechnol 62(4):421–422

Voitsekhovskaia I, Paulus C, Dahlem C, Rebets Y, Nadmid S, Zapp J et al (2020) Baikalomycins A-C, new aquayamycin-type angucyclines isolated from lake Baikal derived *Streptomyces* sp. IB201691-2A. Microorganisms 8:680. https://doi.org/10.3390/microorganisms8050680

Vojnovic S, Aleksic I, Ilic-Tomic T et al (2024) *Bacillus* and *Streptomyces* spp. as hosts for production of industrially relevant enzymes. Appl Microbiol Biotechnol 108:185. https://doi.org/10.1007/s00253-023-12900-x

Walter T, Al Medani N, Burgardt A, Cankar K, Ferrer L, Kerbs A, Lee JH, Mindt M, Risse JM, Wendisch VF (2020) Fermentative N-Methylanthranilate production by engineered *Corynebacterium glutamicum*. Microorganisms 8:866. https://doi.org/10.3390/microorganisms8060866

Williams PG (2009) Panning for chemical gold: marine bacteria as a source of new therapeutics. Trends Biotechnol 27:45–52

Woo PC, Tam EW, Chong KT, Cai JJ, Tung ET, Ngan AH, Lau SK, Yuen KY (2010) High diversity of polyketide synthase genes and the melanin biosynthesis gene cluster in *Penicillium marneffei*. FEBS J 277(18):3750–3758. https://doi.org/10.1111/j.1742-4658.2010.07776.x

Woo SM, Uyeh DD, Kim J, Hong DH, Park T, Ha YS (2019) A study on the optimal fermentation conditions for mixed by-products in livestock feed production. Eng Agric Environ Food 12(4):470–476. ISSN 1881–8366

Xu YY, Yao GX, Liu PX, Zhao J, Wang XL, Sun JM, Qian YZ (2019) Review on the application of metabolomic approaches to investigate and analysis the nutrition and quality of agro-products[J]. Sci Agric Sin 52(18):3163–3176

Xu D, Xue M, Shen Z, Jia X, Hou X, Lai D, Zhou L (2021) Phytotoxic secondary metabolites from fungi. Toxins(Basel) 13(4):261. https://doi.org/10.3390/toxins13040261

Zhang Z, Li Y, Zhang J, Peng N, Liang Y, Zhao S (2020) High-titer lactic acid production by *Pediococcus acidilactici* PA204 from corn stover through fed-batch simultaneous saccharification and fermentation. Microorganisms 8:1491. https://doi.org/10.3390/microorganisms8101491

Ziemons S, Koutsantas K, Becker K, Dahlmann T, Kuck U (2017) Penicillin production in industrial strain *Penicillium chrysogenum* P2niaD18 is not dependent on the copy number of biosynthesis genes. BMC Biotechnol 17:16. https://doi.org/10.1186/s12896-017-0335-8

# Probiotics 9

**Abstract**

Probiotics +are live and viable microorganisms that improve health with their consumption. Many food products are prepared by using probiotic bacteria, fungi, or yeasts. Many products have been discussed in this chapter. The human mankind has been consuming many fermented food products since time immemorial. However, the use of these microorganisms at a large scale or on an industrial scale needs optimization of many physical and chemical parameters to make the process successful. For this, one needs to study the probiotic microorganisms from isolation to characterization and then prove their efficacy for consumption. This chapter deals with many probiotic products and various parameters for their effectiveness. With recent advances, scientists have worked on various aspects to improve the microbial strains including genetic manipulation or using a combination of microbes. The established strains can be obtained from culture collection centers all around the globe. Optimization of various physical and chemical parameters is very critical. The shelf storage of the product should be studied before taking it on a commercial scale.

**Keywords**

Probiotics · Products · Shelf life · Bacteria · Fungi · Yeasts · Genetic manipulation · Synbiotics · Properties · Optimization

## 9.1 Introduction

The supplementation of beneficial bacteria in any food improves the health of the consumers. These are known as probiotics. Most probiotics are lactic acid bacteria which are commonly added to milk to make yogurt, different types of fermented milk, and many other fermented foods. They benefit the consumers as the digestion

 
R. Agrawal, *Textbook of Industrial Microbiology*,
https://doi.org/10.1007/978-981-97-9582-6_9

becomes better. The immune system has been found to improve. The bioavailability of many nutrients becomes more. The problems of allergies caused due to intolerance of lactose have been observed to reduce to a large extent. Many other allergic reactions have been found to come down, and the risks of cancer have been found to be low. The role of probiotics in improving human health has been well explained in the book by Renu Agrawal (2016). The mechanism of action still needs to be worked out; however, it modifies the gut pH and antagonizes pathogens by producing compounds that kill them. These lactic acid bacteria have been found to replace the pathogenic bacteria in the intestine. Many times, they bind or replace the pathogen receptor sites, which results in taking in the nutrients and using growth factors by these probiotics in place of pathogens. Probiotics are known to stimulate the immunomodulatory cells and are known to produce lactase. They can provide a safe and cost-effective barrier against microbial infections and are classified under GRAS category. During the process, the growth of the microorganism takes place in a controlled manner. The food components also get converted with the help of enzymes. There are many foods produced by fermentation which are mainly from various fruits and vegetables, poultry, meat, fish, dairy, vegetables, different types of legumes, and cereals. From the last many centuries, this method of fermenting foods is being used in order to preserve the perishable products. The preservation was noted mainly due to the formation of compounds that had the ability to kill the pathogenic microorganisms. After fermentation, the taste and texture of foods have been found to be more acceptable by the consumers. Fermentation also helps in removing the bitter phenolic compounds. Foods can be fermented by two main methods. Many times the beneficial microorganisms are naturally present in the raw material itself. However, at other times, a particular microorganism is added for the food to ferment. There are many fermented foods that are prepared without adding any microbial culture. These are mainly sauerkraut, kimchi, and soy products. Foods that are fermented with the addition of starter cultures are kefir, kombucha, and natto. The starter cultures, which are added to start the fermentation reaction, are obtained from the culture collection centers. Each microorganism culture imparts a characteristic aroma and flavor to the fermented product. Fermented foods are used all over the globe. They are mainly used for improving the health.

It is important to optimize the culture to sustain in the GUT conditions. Work has been done by Renu Agrawal et al. (2010). They have studied the changes in the pattern of many properties in the native *Lactobacillus plantarum* and probiotic culture strain *Lactobacillus plantarum* cfr after it was adapted to the conditions of the gastrointestinal tract. It was found that the probiotic culture had many beneficial functional properties. The usage of probiotic products is very common in western countries. According to international standards, fermented products must contain a minimum of $10^6$ microbial cells in one gram of any product. However, according to the scientific work done (Patel et al. 2023), it has been found that the yields vary in different parts of the world, and the growth phase and age of the culture taken have a big role in the product formation. The time at which the products are analyzed/consumed is also a very critical factor. The food matrix plays an important role for the probiotic strains to survive against the gut conditions (low pH, high bile acids). These bacteria form peptides that are bioactive and they produce polyamines. These compounds are known to improve the heart health. The immunity and metabolic health of the consumer have been found to increase. Most of the fermented foods contain fibers that impart beneficial effects. These are especially helpful in the bowel movement. They are known as prebiotics. Fermentation helps in reducing toxins and antinutrient factors. Fermentation of soybeans helps in reducing phytic acid concentrations (Gupta et al. 2015). In sourdough, the content of fermentable carbohydrates has been found to be very low. This in turn helps the patients to consume these products with improved bowel movements and irritable bowel syndrome (Satish et al. 2022).

## 9.2 Probiotic Products

A. **Nonalcoholic Products**

(a) **Kefir** When the milk is fermented with yeasts along with kefir grains, the fermented product is of a creamy texture and sour taste. It also has effervescence. The Kefir grains have yeasts of two types. The symbiotic yeasts could ferment lactose (*Kluyveromyces marxianus*) and another one, which does not ferment lactose (*Saccharomyces cerevisiae*, *Saccharomyces unisporus*). Along with the yeasts, the grains also have lactic and acetic acid-producing bacteria. The bacteria are present inside a matrix which is made up of polysaccharide and a protein. This is called as kefiran. When fermentation takes place, many compounds and acids are produced giving the product a very characteristic aroma and flavor. These are lactic acids, flavor-generating components (acetaldehyde), ethanol, and carbon dioxide. Kefir can also be produced without milk called as water kefir. The product contains water, sugar, and kefir grains. However, in water kefir, the cultures used are different. Scientists (Prado et al. 2015) have worked to find out the various microorganisms present in kefir grains that include: *Lactobacillus brevis, L. paracasei, L. helveticus, L. kefiranofaciens, L.plantarum, L.kefiri, Lactococcuslactis, Streprotcoccus thermophilus, Acetobacter lovaniensis, Acetobacter orientalis, Saccharomyces cerevisiae, S. unisporus, Candida Kefyr, Kluyveromyces marxianus,* and *Leuconostoc mesenteroides*. After the fermentation, the microbial composition of kefir can change. A predominant *Lactobacillus* species in the kefir grain starter culture is *L. kefiri* of all *Lactobacillus* species in the final fermented beverage (Chong et al. 2023). At the global level, the amount of cells to be added in any fermented product has been fixed for the reliability. According to The Food and Agriculture Organisation (FAO) and also the World Health Organisation (WHO), the kefir grains to be added should have at least $10^7$ colony-forming units (CFU)/g of the micoorganisms. In the final product, a minimum of $10^4$ CFU/g of the yeast cells should be present. Work has been done to study the activity against the pathogens in kefir. Scientists have also studied if any other compounds are present in kefir grains that are responsible for killing the pathogenic organisms. Antimicrobial activity has been shown in fermented Kefir products against many pathogenic microorganisms (Al-Mohammadi et al. 2021). Work has been done (Ledesma-Martínez et al. 2019), which has shown the production of bioactive peptides from casein after fermentation. These had the property to enhance the immunity in animal models. Kefiran has been shown to reduce the cytokine production in mice with asthma (Lee et al. 2007a, b). Many animal studies have proved the antioxidant, lowering of the blood pressure, reduction of tumor cells, lowering of cholesterol, and lowering of glucose in fermented kefir products. When the microorganisms were isolated from kefir and tested for their adherence properties to mucosa, it has been found to effectively adhere to the human enterocyte-like Caco-2 cells. The property of the kefir microorganisms to adhere and colonise in the human gut is beneficial as it improves the gut microbiota. Studies have shown that this also increased the cell number of lactic acid bacteria. These include *Lactobacillus, Lactococcus,* and *Bifidobacterium* concentrations. A reduction in *Proteobacteria* sp. and Enterobacteriaceae concentrations has been shown in animal studies (Dimidi et al. 2019). Kefiran has shown an increase in the stool weight and in moisture content in rats, in a dose-dependent manner imparting a potentially beneficial effect against constipation (Turan et al. 2014). Kefir can be consumed by people with lactose malabsorption as it contains β-galactosidase expressing bacteria (*Kluyveromyces marxianus*). The flatulence problem has been found to be reduced by using kefir milk. Another double-blind RCT investigated the impact of 500 mL/day kefir on *Helicobacter pylori* eradication rates (Bekar et al. 2010). Kefir is the most common fermented food with benefits for lactose malabsorption. *H. pylori* eradication and improvement of the GI health has been found effective with the consumption of probiotics. Fermented foods can modulate many body functions as they have nutritive as well as nonnutritive components. This, in turn, helps in improving the health. In a study (Tamang et al. 2016), microorganisms have been found even in natural fermented foods. These microorganisms are either functional or nonfunctional. Enhancement of the nutrient bioavailability, the flavor, and aroma gets increased due to the presence of functional microorganisms. They also help in increasing the shelf life of the products. The functional microorganisms are known to decompose the toxic components and also factors that do not provide any nutrition. They impart many beneficial properties which improve the human health (Tamang et al. 2009, 2016). Holzapfel and Wood (2014) have shown the presence of many beneficial microorganisms in various fermented products. Tamang (2015) found out that the microorganism, *Bacillus*, was present when legumes were used for food fermentation. When different cheese products were prepared, the presence of various microorganisms was found (*Bifidobacterium, Brachybacterium, Brevibacterium*, and *Propionibacterium*). Bourdichon et al. (2012) worked on fermented meat products and found the presence of *Arthrobacter and Hafnia* species. Lv et al. (2013) worked on alcoholic beverages and found the presence of different yeasts and bacteria (*Candida, Debaryomyces, Geotrichum, Hansenula, Kluyveromyces, Pichia, Rhodotorula, Saccharomyces, Saccharomycopsis, Schizosaccharomyces, Torulopsis, Wickerhamomyces*, and *Zygosaccharomyces*). Species of *Actinomucor, Amylomyces, Aspergillus, Monascus, Mucor, Neurospora, Penicillium, Rhizopus*, and *Ustilago* are present in many fermented foods (Chen et al. 2014). Strains can be improved to enhance the yields of metabolites. Shobharani et al. (2006) worked on the improvement of the culture strains for lactose tolerance in a native strain of *Leuconostoc mesenteroides* sub sp. using mutation technique. They optimized the culture parameters using the method of response surface.

(b) **Koumiss** This is an ancient beverage used by Scythian tribes made from fermented mare's milk. Mare's milk is usually not consumed raw as it has laxative effect. It is prepared by the fermentation of mare's milk using lactic acid bacteria followed by alcoholic fermentation. Afzaal et al. (2021) studied the chemistry of the process and the microorganisms involved in the fermentation. The lactic acid bacteria, *Lactobacillus* bulgaricus, is responsible for acidifying the milk, whereas an yeast, *Saccharomyces lactis*, helps to make the drink carbonated and mildly alcoholic.

(c) **Kombucha** The term was given by a Chinese when the tea was fermented. It was originated in Northeast China (220 BC). As there are many beneficial properties of this product, the sale has been commercialized. Basically, the process consists of aerobic fermentation of black tea or green tea. Sugar is added along with the bacteria and yeast cultures. A term (SCOBY) has been given to specify this fermentation procedure as it has two known cultures of bacteria and yeast that work in a symbiotic manner. Sreeramulu et al. (2000) studied the reasoning of pathogenic bacteria not growing in this product. They found that due to the high concentration of acetic acid and a low pH the pathogenic bacteria were not able to grow. This prevents the growth of pathogenic bacteria like *Helicobacter pylori, Escherichia coli, Salmonella typhimurium,* and *Campylobacter jejuni*. They also found out with the in vitro experiments that kombucha had high antimicrobial activity against pathogens even if it was thermally denatured or was at neutral pH. This proves that there are many other compounds apart from acetic acid which are contributing factors for the antimicrobial activity. There are many bacteria and fungal species that constitute the SCOBY. These mainly consist of acetic acid bacteria (*Acetobacter, Gluconobacter*), lactic acid bacteria (*Lactobacillus, Lactococcus*), and yeasts (*Saccharomyces, Zygosaccharomyces*). Studies utilizing the high-throughput sequencing analysis have demonstrated that following fermentation, *Candida* and *Zygosaccharomyces* genera are the predominant yeasts in kombucha, and *Komagataeibacter, Lyngbya, Gluconobacter, Lactobacilli*, and *Bifidobacteria* are the most abundant bacteria genera. Kombucha product has shown many useful properties that improve the health. With regular consumption, it has been shown to reduce the glycemic index, decrease the free oxygen radicals, reduce obesity, and reduce toxicity in the kidneys. Even the levels of cholesterol and gut ulcers were able to come down. The product showed the formation of d-saccharic acid-1,4-lactone (DSL). This compound has beneficial effects. The formation of polyphenols and flavonoids in the product helps in killing many pathogenic microorganisms.

(d) **Sauerkraut** Since the fourth century BC, the fermented cabbage has been consumed by humans. The consumption of this product has been in Europe, Asia, and the United States (Di Cagno et al. 2016). To make this product, the basic raw material is shredded cabbage. To this, salt (2.3–3.0%) is added. Zabat et al. (2018) found the presence of many culture strains (*Leuconostoc* spp., *Lactobacillus* spp., and *Pediococcus* spp.) in spontaneous fermentation. Anita Christie et al. (2021) found many more culture strains in the homemade and also in the commercial product. They found the presence of *Bifidobacterium dentium, Enterococcus faecalis, Lactobacillus casei, Lactobacillus delbrueckii, Staphylococcus epidermidis, Lactobacillus sakei, Lactobacillus curvatus, Lactobacillus plantarum, Lactobacillus brevis, Weissella confusa, Lactococcus lactis*, and *Enterobacteriaceae*. Some of the *Lactobacillus* species isolated from sauerkraut have demonstrated the probiotic potential which has tolerance to low pH, ability to adhere to Caco-2 cells and possess antimicrobial activity against pathogens. In sauerkraut, it is the culture strain of *Lactobacillus paracasei* HD1.7, which produces a bacteriocin. This helps in the preservation of the product (Ge et al. 2009). Experiments were conducted using Wistar rats. Sauerkraut juice was fed to the rats by mouth. The results showed enhanced activity of glutathione S-transferase (GST) and NAD(P)H:quinone oxidoreductase 1 (NQO1). These enzymes are known to detoxify the liver and kidney (Krajka-Kuźniak et al. 2011). Some of the lactic acid bacteria in sauerkraut generates conjugated linoleic acid which has anticarcinogenic and antiatherosclerotic activity in animals (Krajka-Kuźniak et al. 2010). The presence of *Lactobacillus plantarum* P2 in sauerkraut has been found to induce TNF-α and IL-12 expression and prevents adhesion and invasion of *Salmonella enteritidis* (Shahbazi et al. 2021). The biosynthetic pathways have been worked out in the formation of kaempferol, (a flavonoid) isothiocyanates, indole-3-carbinol, goitrin, allyl cyanide, and nitriles (Tolonen et al. 2002). These compounds have many beneficial properties. Kaempferol scavenge the free radicals. Isothiocyanates can kill the pathogenic microorganisms like *E. coli, C. difficile, C. jejuni* and *C. perfringens* (Tolonen et al. 2002). As per the literature, there are some probable risk in case the product is consumed regularly as it might cause cancer in the larynx due to the high salt used in the product preparation.

(e) **Fermented Pickles** Fermentative bacteria produce acids which are necessary for the preservation and also produce flavor compounds. Foods are soaked in an acidic liquid for a sour taste. Traditionally, fermented pickles are very popular in Asia. They have a unique flavor and impart many health benefits. Pickles are usually prepared by using lactic acid bacteria (LAB) and followed by fermentation. In a study on street pickles which was done by natural fermentation, it was found that they mainly had LAB and shows antibiotic resistance to foodborne pathogens. However, many pathogenic microorganisms were also found to be present showing that street food pickle consumption can cause health hazards and make people ill (Banik et al. 2023) and therefore, it is important to prepare fermented food under controlled conditions.

(f) **Fermented Soy Products** (Tempeh, Natto, Miso) Asian countries are very vast. Therefore, each part has a different way to ferment soy. The common products made using soy are tempeh, natto, miso, sufu, douche, soy sauce, and doenjang. Each one is explained as below.

**Tempeh** The product has been produced traditionally in Indonesia. The soybeans are initially boiled and then the outer cover is removed. This is fermented by using a starter culture of *Rhizopus oligoporus* fungal species. The fermentation is done around 30 °C for a period of 35–37 h (Dwiatmaka et al. 2021). It is formed into a white cake which is soft and has a chewable texture. The flavor more or less resembles as mushroom. The lactic acid bacteria used are *Enterococcus faecium* and *Rhizopus* filamentous fungi. Studies have shown that fermentation of soybeans has lower concentration of protease inhibitors, phytic acid and phenols, and antinutritional factors that are high in raw soybeans. Fermented soy products have anticarcinogenic, antidiabetic, antioxidant, antiinflammatory and antihyperlipidemic properties.

**Natto** Traditionally, this is prepared by fermenting soybean which is commonly eaten in Japan. It is produced by the fermentation of cooked yellow soybeans with *Bacillus subtilis* var. natto. It produces many bioactive compounds and enzymes like nattokinase, bacillopeptidase F, vitamin $K_2$, and dipicolinic acid (Afzaal et al. 2022). Isoflavone, genistein, are high in concentration as compared to unfermented soy products. *Bacillus subtilis* var. natto produces an enzyme, namely, nattokinase. It is classified under the family subtilisin. Due to the fibrinolytic activity of this enzyme, and being a tissue plasminogen activator, the utility of this enzyme is very high. The enzyme also helps in reducing the aggregation of platelets. Chen et al. (2018) did clinical studies and found that the blood pressure of patients was reduced by this enzyme. The consumption of miso soup with natto continuously for a period of 2 weeks has helped in killing the pathogenic bacteria like *Bacilli*, *Clostridia,* and Enterobacteriaceae. When the stools were tested, the levels of short-chain fatty acids were found to be higher in the stool. The concentration of ammonia and sulfide was lower than the control group.

**Miso** Consumption of Miso soup is very common in Japan. The soybeans are fermented in the form of a paste to make miso soup. The soybeans are fermented, and a mold *Aspergillus oryzae* culture is used as the starter organism. Many times, a yeast, namely, *Saccharomyces* *cerevisiae* and lactic acid bacteria are used to obtain Miso. In the natural fermentation process, many organisms have been found (*Bacillus subtilis, Bacillus amyloliquefaciens, Staphylococcus gallinarum,* and *Staphylococcus kloosii*). *Bacillus* species survive till the end. On analysis, it has been found that most of the samples contain the lactic acid bacteria, namely, *Lactococcus* sp. GM005. This has the ability to produce a bacteriocin. This bacteriocin is known to kill many of the pathogenic bacteria. Antitumor activity is well pronounced, and it helps to lower the rate of colon and prostate cancer. Soybeans produce isoflavone; genistein and daidzein. The work of Spagnuolo et al. (2015) in an in vitro study has shown that these isoflavones were able to prevent cancer by reducing cancer cell migration. The content of genistein and daidzein has been found to be higher in fermented soybean products (miso and natto) as compared to soy milk and tofu (Sharma 2013).

(g) **Kimchi** In Korea, vegetables are salted and fermented. It is made up of Chinese cabbage, carrot, apple, pear, shrimp, and/or radish which are fermented along with various flavoring ingredients (a spice mixture of chilli, pepper, garlic, onion, and ginger in a known concentration) and seasonings (addition of soybean sauce, sesame seed, and salt to taste). A detailed review has been written by De Di Cagno et al. (2016) on fermented vegetables and other products. The vegetables and fruits are brined and drained, and spices are added which are then mixed with the cabbage. This is kept for a few days for fermentation to take place with the help of microorganisms present naturally. For a commercial product, the specific culture strain is obtained from the culture collection center and added (Yun et al. 2021). According to the work of Kim et al. (2005), there are many different bacterial species in the product (*Leuconostoc*, *Lactobacillus*, *Pseudomonas*, *Pantoea,* and *Weissella* genera). However, the fermentation kick starts using *Leuconostoc* sp. culture strain. After 3 days of fermentation, other microorganisms like *Leuconostoc gasicomitatum* and *Leuconostoc gelidum* also start growing. When garlic is added in high quantities, the *Lactobacillus* culture grows at a faster rate. However, addition of red pepper powder, enhances the growth of *Weissella*. This brings down the growth of both *Leuconostoc* and *Lactobacillus* cultures. Apart from these, there are many other archaea (*Halococcus, Natronococcus*) and yeasts (*Saccharomyces, Candida, Trichosporon*) genera that have been found to be present in kimchi. An animal study showed that consumption of kimchi which contains *Leuconostoc mesenteroides* DRC has been helpful in reducing obesity in mice (Meizi et al. 2015). Kimchi is beneficial medically as it has the ability to reduce serum total cholesterol, triglycerides, low-density lipoprotein cholesterol levels, and atherogenic index as studied in rat models (Jo 2015). In a human study, it was found that consumption of kimchi, which was fermented for 8 weeks, can change the expression of genes that are related to metabolic pathways and immunity (Jalili et al. 2023). The work of Lee et al. (2023) has shown that the tumors can be reduced by the consumption of the product on a regular basis. When it was eaten for 4 weeks (300 g/day) the levels of *Lactobacillus* and *Leuconostoc* increased in the stool along with a decrease in the stool pH.

(h) **Soy sauce** In village homes, the soybeans are fermented along with salt either naturally or by adding the cultures (*Aspergillus oryzae* or *Aspergillus sojae*) of molds. The characteristics of the sauce depend on the raw material used, the method and time taken for fermentation, the characteristics of the mold, or any other supplementation done during the process. The chemical composition of the soy sauce helps to give the characteristic taste and flavor to the product.

(i) **Oncom** Molds are used in the preparation of oncom products. Traditionally, the product is commonly used in West Java, Indonesia. Oncom are of two types, red and black oncom. *Neurospora intermedia* var *oncomensis* microorganism is used in the preparation of Red oncom, the only food from the organism for human consumption. It is a healthful food as it is known to reduce cholesterol levels. It also forms isoflavones and carotenoids. These are the bioactive compounds that have high antioxidant activities. They have the ability to quench the free radicals. These qualities are absent in the unfermented ones. Black oncom is prepared when the fermentation is made using *Rhizopus oligosporus*. Andayani et al. (2020) have studied the different preparations as *gorengan* fritters, seasoning with spices, roasting, mixing the oncoms with steamed rice, known as nasi tutug ontjom, and cooking the product in banana leaves.

(j) **Hamanatto** For the preparation of this product, the cooked soybeans are inoculated with *Aspergillus oryzae*. This is then soaked either in brine or soy sauce. After a specific time period it is removed and then dried. The dried product is called as soy nugget. The product is a little sticky on the outside; therefore, it is different from ordinary natto. The product has taken different names depending on the region they are produced. The common products in Japan, are daitokuzi natto, diyohukuzi natto, and hamanatto. As it is soaked in salt in the beginning, it is salty and is black in color. China has a similar product which is called as douchi (Afzaal et al. 2022).

(k) **Sufu** This product is made by solid-state fermentation using fungal cultures. This is then kept in brine solution which contains salt and alcohol for long duration. The product has different colors and flavors. Sufu has been classified into four types, red, white, grey, and others which depend on color of the product. Different colors are obtained due to different ingredients in the mixtures used for dressing the product. When the product is made after 3–6 months, the proteins get hydrolyzed. In China, the product resembles cheese which is called as Spehtze. This is also salted and is matured in brine. The work (Han et al. 2001) has shown that after the red sufu product is ripened for 80 days the Total (TAA) and free amino acid (FAA) content increases many folds. However, in white sufu, the levels of FAA are much more. The essential amino acid content is so high that it is almost equivalent to eggs/cow's milk. The amino acid analysis has shown to produce high amounts of glutamic acid, leucine, aspartic acid, alanine, phenylalanine, and lysine.

(l) **Idli** It is a savory rice cake, which has the origin from South India and Sri Lanka. The batter is steamed which basically contains fermented black lentils and rice. The work done (Shivangi et al. 2020) has shown that the strains were tolerant to GIT tract conditions as tolerance to strong acid and bile was found good. They were able to travel the digestive tract and had cell surface hydrophobicity. They are found to adhere to the human colon adenocarcinoma (HT 29) cell line. Among the biological properties, the culture strains could bind to the cholesterol and were thermostable. Due to the presence of an enzyme β-galactosidase, the lactose gets utilized easily. This property makes the product very useful as it can be consumed by the lactose-intolerant patients. The presence of proteolytic, amylolytic, and lipolytic activity in the culture strains, the proteins, starch, and lipids get digested easily. The biofilm formed by pathogen *Pseudomonas aeruginosa* was also inhibited to a large extent with the use of cell-free supernatant. Intact cells also showed significant DPPH (41%), hydroxyl (31%), radical scavenging activity, and lipid peroxidation inhibition (20.38%). Renu Agrawal et al. (2000) studied the flavor profile of idli batter prepared for defined microbial starter cultures.

(m) **ANG-KAK** The product is reddish purple in color. It is prepared by fermenting rice, with the mold *Monascus purpureus*. Traditionally, it is used in Chinese medicine, which has been produced by solid-state fermentation of cooked non-glutinous rice (*Orizae sativa* L. Gramineae) with *Monascus purpureus*, *M. ruber*, *M. anka*, and *M. pilosus*. The food product is known for improving digestion, maintaining the blood circulation and for the treatment of muscle bruising and dysentery. The old Chinese pharmacopoeia (Ben-Taso-Gum-Mu) has given the manufacturing process for angkak. The therapeutic properties of the product have also been explained well.

A secondary metabolite, namely, lovastatin (which is a HMG CoA reductase inhibitor), is formed by the product. It is also called as monacolin K. This is known to lower the blood lipid levels in animal models and humans. Another metabolite is γ-aminobutyric acid (GABA). This compound has the ability to lower the blood pressure. Bioactive compound dimerumic acid can quench the free radicals and monascin can reduce the inflammation. When a co-culture of *Monascus purpureus* MTCC 369 and *Monascus ruber* MTCC 1880 are used in making the product, then they work in a symbiotic manner, resulting in the reduction of blood lipid levels to a large extent (Panda et al. 2010).

(n) **Fermented Sausages** In this method, the sausages are partially or completely dependent on fermentative action of certain types of bacteria. The chopped/ground meat is salted to remove moisture. The beneficial bacteria break down the sugars into flavor molecules. The product quality and the desired flavor are affected by the microflora. Vuyst et al. (2008) have shown the sustainability of the probiotic bacteria in the sausages. The bacteria with probiotic properties are able to survive in the product as the sausages are not heated at very high temperatures. However, some amount of bacterial viability goes down due to the high amount of salt used during curing. Probiotic strains in sausages are used either from the existing probiotic bacteria present in the meat products or are obtained from the culture collection center.

(o) **Fermented Fish Products** Fermented fish is consumed in Europe as a staple food. In Greek and Rome, the product is known as garum. The product is like pasta which has a very strong smell. During the process of fermentation, many flavor compounds are formed. This happens as the organic molecules get converted into secondary metabolites with the help of enzymes. The common products are fish sauce and shrimp paste.

(p) **Sourdough Bread** Sourdough is a very old and a traditional method in which the cereal flour is fermented. Fermentation makes the quality of the bread better. During the process, many nutritionally important and flavor compounds are formed. Scientists have studied the microbiota in the fermented dough which majorly contained lactic acid bacteria and yeasts. These also help in improving the sensory qualities. Sourdough fermentation can utilize the flours of wheat, rye as well as other cereals and water. Based on the inoculum and technique, the sourdough fermentation has been classified into three (types I, II, and III) and their technology process as types 0, II, and III.

In type I, the microorganisms used are present in the dough itself. Therefore, the fermentation is spontaneous and utilizes back-slopping for the microorganism to grow in the sourdough. It is important to optimize different parameters like temperature, cell number in the culture to be used, and the time of back-slopping. In type II inoculums, a starter culture is added to sourdough fermentation. The culture is usually obtained from any culture collection center depending on the need of the product. Type III is a hybrid which is a combination of type I and type II. Most of the bakeries and industries use the Type III method.

The different types of fermentation for sourdough bread are:

1. The simple type 0: For the fermentation, initially a mixture of flour and water is taken. The fermentation is done for a less time. It has been found that this flour forms many beneficial compounds and organic acids once the pH is ~4. As the time of fermentation is not enough to produce other characteristic products of sourdough, it is known as sponge dough and not a true sourdough.
2. In type II, the starter culture initiates the fermentation process. The process forms a sourdough in liquid form under optimized physical and chemical parameters. This method has been used traditionally and is spontaneous with a mixture of flour and water at an ambient temperature for 24 h. The quality of the products becomes better.
3. In type III, a dried and lyophilized culture is used. These are mainly used in industries for stable sourdough. The temperature is kept above 30 °C with a fermentation time between 24 and 72 h.

During the fermentation process, many metabolites are produced as acids, alcohols, aldehydes, esters, and ketones. The wheat flour consists of starch, water, proteins, non-starch polysaccharides, lipids, and soluble carbohydrates. Among the carbohydrates, it can be maltose, sucrose, and/or glucose. The flour has enzymes like amylase and glucoamylase which help in the degradation of starch into glucose and maltose. This is done by fermenting the flour using lactic acid bacteria. Yeasts help in the decomposition of maltose as it produces maltase enzymes.

Fermentation increases the concentration of glucose as the starch gets metabolized with the help of LAB and yeasts. There are two major fermentation routes which include, lactic fermentation and alcoholic fermentation. In lactic fermentation, the pyruvate molecules get reduced to lactic acid (homolactic fermentation). This pathway is usually followed by lactic acid bacteria (*Streptococcus*, *Lactobacillus*, and *Enterococcus*). Whereas alcoholic fermentation uses yeast and forms ethanol and $CO_2$ after the reduction of pyruvate. The gas increases the volume. During the fermentation, many flavor compounds are produced which give a unique flavor and aroma to the bread. The yeast *S. cerevisiae* produces various compounds like ethanol, 2,3-butanedione, 2-methyl-1-propanol, 3-methyl-1-butanol, and phenyl ethyl alcohol. Koistinen et al. (2018) have done the metabolic profiling of sourdough. More than hundred (118) bioactive compounds were produced. However, different parameters play a major role. As the temperature goes high during baking, many cells die which forms cell debris. These are known as postbiotics. These impart beneficial properties to the product (Pérez et al. 2022). *Lactiplantibacillus plantarum* strain ITM21B has been used to prolong the shelf life of bread as it produces many antimicrobial compounds like lactic, acetic, phenyl lactic (PLA), and hydroxy-phenyl lactic (OH-PLA) acids (Di Biase et al. 2022). GABA is a neurotransmitter that is helpful in regulating the blood pressure and can combat depression. It is also known to have diuretic and antioxidant activities. This versatile compound has been found to be produced in sourdough. It is formed from L-glutamate by using the enzyme glutamate decarboxylase (Kim et al. 2022). Most of the microorganisms in traditional sourdough significantly produced phenolic compounds which had antioxidant properties. This makes the consumption of bread in improving human health. Many exopolysaccharides (EPS) are produced by LAB in sourdough fermentation (β-glucan, dextran, and inulin). These are mainly a prebiotic homopolysaccharide that imparts health benefits. These are associated in stabilizing cholesterol levels and have anti-inflammatory properties.

The starter culture is added to the base ingredients for fermentation. This step is known as back-slopping. According to the work of Pasquale (2011), it helps to make the quality, texture, flavor, nutritional content, and shelf-life of the bread better.

The role of lactic acid bacteria is to reduce the pH and modulate the cereal enzymes and solubility of the substrate. These parameters help in the growth of microorganisms like *Lactobacillus, Leuconostoc, Weissella, Pediococcus,* and *Streptococcus* genera (Pérez-Alvarado et al. 2022). This process of making bread can lower the nondigestible oligosaccharides, fructans, and raffinose in the bread. This helps the patients with IBS to consume this bread. Sourdough and its constituent microorganisms have antimicrobial, antihypertensive, and cholesterol-lowering properties (Gabriele et al. 2023). The specific sourdough microorganisms form enzymes (proteases) that hydrolyze proteins (Di Cagno et al. 2002). All the gluten is not hydrolyzed during fermentation.

(q) **Cheese** It is a concentrated form of milk protein, namely, casein and fat of milk. In the making of cheese, LAB, milk clotting compound, and NaCl are utilized. Cottage cheese, Cheddar, Swiss cheese, Brick cheese, Blue cheese, mold-ripened cheese, Camembert cheese, and Colby cheese are different types of cheese obtained after fermentation of probiotic cultures (Emmons et al. 2010). Shobharani and Renu Agrawal (2011) isolated and characterized a lactic acid bacteria from cheddar cheese which was a potent probiotic strain. The infection of *Staphylococcus* is common during making of cheese. This can be prevented by using probiotics. The pH should not be taken above 5.8. Thermal destruction at 60 °C can be done. Probiotics also help in eliminating *E. coli* infection.

The microorganisms are specific for each type of cheese. For cheddar, *Streptococcus* cremoris; for Swiss cheese, *Lb. bulgaricus* and *St. thermophilus* are used. Then a coagulant, which is an enzyme, is added. This splits casein into carbohydrate fraction and insoluble fraction. Hydraulic pressure is used in pressing the curd. Ripening is done at 4 °C and 85% relative humidity. As the cheese ripens, the cultures start dying after 2–8 weeks. The product releases proteolytic and lipolytic enzymes. This in fact changes the flavor and texture of the product. The stiffness and elasticity of the product goes down.

The flavor in cheddar cheese is due to many compounds. These are not only carbonyl but also sulfur and nitrogen compounds.

During packing precaution must be taken against mold spores.

In swiss cheese, there are two fermentations. These include lactic acid and propionic acid fermentation. Gas gets released, which makes hole-like structures in the product. To the pasteurized milk cultures are added at 32 °C. After the coagulant is added, the curd is cooked by raising the temperature slowly uptill 50 °C. The product is dipped in NaCl solution (23%). Cheese can be kept in brine upto 2 days. This helps the salt to get absorbed in the product. This also helps in reducing any contamination. Usually, the wrapping material should be impermeable to liquids. In this product, lipolysis takes place due to bacterial metabolism. Many volatile compounds are formed. Due to contamination of other microorganisms, it is possible to have a change in texture, color, and quality.

There can be diseases like Salmonellosis due to Salmonella and due to *Staphylococcus aureus* microorganisms. The symptoms are headache, chills, and green watery stool. Infection of *E.coli* can cause diarrhea. Presence of Coliforms can produce gas. However, the use of probiotics reduces these infections. The contamination of *Clostridium* spp. is extremely dangerous as it can even be fatal.

Infection of *Aspergillus flavus* and other species produces aflatoxin. This can deteriorate the health. This could come in the milk via the feed of cows. The researcher/industrialist should look into that. It has been found to be stable even during processing.

(r) **Yogurt** It is a milk product prepared after the fermentation. It possesses all the nutrients of milk. Many varieties of yogurt are available. The product can vary in fat content. From low to high fat (Greek yogurt) with a fat content of 8 g/100 g. The terminology is specific to the product which is fermented with culture strains of *Lactobacillus* bulgaricus and *Streptococcus thermophilus*. They contain calcium, vitamin D, and potassium. Many consumers are not able to digest milk due to the presence of lactose. However, yogurt can decompose lactose easily. They increase the bone mineral acquisition, weight management, heart health, metabolic health, digestion, and immune health (Anastasios et al. 2023). Culturing the two organisms together imparts a symbiotic effect. A ratio of 1:1 has been found good by the workers. With 2% inoculum each, at 44 °C for 2–3 h has been found optimum.

Fruit style flavored yogurts This product is usually prepared with the addition of flavorings to the plain yogurt.

Western Sundae style yogurt In this preparation, a fruit puree is layered at the bottom of the container which is usually 15–18% of the total weight of yogurt. Then it is sealed and incubated for an optimized period of time.

Many times, the bacteriophages attack slows down the acid production in lactic acid bacteria. They can easily lyse the beneficial bacteria in the system. It is well known to attack the curd and yogurt cultures. This bacteriophage attack can be prevented if chlorine (200–300 ppm) is used on processing equipments. The culture rooms can also be fogged to prevent bacteriophage infection. Inoculum size and the time of incubation should be optimized for maximum product as the acid content can increase if it is high during processing.

**Frozen Yogurt** It is a frozen product that contains the same basic ingredients as ice cream, along with live bacterial cultures as that of yogurt. Usually for this product, around 2% of milk fat is taken with 13% nonfat milk solids. Very little gelatin is added with corn syrup solids (24–26 DE). The mixture is pasteurized and homogenized. This is cooled to 44 °C. The yogurt cultures are inoculated. This is then frozen. For the good quality of product, it is important to optimize all the parameters. Many times, ice crystals are formed due to power fluctuations as they vary the temperature in the product during storage.

(s) **Butter Milk** After the preparation of butter, it is churned. The liquid which is left behind after churning is the butter milk product. Before homogenization of the milk, the cream and milk are separated. The naturally occurring lactic acid bacteria in the milk help in fermentation and in the butter churning process. The harmful microorganisms do not grow in this acidic environment. This in turn increases the shelf life of the product. It is a fermented dairy drink. In recent times, the buttermilk is prepared by using defined cultures of *Lactococcus lactis* or *Lactobacillus bulgaricus* along with the culture of *Leuconostoc citrovorum* which simulates the naturally occurring bacteria (Barukčić et al. 2019). The processing steps involve pasteurization, homogenization, and inoculation of the culture. The raw material in the preparation of butter milk can be skim milk/low-fat milk/cream/nonfat dry milk, culture, and salt. The flavor is usually increased by adding sodium citrate (0.20%). Sometimes, butter flakes or particles are supplemented to the chilled product. However, care should be taken for acidity during the process, temperature, and contact with metal should be avoided so that metallic taste does not come into the product and good flavor is obtained in the product.

(t) **Cultured Cream** It is a double cream obtained by a fermentation process during which the live bacteria convert the milk sugars into lactic acid. Many distinct aroma compounds are produced due to fermentation. The starter cultures in cultured cream do not have the same two starter bacteria which are recommended as per the FDA for yogurt (Ekinci et al. 2008). The product has around 18% fat. The fermentation is done using lactic acid bacteria. It has a buttery aroma. It is usually added as toppings in many food items. Homogenization is done 17 kPa at 71 °C. The parameters for processing should be optimized as the acidity, agitation speed, heat, fat content, and rennet addition affect the quality of the product. When rennet is added in small quantities, then it helps in thickening of the product.

Sometimes, instead of milk fat either oils or other fats are used. Then the product is called as artificial cultured cream.

(u) **Sour Cream Dip** Cultured sour cream is made by adding a culture of *Streptococcus lactis*. For pasteurization, the cream is incubated at 72 °F for 8–12 h. The organism produces lactic acid which helps the proteins to coagulate. It thickens the cream and gives a sour flavor (Feryal et al. 2022). In this product formation, seasonings are blended with sour cream. To the cream, skim milk and stabilizer are added which is then pasteurized and homogenized. To this, cultures are added after cooling. Rennet is also added. This is kept in the ripening tank for a period of 12–14 h. The final product is stored at refrigerated temperature for 2–3 weeks.

(v) **Probiotic Salad Dressing** Probiotic cultures/probiotic yogurt is added to a mixture of oil, vinegar, herbs, and other flavorings which are used as a salad dressing (Akcicek et al. 2021).

Acidophilus Milk This product forms the biggest segment as functional food around the globe. The culture used to prepare acidophilus milk is *L. acidophilus*. They help in gut-associated functions. The product helps in preventing many skin diseases. It is also known to prevent secretion of the conjunctiva, infections in vagina, mouth, and nose.

B. **Alcoholic Fermentation** Alcoholic fermentations are carried out using different microorganisms like yeasts, fungi, and bacteria. During the alcoholic fermentation, pyruvate is formed by the yeast through the EMP pathway. *Zymomonas* (bacteria), however, uses the ED pathway and not the EMP pathway.

- **Wine, Brandy, and Champagne** Grapes are fermented to produce wine which is an alcoholic beverage. Wine can also be prepared using fruits, rice, and other grains. When wine is distilled, another alcoholic product is made known as Brandy. Grapes grown in each area are very specific which gives a characteristic taste to the product on fermentation, for example; when the grapes grown in the Champagne region of France are fermented, then it produces a sparkling wine known as Champagne. Fermentation with microorganism *Saccharomyces cerevisiae* or non-*Saccharomyces* yeasts along with lactic acid bacteria produces a particular type of product after fermentation. Each product has a specific aroma profile. The growth of pathogenic microorganisms can be inhibited if prevention policies are followed, advanced microbial assessment system is in place, and the antimicrobial agents like probiotics are supplemented into the food or incorporated into the packaging films (Madni et al. 2019). Wine making has been explained in detail in Chap. 7 of this book.

Wine quality can deteriorate with undesirable bacteria. Lactic acid bacteria can spoil the product. While processing the amount of sugar is an important factor in the system. Variable amounts of carbon dioxide, ethanol, and acids can be formed. The acidity can increase with high lactic acid in the system. Presence of diacetyl can impart an off flavor. Usually in Swiss wines, the taste is good as diacetyl content is within limits (0.78 mg/l). When other bacteria enter during the process (acetic acid bacteria), then the product gets contaminated and spoils the quality of wine. This has been observed when Acetobacter and Gluconobacter are present. In the presence of acetic acid bacteria, the ethanol produced gets oxidized to acetic acid. On esterification, it can form ethyl acetate.

If the temperature goes above 38 °C the yeats can be inactivated. Therefore, the fermentation is usually done below this temperature. This also prevents the extraction of tannins. Color can be extracted by short time high-temperature heating of the crushed grapes.

**Champagne** When a dry white wine undergoes a secondary fermentation which is done under pressure after the addition of sucrose, then it produces champagne. The yeasts get collected near the neck of the bottle which gets frozen there. It retains carbon dioxide.

Carbonation is done when wine is impregnated with carbon dioxide. However, there is no secondary fermentation. Around 400 to 600 volatile compounds are attributed to the aroma of wine. These include different classes like alcohols, acids, esters, aldehydes, ketones, terpenes, and many more. These come either from the grapes or are produced during the fermentation with yeats. In case, the higher alcohols are formed, the quality of wine goes down. This should be avoided. These could be propanol or butanol.

**Brandy** When wines are distilled then the product brandy is produced. The starting material is grapes. Mostly the white wine is used for distillation. The quality of brandy is better when prepared under low pressure and without the use of sulfur dioxide. The fermentation temperature should also be kept low so that higher alcohols, acetals and fusel oils are not formed. Therefore, it is important to optimize all the parameters before taking up the process.

- **Beer** For the production of beer, the process of fermentation is done using a specific culture strain of *Saccharomyces cerevisiae* (ale-type beers) and *Saccharomyces pastorianus* (lager-type beers). They convert large amounts of sugar, basically from barley, which is present in the medium into ethanol. In sour beer, bacteria have a role in acidifying the drink which imparts a characteristic flavor and aroma. During fermentation, many volatile compounds are produced. These are responsible for the characteristic taste and aroma in beer. In lambic beers, the yeast *Brettanomyces bruxellensis* is responsible for producing volatile compounds for the aroma. Many studies are on the way to understand the aromatic profile of beers using a particular yeast organism through their metabolism and biosynthetic pathways. Study has been done on probiotic beer produced by using *Saccharomyces boulardii* as an alternative to conventional brewer's yeast. The strain was grown in worts along with hops. It was also supplemented with alpha-acid with a final cell concentration of $1 \times 10^8$ cells $mL^{-1}$. The probiotic beer produced in this manner was found to be less cloudy, less alcoholic and healthier. As the pH in the product is kept low, the risk of spoilage is reduced to a large extent. The product contained 30 volatile compounds, which resulted in imparting a unique taste and aroma to the product.

The microorganism used and the condition of the organism decides the quality of beer to a large extent. The pure culture of yeast should be used in the process which should be monitored time and again during the process. The contamination of wild yeast can be dangerous. They may eat away the nutrients supplied and grow profusely. This spoils the beer quality. There are different types of beer. These are differentiated based on the type of fermentation. Bottom yeasts flocculate and settle down whereas in top fermented beers the top yeast does not settle and can be discarded easily.

Hops are added during the fermentation to impart characteristic aroma to the beer. The top-fermented beer is also known as Ales. These have different colors depending on the amount of hops added during the fermentation. Weissbier is prepared using malted wheat and is light. The basic barley is not used in this product. In Lambic beer, a combination of 60% malted barley and 40% raw wheat is used for fermentation.

The basic steps in the process include malting of barley, milling, mashing, kettle boiling, and addition of hops. A mash is prepared by mixing ground malt with water. Kettle boiling is important as the wort is sterilized and in the process many enzymes can denature. Hops are also added during this step. Due to the caramelization of sugars the color of the wort becomes dark. Also there is oxidation of tannins from the hops. This dark color is preferred by the consumers.

The wort is clarified and aerated to which the culture is added in pure form. This allows the fermentation process to take place. The culture in log phase is metabolically steady and is taken for fermentation. The temperature, pH, and time of fermentation are important parameters to be monitored. The nutrient supply and check on the purity of culture during the process are very vital. All these decide the quality of the product.

- **Fermented Ragi** Work has been done on fermented ragi (Basappa et al. 1997). The nutritional composition of fermented ragi (chang) was studied using Phab and defined starter cultures as compared to unfermented ragi (*Eleucine caracanis G*). It was found that the fermented product produced many beneficial and healthful compounds.
- **Paddy Liquor** Basappa and Renu Agrawal (1996) studied the simultaneous saccharification and fermentation of paddy malt mash to ethanol. The work was done using two cultures, namely; *Saccharomyces cerevisiae* PHO3 and *Zymomonas mobilis* ZM4. The symbiotic effect of both the cultures was studied for the activity against pathogens. Renu Agrawal and Basappa (1996) also worked on the yield of ethanol produced using Penicillin G in a fermentation process with mixed cultures of *Saccharomyces cerevisiae* and *Zymomonas mobilis* ZM4. It was found that the alcohol yield increased many folds.

## 9.3 Properties of Probiotics

- **Antimicrobial Properties** Many species of LAB produce antimicrobial compounds like bacteriocin and nisin (Grosu Tudor and Zamfir 2013). Renu Agrawal and Shylaja (2011) isolated an antishigella protein from *Pediococcus pentosaceous* MTCC 5151.
- **Antioxidant Activity** Many scientists (Liu and Pan 2010; Abubakr et al. 2012) have studied the antioxidant activities in fermented foods. The activity can be analyzed quantitatively using; 1,1-diphenyl-2-picryl hydrazyl (DPPH) radical scavenging activity, 2,2′-azino-bis (3-ethylbenzo-thiazoline-6-sulfonic acid; ABTS) radical scavenging activity, total phenol content (TPC) estimation, and reducing power assay. It has been found that most of the Asian fermented soybean foods have many beneficial properties. Among them, the antioxidant properties of product natto obtained from *Bacillus* in Japan are very famous and are formed by the fermentation of soybean (Ping et al. 2012; Shon et al. 2007). Shin and Jeong (2015) worked on the fermented soybean foods in Korea which are known as chungkokjang and jang. These have high antioxidant properties. In China, the fermented soybean product is called Douchi (Wang et al. 2007a). In India and Nepal, work has been done by Moktan et al. (2008), Chettri and Tamang (2014), and Tamang (2015) on the fermented soybean food utilizing the microorganism *Bacillus* spp. The product is named as kinema in Nepal and

bekang and tungrymbai in India. Dajanta et al. (2013) worked on the fermented soybean food of Thailand using *Bacillus*. The food is named as thua nao. Using mold, a fermented soybean food is commonly prepared in Indonesia which is known as tempe (Nurrahman et al. 2013). All these foods obtained after the fermentation of soybean have an excellent antioxidant activities. Products like kimchi (Park et al. 2011) and yogurt (Sabeena et al. 2010) have shown to have high antioxidant activities. Vanaja et al. (2010) studied the therapeutically important volatile compounds in probiotic culture strain of *Lb. plantarum* LB cfr. Shobharani and Renu Agrawal (2007) studied the therapeutic importance of volatile compounds produced by *Leuconostoc paramesenteroides*.

- **Bioactive Peptide Production** Many microorganisms produce proteolytic enzymes during fermentation process. These degrade the proteins into peptides which possess bioactivity (De Mejia and Dia 2010). Work has been done by many scientists on the bioactive peptides. It has been found that these peptides have some specific function. Qian et al. (2011) have proved that the bioactive peptides can enhance immunity. Work by Phelan and Kerins (2011) has shown that these peptides can reduce blood pressure and antithrombic properties have been studied by Singh et al. (2014). In another study done by Nagai and Tamang (2010), it has been found that when the proteins in *Bacillus* are hydrolyzed, with the help of enzymes, then peptides and amino acids are produced which contribute in healthful properties. Literature studies have shown that many fermented milk products like kefir (Quiros et al. 2005), koumiss (Chen et al. 2010), yogurt (Papadimitriou et al. 2007), fermented camel milk (Moslehishad et al. 2013), cheese (Meyer et al. 2009), and fermented fish products (Ichimura et al. 2003) have the potential to form angiotensin (ACE). This is very important in kidney function and in keeping the blood pressure regulated. It helps in improving the cardiovascular diseases. A protein was isolated (Vanitha and Renu Agrawal 2012) from the probiotic culture strain, *Leuconostoc mesenteroides*, which showed the potential to kill *V. cholerae* and is helpful in preventing cholera. In another study, Renu Agrawal and Shylaja (2011) isolated an antishigella protein from *Pediococcus pentosaceous* MTCC 5151.
- **Production of Enzymes** Many enzymes are produced during the process of food fermentation. These are produced by the microorganisms supplemented for fermentation. The complex compounds get broken down into simple biomolecules which are helpful in many biological activities. Most of the Asian fermented soybean foods are produced by *Bacillus* spp. (Tamang and Nikkuni 1996; Chettri and Tamang 2014). These are known to have high levels of the enzymes which help in the production of bioactive compounds. Studies have shown the production of proteinase enzymes which help to breakdown complex proteins; amylases, which degrade the complex starch molecules and cellulases which have the property to decompose cellulose. Many fungi are used in the making of these fermented foods and beverages. The most commonly used organisms are *Actinomucor, Amylomyces, Aspergillus, Monascus, Mucor, Neurospora*, and *Rhizopus*. They have the ability to produce many enzymes especially to degrade the carbohydrates. The common enzymes produced are $\alpha$- amylase, amylogluco-

sidase, maltase, invertase, pectinase, ß-galactosidase, cellulase, hemicellulase, acid and alkaline proteases and lipases (Nout and Aidoo 2002). The work done by Suganuma et al. (2007) have shown the production of amylases by *Aspergillus oryzae* (Taka-amylaseA,TAA). This enzyme is widely used by the industries to degrade amylose. The amylolytic starters are available in many forms. It can be in a dry form, solid, or cake like. These are being used by the alcohol industries. When a consortium of microorganisms are used (*Saccharomycopsis fibuligera, S. capsularis* and *Pichia burtonii*), it is able to degrade high amylose easily (Tsuyoshi et al. 2005; Tamang et al. 2007). Fibrinolytic activity has been found in *Bacillus subtilis* subsp. *natto* (Kotb 2014). Work has been done on the production of fibrinolytic enzymes by microorganisms. Scientists have found that many microbial fermented foods produce these enzymes and the organisms involved are mainly; *B. subtilis* and *B. amyloliquefaciens* (Chang et al. 2012; Zeng et al. 2013), *Vagococcus carniphilus, V. lutrae, Enterococcus faecalis, E. faecium, E. gallinarum,* and *P. acidilactici* (Singh et al. 2014). Work has been done on fish sauce fermentation (Montriwong et al. 2012) using the microorganism *Virgibacillus halodenitrificans* SK1-3-7, which has shown the potential to produce fibrolytic enzymes

- **Increase in Isoflavones, Saponin, and Production of PGA** Isoflavone compounds as daidzein, genistein and glyciteinare are present in soybean. They exist in four chemical forms such as aglycones, β-glucoside, acetylglucoside, and malonylglucoside (Kudou et al. 1991). During fermentation, in products like sufu and douchi of China, it has been seen that the aglycones get hydrolyzed into isoflavone glucosides (Wang et al. 2007b; Yin et al. 2007). The hydrolyzed isoflavone glucosides have been found to be present in products like miso and natto of Japan (Chiou and Cheng 2001), in chungkokjang and doenjang products of Korea (Lee et al. 2007a, b), in tempe product of Indonesia (Lu et al. 2009), and in thua nao which is a product of Thailand (Dajanta et al. 2013). Nakajima et al. (2005) found the formation of many isoflavone compounds during the making of a product known as tempe. The study observed an increase in the content of Factor-II and aglycones. These isoflavones (studied in doenjang product) in turn increase and activate the receptor LDL-C. This prevents the vascular diseases and is helpful in human health (Kwak et al. 2012). The Triterpenoid glycosides are of two types; Group A and DDMP (2,3-dihydro-2,5-dihydroxy-6-metyl-4H-pyran-4-one; Paucar-Menacho et al. 2010). These are found in soybeans. All saponins are known to improve human health especially in maintaining the cholesterol levels (Murata et al. 2006). The colon cancer cell multiplication can be prevented (Ellington et al. 2006), and lipids can be prevented for any peroxidation reaction (Ishii and Tanizawa 2006). The Natto product has high saponin content due to the fermentation by *Bacillus natto* bacteria (Yanagisawa and Sumi 2005). In Kinema, the Group B saponins have been found to be high, which impart health to the consumers (Omizu et al. 2011). Studies have shown that in fermented soybean foods, the strains of *Bacillus* spp. produce poly-glutamic acid (PGA). These are biodegradable, can be dissolved in water, and are not toxic to humans (Yoon et al. 2000). Strains of *B. subtilis* and *B. licheniformis* produce γ-PGA (Stanley and Lazazzera 2005). Kwak et al. (2014) have shown the cancer prevention potential of kimchi lactic acid bacteria.

- **Degradation of Anti-nutritive Compounds** The preparation of fermented products follows particular steps. Starting from peeling the vegetable, it is then washed and grated. Before roasting, the water is removed. It has been worked out in the processing of fermented cassava products of Africa for the preparation of gari and fufu products. During these steps, it has been found that the residual cyanide content in the products comes down (Babalola 2014). Bitterness is present in most of the cassava tubers due to the presence of glycosides, namely, linamarin and lotaustralin. These are known to have cyanide. It has been found that these toxic compounds can be removed with the supplementation of probiotic species of *Leuconostoc, Lactobacillus*, and *Streptococcus*. These probiotic organisms help in producing hydrocyanic acid (HCN). The boiling point of hydrocyanic acid is low and during toasting it goes out from the pulp. In this way, the product becomes very safe to be consumed (Bamidele et al. 2015). Degradation of antinutritive compounds has been reported by *B. subtilis* in the product called as kinema. Work has been done on the fermentation of idli product. It has been found that fermentation with defined probiotic cultures reduces the levels of phytic acid (Reddy and Salunkhe 1980).

## 9.4 Fermented Foods to Combat Diseases and Promote Health

Longevity of people in Japan has been inferred due to the consumption of traditional and cultural foods like natto, miso, tofu, shoyu, and fermented vegetables. All these products are cholesterol free and have low fat and high bioactive compounds (Willcox et al. 2004). Korean kimchi has been claimed to possess health-promoting benefits along with antiaging properties (Han et al. 2015). Natto has several health benefits such as high contents of nattokinase, isoflavones, saponins, vitamin K, and unsaturated fatty acids are very high, and it has immunomodulating activities due to *B. subtilis* (Nagai 2015). Kinema has health-promoting benefits (Tamang 2015). Inhibition of 1,2 dimethylhydrazine induced genotoxicity in rats with the administration of probiotics (Arvind et al. 2009). Lactic acid produced in kimchi may prevent fat accumulation and help in weight loss (Park et al. 2008). Dinesh and Renu Agrawal (2008) prepared a probiotic fermented milk with a probiotic culture of *Pediococcus pentosaceous* MTCC 5151 and found many healthful compounds in the product.

**Nutrient Synthesis**

Scientists have found that the substrates get enriched with vitamins, essential amino acids, and bioactive compounds during food fermentations (Thapa and Tamang 2015). Fermented soybean foods are known as tempe in Indonesia. When the soybeans were inoculated with mold (*Rhizopus oligosporus*), then the food product has been found to have high contents of folic acid, niacin, riboflavin, nicotinamide, and pyridoxine (Astuti 2015). Products fermented with *Citrobacter freundii* have shown to have high levels of vitamin $B_{12}$ (Okada 1989; Keuth and Bisping 1994). For the preparation of Idli batter, a combination of rice and legume is taken. This is fermented with yeast and lactic acid bacteria. This food is commonly eaten in India and Sri Lanka. Studies have shown high levels of thiamine, riboflavin, methionine and vitamin $B_{12}$ (Ghosh and Chattopadhyay 2011). Pulque is an alcoholic drink of Mexico, which is made from a cactus plant. Many beneficial compounds at high levels are known to be present in this alcoholic product. The work of Ramrez et al. (2004) has shown the formation of vitamins B complex and C, lysine, and tryptophan along with high iron content. Asian fermented foods with probiotic *Bacillus* spp. have been found to be rich in riboflavin and niacin contents (Kim and Hahm 2002; Nagai 2015). Riboflavin and folic acid are synthesized in kimchi by *L. mesenteroides* and *L. sakei* (Jung et al. 2013). Enhanced amounts of free amino acids have been found in fermented soybean foods. Vanajakshi et al. (2015) found many beneficial effects of probiotic cultures on the fermentation of Moringa leaf extract. Therefore, the product can be consumed for improving the health.

**Prevention of Hypertension and Heart Disease**

Many probiotic products have antihypertensive properties. This has been studied using animal models and clinical trials (Sipola et al. 2002). Consumption of fermented milks or probiotic bacteria (Agerholm-Larsen et al. 2000) and fermented soybean foods (Chen et al. 2012) helps in cardiovascular diseases. Chen et al (2018) have explained the use of Nattokinase as a promising alternative in the prevention and treatment of cardiovascular diseases. According to Anderson (2003), the serum LDL-cholesterol levels can be brought down with the consumption of fermented whole-grain foods. It also helps in reducing the blood pressure, in controlling heart diseases, and in maintaining the insulin levels and homocysteinaemia. Reduction of cholesterol levels has been documented by consuming products like tempe (Su et al. 2023), fermented soybean foods (Lee 2004), and kefir (Otles et al. 2003). Calpis, the Japanese fermented sour milk, has been found to contain two peptides VPP and IPP. These peptides have hypotensive properties (Nakamura et al. 1996). *L. helvetius* in fermented milk reduces the elevated blood pressure (Aihara et al. 2005; Shah 2015). *Monascus purpureus* is used in the fermentation of red rice of China locally called *angkak*. This prevents the formation of cholesterol as it blocks a key enzyme, HMG-CoA reductase due to the presence of mevinolin citrinin (Pattanagul et al. 2008). Consumption of fermented tea prevents many heart diseases (Mo et al. 2008). Some Asian fermented soybean foods have antihypertensive properties as observed in *natto* (Nagai 2015) and tempe (Astuti 2015). Isoflavone in *doenjang*, mold-fermented soybean food of Korea, plays an important role in preventing cardiovascular diseases (Tamang et al. 2016). Consumption of fermented whole grains protects from heart diseases and diabetes (Anderson et al. 2003). Red wines are rich in polyphenols which inhibit the peroxidation of lipids (Feher et al. 2007). It has been found that when Korean fermented soybean foods are consumed on a regular basis, it reduces the blood pressure and diabetes (Jung et al. 2014). The work of Lim et al. (2014) has also confirmed that the consumption of fermented foods can reduce cholesterol levels. Ogunremi et al. (2015) have shown the probiotic potential of yeasts isolated from some cereal based Nigerian traditional fermented food products.

Omotade et al. (2015) studied the functional properties by feeding the rats with a cereal-mix produced with probiotic yeast. It was found that the lipid levels in rats came down drastically and the capacity to quench the free oxygen radicals was enhanced many folds. Ogunremi et al. (2015b) developed a cereal based functional food using a cereal mix substrate which was fermented with probiotic strain of Pichia.

**Gastrointestinal Disorders and Protection**

Probiotics can reduce many of the digestive tract disorders (Verna and Lucak 2010). It has been found that many *Lactobacillus* strains help in improving the inflammatory bowel disease, paucitis, and colitis caused due to ulcer formation (Li et al. 2023). The probiotic culture, *L. rhamnosus* GG, has been able to treat acute diarrhea (Sanklecha et al. 2022) and consumption of fermented milk with *L. helveticus* culture improves the cognitive function (Young et al. 2014), immunomodulation (Granier et al. 2013) and inhibits diarrhea (Balamurugan et al. 2014). High antishigella activity (dysentery) was found in a study with probiotic lactic acid bacteria (*Pediococcus pentosaceus*) (Srividya et al. 2015).

**Antiallergic Properties**

*Lactobacillus kefiranofaciens* M1 in *kefir* has been found to have antiallergic properties (Hong et al. 2010). Kimchi has many beneficial properties. A lactic acid bacteria, *Lactobacillus* species, has been isolated from kimchi. This can balance Th1/Th2. This probiotic organism has the ability to produce IL-12 and IFN-γ in high amounts which can reduce atopic dermatitis and food allergies (Won et al. 2011). Fermented fish oil, which is rich in Omega-3 polyunsaturated fatty acids, is helpful in reducing allergies (Han et al. 2012).

**Diabetes and Osteoporosis Protection**

Scientists have found that consumption of probiotic dahi can reduce the intolerance of glucose, can bring down glycemia, and regulate insulin, oxidative activity, and dyslipidemia. This ultimately can reduce and regulate diabetic conditions (Yadav et al. 2007). When the fermented products are eaten every day, it helps in controlling diabetics (Tolhurst et al. 2012). Natto product contains Vitamin $K_2$, which is known to stimulate the formation of bones and osteoporosis can be prevented (Yanagisawa and Sumi 2005). Many minerals are present in yogurt like magnesium, calcium, phosphorus, and potassium. It also contains a large amount of proteins. These help in the formation of healthy bones (Rizzoli 2014).

**Protection of Lactose Malabsorption**

The common organisms utilized in the preparation of yogurt are *L. delbrueckii* subsp. *bulgaricus* and *S.thermophilus*. The enzyme ß-D-galactosidase is in large amounts in these probiotic cultures. Due to this particular enzyme, the galactose can be hydrolyzed easily (Shah 2007). Therefore, it is generally observed that using fresh yogurt (with live cultures) improves the digestion of lactose (Oskar et al. 2004). Lactose easily gets digested with Kefir as it contains the enzyme β-galactosidase (Hertzler and Clancy 2003). Major antimicrobial activity of some probiotic cultures has been mentioned (Table 9.1).

**Table 9.1** Major antimicrobial activity of some probiotic cultures

| Probiotic culture | Antimicrobial compound | References |
|---|---|---|
| *Lactococcus lactis* | Nisin, Lactostrepsin, Lactocin, Lacticin | Bravo et al. (2009) |
| *L. acidophilus* | Acidolin, Acidophilin, Lactocidin, Lactocin B | De Vuyst (1994) |
| *Pediococcus pentosaceous, P. acidilactis* | Pediocin | Papagianni and Anastasiadou (2009) |
| *S. thermophilus* | Streptophilin | Ravinder et al. (2012) |

**Anticarcinogenic Properties**

Activity against cancer has been well documented by using many LAB (lactic acid bacteria). Antimutagenic property has also been worked out (Ahmad et al. 2018). Consumption of kefir has been found to be effective in the treatment of cancer (Yanping et al. 2009). Sauerkraut contains *s*-methylmethionine that helps in bringing down the risk of tumourigenesis in the stomach (Liang et al. 2023). Fermented milk (which contains live cells of *L. acidophilus*) has been found to decrease the amounts of ß-glucuronidase, azoreductase, and nitroreductase enzymes which are cancer-causing enzymes. Inhibition of these procarcinogens and simultaneous activation of the immune system has been found in the consumers which took probiotic products made with *L. acidophilus* (Macouzet et al. 2009). Kimchi, when prepared using the cultures of *W.cibaria*, and *L. plantarum,* has shown to inhibit the multiplication of cancer cells (Kwak et al. 2014). Studies have shown that use of probiotic yogurt in food can reduce cancer in the urinary bladder, in the digestive tract regions especially colon, and in the cervical region (Chandan and Kilara 2013).

According to Goldin and Gorbach (1980), use of *L.acidophilus* in diet could reduce colon tumors in rats. Cancer-preventing attributes of probiotics have been discussed well in the literature (Kumar et al. 2010). Hepatitis B virus is the most common organism to cause cancer. The presence of Aflatoxin $B_1$ (AFB1) makes the genetic changes in the p53 tumor suppressor gene and in turn promotes cancer by protooncogenes. This harmful aflatoxin can be bound by many probiotic bacteria strains. In vivo experiments have shown to neutralize the toxin AFB1.This results in reducing the bioabsorption of the toxin from the gut (Haskard et al. 2000).

Cultures of *Bifidobacterium longum* have also depicted the antitumor activity on colonic mucosa of rats by reducing the level of ras-p21. It has shown to inhibit the proliferation of cells (Abbas and Dutta 2009). Use of *Lactobacillus* GG has shown the up regulation of 334 genes and downregulation of 92 genes. The genes responsible in immune system and inflammation also get involved. The growth factor-beta (TGF-β) gets transformed and there is a change in tumor necrosis factor (TNF), cytokines, nitric oxide synthase 1, defensin alpha-1, apoptosis, cell growth, and cell differentiation (cyclins and caspases, oncogenes).

Probiotics are known to reduce the fecal concentrations of cancer-causing enzymes (glycosidase, β-glucuronidase, azoreductase, and nitroreductase) and secondary bile salts. The probiotics have shown to bring down the absorption of harmful mutagens which are responsible to cause colon carcinogenesis (Rafter 1995). Consumption of probiotic products using *Lactobacillus acidophilus* and *L.casei* have shown to inhibit the cancer-causing enzymes (Lidbeck et al. 1991).

The mechanisms followed by the probiotic cultures include enhancing the immune response in the host, they can also change the metabolic activity of the other microbes present in the intestine. They can bind with the carcinogens and can also degrade them and can produce many compounds without allowing mutagenesis and without altering the physiochemical conditions in the colon (Hirayama and Rafter 2000). When prebiotic inulin is used along with two probiotic cultures, *Lactobacillus rhamnosus* and *Bifidobacterium lactis*, tumor formation can be reduced. This has been confirmed by Femia et al. (2002) in rat models in which colon cancer was induced using azoxymethane (AOM).

**Immunologic Enhancement**

Many of the probiotic strains have shown to increase the immune functions in the host. The work done by scientists (Mazziotta et al. 2023) has verified the use of *B. longum, L. acidophilus, L. casei subsp. rhamnosum,* and *Lactobacillus helveticus* for increasing immunity. Increase in phagocytic activity of granulocytes, cytokine excretion in lymphocytes, and increased immunoglobulin-secreting cells in blood have been found with probiotic supplementation. An inflammatory immune response produced cytokine-activated monocytes and macrophages, that release cytotoxic molecules and are capable of lysing tumor cells in vitro (Philip and Epstein 1986). Enhancing immunity in the elderly with probiotics is a safe proposition (Gill and Rutherfurd 2001). They stimulate both nonspecific immunity and specific immunity which shows its potential for therapeutic and prophylactic applications to treat infections and carcinogenesis.

**Increase in Short-chain Fatty Acid Production**

With the consumption of probiotic foods, the intestine receives high number of probiotics. Due to this, the short-chain fatty acid (SCFA) production gets stimulated (Markowiak-Kopeć and Slizewska 2020). Studies have been conducted on the stressed cells of *Leuconostoc mesenteroides* sub sp. Dextranicum to know the fatty acid content in the cellular membrane. Many beneficial fatty acids have been found as compared to the control samples (Shobharani and Renu Agrawal 2008).

**Antiatherogenic and Cholesterol-Lowering Properties of Probiotics**

Probiotic fermented milk products have shown great promise in lowering cholesterol levels. Studies have shown that plasma cholesterol levels can come down by almost 18% after feeding fermented milk. Mann (1977) has found out that when yogurt is consumed in large quantities, then the level of cholesterol goes down which helps in the problem of cholesterolemia in human beings. Supplementation of *L. acidophilus* strains in food can help in reducing the levels of plasma cholesterol. This microbial strain has the ability to assimilate cholesterol and deconjugate the bile salts. It also binds the cholesterol to the bacterial cell walls which ultimately reduces the biosynthesis of cholesterol (Pereira and Gibson 2002).

**Reduction in Diabetes and Obesity with Probiotics**

It is the microorganisms present in the gut which are responsible for body weight and insulin content. However, many studies suggest that consumption of probiotics help in decreasing the obesity and diabetes as they improve the gut microflora. Strains of *Lactobacilli* and *Bifidobacteria* have shown beneficial effects to reduce obesity and type-2 diabetes (Koutnikova et al. 2019).

**Anticariogenic properties and oral microbial dysbiosis** For dental diseases especially like dental caries, it is very important to understand the oral microbiome. There are many ecological approaches which give corrective measures toward this problem. It helps in correcting the oral microbial dysbiosis with an approach of a symbiotic benefit in preventing dental caries. In a study, the role of a prebiotic, Galacto-oligosaccharide (GOS), was taken up for the growth of probiotic culture. The oral infection is mainly due to the pathogenic microorganisms, namely, *Streptococcus mutans* and *Candida albicans*. An in vitro study was conducted to check the inhibitory effect on the growth of *Streptococcus mutans* and *Candida albicans* with the use of probiotics. Using the technique of real-time PCR, the expression levels of virulent genes in *C. albicans* and *S. mutans* were studied. The work of Zeng et al. (2023) found that GOS was able to inhibit the growth of *C. albicans* and *S. mutans* to a large extent and at the same time the growth of *L. plantarum* was not affected. Supplementation of GOS also improved the anti-fungal activity in the presence of *L. plantarum*.The supplementation of GOS helped in reducing the disease-causing genes (HWP1, ECE1, atpD, and eno genes) to a large extent. Ogunremi et al. (2015) studied the probiotic potentials of yeasts which were isolated from some cereal-based Nigerian traditional fermented food products and found many beneficial properties.

**Other Advantages of Probiotics**

The presence/absence of a particular enzyme is responsible to prevent/increase any disease. For example, it is the lactase enzyme that helps in the decomposition of lactose, it also regulates many allergic diseases. They produce antiinflammatory cytokines, transfer growth factor-β, possess prophylactic and therapeutic properties, prevent vaginal or urinary tract infections, reduce ulcers and intestinal tract infections, increase the nutritional value, maintain mucosal integrity, reduce catabolic products, stimulate the repair mechanism of cells, break down and rebuild hormones, and provide relief for anxiety and depression. Probiotics have shown to reduce seizures, anxiety, and depression in WAG/Rij rats by increasing neurotrophic factors and decreasing proinflammatory cytokines. Probiotics have the ability to detoxify mycotoxins.

## 9.5 Disadvantages of Probiotic Foods

The biogenic amines of low molecular weight bring a disadvantage to the use of probiotic products as they may cause migraine and tachycardia (Saha et al. 2024). These are mainly produced by Enterobacteriaceae and enterococci bacteria (Pircher et al. 2007). At higher levels, amines like histamine and tyramine (>100 mg/kg) can become harmful to human health (Rauscher-Gabernig et al. 2009). Rabie et al. (2011) have shown that fermentation of cabbage with certain lactic starters such as *L. casei* subsp. *casei, L. plantarum*, and *L. curvatus* can reduce the biogenic amine content in sauerkraut.

**Probiotic Activity of Microbial Strains**

The strains should be live, should be growing, and should be able to colonize and survive the immune system. It is important that the strain should not cause any disease or allergy. Nor should the probiotic culture used should mutate or cause cancer (Nagpal et al. 2012). In case the probiotic organism are of human origin, then they can easily survive, and have the ability to grow in the intestine. They are able to tolerate low pH and high concentration of bile acids. In foods, when the probiotic culture is supplemented, then the culture strain should be compatible with the food. It is vital to check the viability of the culture from time to time during the manufacturing process, as the probiotic culture could provide the beneficial properties in the live form only. Moreover, the culture must stay live till the end of storage period in the product.

## 9.6 Molecular Characterization of Probiotics

With the recent developments in molecular sciences, it has become easy and rapid to obtain the phylogenetic information by the technique of 16S rRNA. This allows the characterization of the whole human gut microbiota. It can monitor specific strains. Other molecular techniques like bile salt hydrolase (BSH), mucus binding protein (mub), and fibronectin-binding protein (fbp) help in the identification of probiotic marker genes.

Bile salt hydrolase (BSH) gene This particular enzyme has the ability to hydrolyze different bile acids. It can convert the glycine or taurine conjugated bile acids into amino acids and deconjugated bile acid. The probiotic strains hydrolyze the bile salts; therefore, the strain is able to survive in the GI tract which is used as a potential marker for the screening of probiotic strains (FAO/WHO 2002). It has been worked out that the presence of Mucus binding protein (Mub), Fibronectin binding protein (FbpA), and surface layer protein (SlpA) help the probiotic strain, *L. acidophilus* NCFM, to adhere to the Caco-2 cells in vitro.

In recent times there are many probiotic foods available commercially in the market by different companies. The most common ones are yoghurt, frozen fermented dairy desserts, spray-dried milk powder, cheeses, ice cream, freeze-dried yoghurt, and fruit juices which deliver probiotics efficiently. For a successful

**Table 9.2** Major uses with probiotic cultures

| Probiotic culture | Uses | Companies producing the product |
|---|---|---|
| *Pediococcus* spp. | Controls vibriosis | Eon Pharmaceuticals |
| *Bacillus subtilis* | Controls gas formation | ACI animal health |
| *Rhodococcus* sp. | Antimicrobial activity | Rals Agro Chemicals |
| *Bacillus* | Protect marine life and protects from pathogens in water | CP aquaculture |

probiotic product in the market each parameter is important starting from the selection of a probiotic strain, food processing conditions, product matrix, taste and texture of the product, packaging, and other environmental conditions (Table 9.2).

## 9.7 Safety Regulations

When probiotics are used in food, it is important to study and establish efficacy, safety, labeling, and other claims at the international level. Good surveillance systems must be set in place to record and analyze the health benefits/adverse effects of the microbial strain/s after using them for a long time. The probiotic foods should be categorized under functional foods. The US Food and Drug Administration (FDA) has defined four food categories: conventional foods, foods for special dietary use, medical foods, and dietary supplements. Probiotic foods can fit into any legislative consideration. However, each country forms their own government regulations to assess the safety of probiotic foods. To check the safety of probiotics many factors should be considered. The culture should be nonpathogenic, noninfective, and nonvirulent. All these features will depend on the metabolic activity and the intrinsic properties of the microorganisms. Globally, a set of standard methods have been set in order to assess the safety of lactic acid bacteria (Gueimonde et al. 2006). Another study was done by Salminen and Marteau (1997) on the assessment of safety of probiotics. They have proposed the intrinsic properties, pharmacokinetics, and interactive parameters between the host and the probiotics. An accurate enumeration of the probiotic bacteria in food products is necessary for proper labeling and handling procedures. Also a proper check should be done on the viability and activity of probiotics after processing, handling, and storage of probiotic foods. Indian Council of Medical Research and Department of Biotechnology, Government of India have taken initiatives to formulate guidelines for the regulation of probiotic products in the country (ICMR-DBT 2011). These define various parameters required for a product/strain to be termed as probiotic. The parameters include the identification of the strain, screening of probiotic characteristics, in vivo animal and human studies to establish efficacy, requirements for labeling the products with strain specification, viable numbers at the end of the shelf life, storage conditions, and other details. These help the consumers toward safety of the product and the consumption of the probiotic product. It is very important to validate a product for a health claim before using it. Clinical trials are very vital. For regulations to be in place, it is important to have evidence for any health-related claims. The US Food and Drug Administration (FDA) is authorized for regulating any health claims on the food labels as per the Nutrition Labelling and Education Act (Rogeaux 2010). According to the procedure, the labels should have the information of specific nutrients present in the food along with the information of a particular disease or health-related condition that could appear/happen on the consumption of that particular food.

## 9.8 Genetic Manipulation of Probiotics

A lot of research has been done to prove that by genetic modification the probiotic properties can be enhanced (Steidler 2003). Consumers usually have resistance towards genetically modified organisms unless it is proved that they do not cause any harm to the human health. The use of genetically modified *Lactococcus lactis* in mice has been studied. This can help in delivering cytokine IL-10 to the mucosa of intestine which can prevent colitis. Braat et al. (2006) constructed a transgenic bacteria with *L. lactis* which could produce human IL-10. This was found to treat Crohn's disease.

**Use of Synbiotics**

When prebiotics and probiotics are used in combination, then it is known as synbiotic. A prebiotic is usually a fibre. It is a food ingredient which can not be digested. A prebiotic is known to increase the growth of the probiotic cultures in the digestive tract especially in the colon region. As the probiotics multiply at a faster rate, the benefits to the consumer also increases. This helps in improving the host health (Davani-Davari et al. 2019). The use of synbiotics also enhances the rate of survival of the probiotics which ultimately gets adhered to the mucosal membrane. They also help in the growth of other probiotic strains present in the digestive tract. The European dairy drink and yoghurt manufacturers such as Aktifit (Emmi, Switzerland), Proghurt (Ja Naturlich Naturprodukte, Austria), Vifit (Belgium, UK), and Fysiq (the Netherlands) use synbiotics. The combination of *Bifidobacterium* and oligofructose has been found to synergistically improve colon carcinogenesis in rats (Gallaher and Khil 1999). Rayes et al. (2005) conducted clinical study in which it was found that a synbiotic of *Pediococcus pentoseceus, Leuconostoc mesenteroides, Lactobacillus paracasei* and *L. plantarum* along with four fermentable fibres, namely, β-glucan, inulin, pectin, and resistant starch were able to reduce the infections after the operations were done. The infection rate of 48% without the use of synbiotics, came down to 13% when the synbiotics were used. Organisms like *Bifidobacteria* and *Lactobacilli* are usually used along with the prebiotics.

Work with anaerobic bacteria and keeping the shelf life is a challenging task. However, the work done by Qiu et al. (2020) found that the shelf life of anaerobic bacteria could be increased by using polymeric encapsulation. Probiotics can also help in improving the muscle mass and fatigue (Prokopidis et al. 2023). Work of VidyaLaxme et al. (2012) has found that when two cultures (namely, probiotic

*Leuconostoc mesenteroides* and *Bacillus subtilis*) were used synergistically, with basic raw material as ragi (*Eleucine corocana*) malt, then the activity against pathogenic bacteria, *Vibrio cholerae*, increased many folds. Many other beneficial properties have also been found with higher yields as compared to the monocultures.

**Role of Fermented Foods**

In a fermentation process bacteria and yeast break down the sugars present in the system. They improve food preservation and beneficial bacteria (probiotics) in the gut. Fermented foods are defined as foods or beverages produced through controlled microbial growth, and the conversion of food components through enzymatic action. Many foods undergo fermentation, including meat and fish, dairy, vegetables, soybeans, other legumes, cereals, and fruits. There are several variables in the fermentation processes. These include the microorganisms, the nutritional ingredients, and the environmental conditions, giving rise to thousands of different variations of fermented foods. Historically, food fermentation was a method of preservation. Food fermentation can produce antimicrobial metabolites (organic acids, ethanol, and bacteriocins) which could reduce the risk of contamination with pathogenic microorganisms. Agrawal and Basappa (1994) worked on the fermentation of paddy malt mash to ethanol by mixed cultures of Saccharomyces cerevisiae and Zymomonas mobilis ZM4 with Penicillin G. Agrawal and Basappa (1996) studied the role of antimicrobial agents in simultaneous saccharification and fermentation of paddy malt mash to ethanol by mixed cultures of Saccharomyces cerevisiae PHO3 and Zymomonas mobilis ZM4. This has also shown to enhance the sensory qualities (taste and texture). Foods can be fermented naturally (wild ferments or spontaneous ferments) with the microorganisms present naturally in the raw food or processing environment. Sauerkraut, kimchi, and certain soy-fermented products are examples of these products. Foods can be fermented by adding starter cultures. These are culture-dependent fermentations like kefir, kombucha, and natto.

In another method small amount of culture is added from a previously fermented batch known as backslopping as in sourdough bread. Starters used for the initiation of fermentation can be either natural (backslopping), or selected commercial starters. Fermented foods are very much in demand all around the globe due to their health benefits and for improving the gastrointestinal health. The fermented products must contain a minimum of $10^6$ microbial cells per gram. For the survival of probiotic strains the food matrix matters due to the buffering and protective effect against the gut conditions (low pH, bile acids). Gayathri and Renu Agrawal (2012) studied the antioxidant activity and fatty acid profile of fermented milk prepared by *Pediococcus pentosaceus*.

## 9.9 Commercial Probiotic Organisms Involved/Starter Cultures in Industries

The formation of bioactive peptides and polyamines takes place in lactic acid bacteria. These compounds are responsible for improving the heart health, immunity, and metabolic health. Prebiotics and vitamins present in fermented foods help in these healthful benefits. Fermentation reduces the level of toxins and anti-nutrients which help the patients against bowel disorders such as irritable bowel syndrome. The most common lactic acid bacteria (LAB) used in commercial products are *Enterococcus*, *Lactobacillus*, *Lactococcus*, *Leuconostoc*, *Pediococcus*, and *Weissella*. Cheeses are found to be enriched with species of *Bacillus*, *Bifidobacterium*, *Brachybacterium*, *Brevibacterium*, and *Propionibacterium*. However, the fermented meat products were found to have species of *Arthrobacter* and *Hafnia*. Yeasts have been isolated mainly from alcoholic beverages and nonfood mixed amylolytic starters (*Candida*, *Debaryomyces*, *Geotrichum*, *Hansenula*, *Kluyveromyces*, *Pichia*, *Rhodotorula*, *Saccharomyces*, *Saccharomycopsis*, *Schizosaccharomyces*, *Torulopsis*, *Wickerhamomyces*, and *Zygosaccharomyces*). Many fermented foods also contain species of *Actinomucor*, *Amylomyces*, *Aspergillus*, *Monascus*, *Mucor*, *Neurospora*, *Penicillium*, *Rhizopus*, and *Ustilago*. The commercial probiotic cultures are available in global markets. *Bacillus coagulans* BC30 marketed by Ganeden Biotech, Inc., Cleveland, Ohio, USA; *Lactobacillus acidophilus* NCFM, *Lactobacillus rhamnosus* HN001 (DR20) and *Bifidobacterium lactis* HN019 (DR10) marketed by Danisco (Madison, WI, USA; https://www.ncbi.nlm.nih.gov/pmc/articles/PMC4844621//), *L. casei* strain Shirota and *B. breve* strain Yakult marketed by Yakult (Tokyo, Japan), *L. fermentum* VRI003 (PCC) marketed by Probiomics (Eveleigh, NSW, Australia; https://www.ncbi.nlm.nih.gov/pmc/articles/PMC4844621/), *L. rhamnosus* R0011 marketed by Institut Rosell (Montreal, QC, Canada), *Streptococcus oralis* KJ3 marketed by Oragenics, Inc. (Alachua, FL, USA; https://core.ac.uk/download/pdf/82889782.pdf), and *Saccharomyces cerevisiae* (boulardii) marketed by Biocodex (Creswell, OR, USA; US Probiotics Home, 2011; https://core.ac.uk/download/pdf/82889782.pdf).

**Shelf life extension of products by cultures** Biopreservation is an important method to enhance the microbiological safety and quality of fermented food products along with extending their shelf lives. Application of probiotics as natural antimicrobials and protective cultures has been found effective as they produce compounds that can kill pathogenic organisms (bacteriocins, organic acids, and hydrogen peroxide). These include LAB (*Lactococcus*, *Lactobacillus*, *Pediococcus*, *Leuconostoc*, *Carnobacterium*, *Enterococcus*, and *Streptococcus*) and *Bifidobacterium*. They inhibit pathogenic bacteria which usually include; *Salmonella* spp., *Escherichia coli* O157:H7, *Staphylococcus aureus*, and *Listeria monocytogenes*. Thus the shelf life of the products gets enhanced. Many of the cultured dairy products (yogurt, sour cream, kefir, quark, and cottage cheese) have a low a pH. This is usually less than 4.6. Due to this, the pathogenic bacteria are unable to grow and in turn make the products safe for a longer duration. However, the product requires safety procedures against yeast and mold growth. As per the international rules, the yogurt products produced under good manufacturing practice (GMP) conditions should contain <10 yeast colony-forming units (CFU)/gram. The spoilage can be seen visibly as the cell number becomes more (>$10^5$–$10^6$ CFU/gram). To avoid infection by molds, their cell number should be <1 spore/gram. As soon as the cell number becomes 1–5/gram of the product, the infection becomes visible and the product can not be sent to the market. This infection also reduces the shelf life of the product. Freeze drying has been found to be a very effective method in enhancing the shelf life of probiotic products. The work of Shobharani and Renu Agrawal (2010) used freeze drying for probiotic *Leuconostoc mesenteroides* MTCC 5209 and they were able to enhance the cell stability and viability for a much longer time. Use of oxygen scavengers has a role in improving the stability and viability of probiotic lactic acid bacteria (Surya et al. 2007). Sudha et al. (2006) induced the conditions as that of the gastrointestinal tract for a local probiotic isolate, *Pediococcus pentosaceous* (MTCC 5151). They studied the stability and viability of a culture without any additional preservation method.

**Preparation and Maintenance of Starter Cultures**

Probiotic lactic acid fermentations are the main contributors in the formation of dairy products. They are mainly produced by using the starter cultures to obtain various textures and aromas. When a starter culture is prepared in bulk, a complex starter medium containing the desired nutrients is needed. This is reconstituted in water and is pasteurized before use. For sterilization procedures, either vat-pasteurization (30 to 45 minutes at approximately 85 to 91 °C) or UHT-pasteurization (99 to 113 °C for 3 seconds to 6 minutes) is done.

As per the literature, the efficient method to preserve the starter culture is by adding a cryoprotectant. For preservation, the culture is taken in skimmed milk along with lactose. This is then frozen below (−) 60 °C using liquid nitrogen. It is then stored in a temperature range of (−) 20 °C to (−) 40 °C. A fermented or cultured dairy product may or may not be probiotic. At the time of using the culture, it can be reconstituted in skimmed milk.

**Causes of Starter Defects and Implementation**

Defects can happen due to the use of unhygienic production practices or sick personnel's. It can also happen if the packaging material is not properly cleaned/sterilized or with the use of improper storage conditions and temperatures. Many factors are responsible to cause defects. These are; if the connection wire to and/or from the starter culture vessel is not tight or if the connection to the starter culture vessel has got corroded or if the battery has got corroded/damaged/worn-out. The activity of the starter culture depends on many factors like the age of the culture, the growth phase of the culture, source of the culture, storage conditions of the culture, the duration of the culture being kept in storage, the incubation temperature, the quality of the raw material being used, presence of bacteriophages if any inhibitors like drugs/sanitizers are present or not.

**Degradation of Antinutritive Compounds**

Probiotic bacteria and yeasts have been found to be effective as they can reduce the levels of anti-nutritional factors (Ana María Romero-Espinoza et al. 2020). Legume seeds are rich in proteins but their daily consumption can result in cardiovascular diseases due to the anti-nutritional factors (ANFs) present in them. The use of probiotics have shown promise in reducing the anti-nutritive compounds to a large extent.

**Health Risk of Fermented Foods**

Biogenic amines have been found in some fermented foods like sauerkraut, fish products, cheese, wine, beer and dry sausages. These amine compounds are harmful and risky for health. The maximum amounts of amine compounds are produced by the bacteria belonging to the family Enterobacteriaceae. When the level (>100 mg/kg) of histamine and tyramine is more, then it is a matter of concern and is unhealthy. Although lactic starters like *L. casei* subsp. *casei, L. plantarum* and *L. curvatus* have shown to reduce the levels of biogenic amines.

**FAQ's**

1. Which microorganisms are commonly used as probiotics?
2. Among pro and prebiotics which are better?
3. What is the role of probiotic in digestive tract?
4. Can probiotics cause abdominal discomfort?
5. How do probiotics stay alive in supplements till they are consumed?
6. Which fermented food has the maximum live cells?
7. Can probiotics be taken along with antibiotics?
8. What are the requirements for microorganisms that are part of probiotics?
9. What are the probioitic foods from yeasts for human consumption?
10. What are the common probiotic foods from lactic acid bacteria?

## References

Abbas T, Dutta A (2009) p21 in cancer: intricate networks and multiple activities. Nat Rev Cancer 9(6):400–414. https://doi.org/10.1038/nrc2657

Abubakr MH, Zaiton I, Mohamed SA, Sharifah NR (2012) Antioxidant activity of lactic acid bacteria (LAB) fermented skim milk as determined by 1,1-diphenyl-2-picrylhydrazyl (DPPH) and ferrous chelating activity (FCA). Afrn J Microbiol Res 6:6358–6364. https://doi.org/10.5897/AJMR12.702. Academic Press. 2013. 623–655. ISBN 9780123971548. https://doi.org/10.1016/B978-0-12-397154-8.00181-0

Afzaal M, Saeed F, Anjum F, Waris N, Husaain M, Ikram A, Ateeq H, Muhammad Anjum F, Suleria H (2021) Nutritional and ethnomedicinal scenario of koumiss: a concurrent review. Food Sci Nutr 9(11):6421–6428. https://doi.org/10.1002/fsn3.2595

Afzaal M, Saeed F, Islam F, Ateeq H, Asghar A, Shah YA, Ofoedu CE, Chacha JS (2022) Nutritional health perspective of natto: a critical review. Biochem Res Int 21:5863887. https://doi.org/10.1155/2022/5863887

Agarwal R (2016) Role of probiotics in improving human health. pub.NIPA. ISBN 8119254686, 9788119254682. 186 pages

Agerholm-Larsen L, Raben A, Haulrik N et al (2000) Effect of 8 week intake of probiotic milk products on risk factors for cardiovascular diseases. Eur J Clin Nutr 54:288–297. https://doi.org/10.1038/sj.ejcn.1600937

Agrawal R, Basappa SC (1994) Fermentation of paddy malt mash to ethanol by mixed cultures of *Saccharomyces cerevisiae* and *Zymomonas mobilis* ZM4 with Penicillin G. J Ferment Bioeng 77:218–220

Agrawal R, Basappa SC (1996) Role of antimicrobial agents in simultaneous saccharification and fermentation of paddy malt mash to ethanol by mixed cultures of *Saccharomyces cerevisiae* PHO3 and *Zymomonas mobilis* ZM4. Biotechnol Lett 18(6):673–678

Agrawal R, Shylaja D (2011) An antishigella protein from *Pediococcus pentosaceous* MTCC 5151. Turk J Biol 36:177–185

Agrawal R, Vanaja G, Gotcheva V, Angelov A (2010) Differences in biochemical and electrophoretic properties in native *Lactobacillus plantarum* and probiotic culture strain *Lactobacillus plantarum* cfr on adaptation to GIT conditions with functional properties. Res J Biotechnol 5(2):67–73

Agrawal R, Rati Rao E.R, Vijayendra S.V.N, Varadaraj M.C, Prasad M.S and KrishnaNand (2000) Flavour profile of idli batter prepared for defined microbial starter cultures. World Journal of Microbiology and Biotechnology. 16: 687–690

Ahmad A, Salik S, Yap WB, Kofli N, Ghazali, R (2018). Mutagenicity and Antimutagenic Activities of Lactic Acid Bacteria (LAB) Isolated from Fermented Durian (Tempoyak). Jurnal Sains Kesihatan Malaysia. 16. https://doi.org/10.17576/JSKM-2018-04.

Aihara K, Kajimoto O, Hirata H, Takahashi R, Nakamura Y (2005) Effect of powdered fermented milk with *Lactobacillus helveticus* on subjects with high-normal blood pressure or mild hypertension. J Am Coll Nutr 24:257–265

Akcicek A, Bozkurt F, Akgül C, Karasu S (2021) Encapsulation of olive pomace extract in rocket seed gum and chia seed gum nanoparticles: Characterization, antioxidant activity and oxidative stability. Foods, 10, 1735.

Al-Mohammadi AR, Ibrahim RA, Moustafa AH, Ismaiel AA, Abou Zeid A, Enan G (2021) Chemical constitution and antimicrobial activity of kefir fermented beverage. Molecules 26(9):2635. https://doi.org/10.3390/molecules26092635

Ana María RE, Sergio OSS, María CVA, Miriam BG, Marco ALV (2020) Effects of fermentation with probiotics on anti nutritional factors and proximate composition of lupin (*Lupinus mutabilis*sweet). LWT 130:109658. ISSN0023-6438. https://doi.org/10.1016/j.lwt.2020.109658

Andayani SN, Lioe HN, Wijaya CH, Ogawa M (2020) Umami fractions obtained from water soluble extracts of red oncom and black oncom— Indonesian fermented soybean and peanut products. J Food Sci 85(3):657–665. https://doi.org/10.1111/1750-3841.14942

Anderson GG, Palermo JJ, Schilling JD, Roth R, Heuser J, Hultgren SJ (2003) Intracellular bacterial biofilm-like pods in urinary tract infections. Sciences 301:105–107

Anita CJ, Andrea SG (2021) Chapter 27 - Therapeutic potential of different probiotic foods. In: Dharumadurai D, Alwarappan S (eds) Advances in probiotics. Academic Press, pp 449–477. ISBN 9780128229095. https://doi.org/10.1016/B978-0-12-822909-5.00027-7

Arvind KS, Nikhlesh S, Pushpalata (2009) Inhibition of 1,2 dimethylhydrazine induced genotoxicity in rats by the administration of probiotic curd. Mol Biol Rep 37:1373–1376. https://doi.org/10.1007/s11033-009-9519-1

Astuti M (2015) In: Tamang JP (ed) Health benefits of tempe in health benefits of fermented foods. CRC Press, New York, pp 371–394

Babalola OO (2014) Cyanide content of commercial gari from different areas of Ekiti State Nigeria. World J Nutr Health 2:58–60

Balamurugan V, Hemadri D, Gajendragad MR, Singh RK, Rahman H (2014) Diagnosis and control of peste des petits ruminants: a comprehensive review. Virus Dis 25(1):39–56. https://doi.org/10.1007/s13337-013-0188-2

Bamidele OP, Fasogbon MB, Oladiran DA, Akande EO (2015) Nutritional composition of fufu analog flour produced from Cassava root (*Manihot esculenta*) and Cocoyam (*Colocasia esculenta*) tuber. Food Sci Nutr 3:597–603. https://doi.org/10.1002/fsn3.250

Banik A, Anjum H, Habib H, Abony M, Begum A, Ahmed Z (2023) Characterization of lactic acid bacteria isolated from street pickles of Dhaka, Bangladesh. Heliyon 9(6):e17508. https://doi.org/10.1016/j.heliyon.2023.e17508

Basappa SC, Somashekar D, Agrawal R, Suma K, Bharathi K (1997) Nutritional composition of fermented ragi (chang) by Phab and defined starter cultures as compared to unfermented ragi (*Eleucine caracanis G*). Int J Food Sci Nutr 48:313–319

Bekar OY, Yusuf GM (2010) Kefir improves the efficacy and tolerability of triple therapy in eradicating *Helicobacter pylori*. J Med Food 14:344–347. https://doi.org/10.1089/jmf.2010.0099

Bourdichon F, Casaregola S, Farrokh C, Frisvad JC, Gerds ML, Hammes WP et al (2012) Food fermentations: microorganisms with technological beneficial use. Int J Food Microbiol 154:87–97

Braat H, Rottiers P, Hommes DW, Huyghebaert N, Remaut E, Remon JP, van Deventer SJ, Neirynck S, Peppelenbosch MP, Steidler L (2006) A phase I trial with transgenic bacteria expressing interleukin-10 in Crohn's disease. Clin Gastroenterol Hepatol 4(6):754–759. https://doi.org/10.1016/j.cgh.2006.03.028

Bravo D, Rodríguez E, Medina M (2009) Nisin and lacticin 481 coproduction by *Lactococcus lactis* strains isolated from raw ewes' milk. J Dairy Sci 92(10):4805–4811. https://doi.org/10.3168/jds.2009-2237

Chandan RC, Kilara A (2013) Manufacturing yogurt and fermented milks, 2nd edn. Wiley, Chichester, p 477

Chang CT, Wang PM, Hung YF, Chung YC (2012) Purification and biochemical properties of a fibrinolytic enzyme from *Bacillus subtilis*—fermented red bean. Food Chem 133:1611–1617

Chen Y, Wang Z, Chen X, Liu Y, Zhang H, Sun T (2010) Identification of angiotensin I-converting enzyme inhibitory peptides from koumiss, a traditional fermented mare's milk. J Dairy Sci 93:884–892

Chen KI, Mei HE, Nan WS, Wen HL, Cheng CC, Kuan CC (2012) Soyfoods and soybean products: from traditional use to modern applications. Appl Microbiol Biotechnol 96:9–22

Chen H, McGowan EM, Ren N, Lal S, Nassif N, Shad-Kaneez F, Qu X, Lin Y (2018) Nattokinase: a promising alternative in prevention and treatment of cardiovascular diseases. Biomark Insights 5(13):1177271918785130. https://doi.org/10.1177/1177271918785130

Chen F, Ludwiczuk A, Wei G, Chen X, Crandall-Stotler B, Bowman JL (2018) Terpenoid Secondary Metabolites in Bryophytes: Chemical Diversity, Biosynthesis and Biological Functions. Critical Reviews in Plant Sciences, 37(2–3), 210–231. https://doi.org/10.1080/07352689.2018.1482397

Chettri R, Tamang JP (2014) Functional properties of tungrymbai and bekang, naturally fermented soybean foods of North East India. Int J Fer Foods 3:87–103. https://doi.org/10.5958/2321-712X.2014.01311.8

Chiou RYY, Cheng SL (2001) Isoflavone transformation during soybean koji preparation and subsequent miso fermentation supplemented with ethanol and NaCl. J Agric Food Chem 49:3656–3660. https://doi.org/10.1021/jf001524l

Chong AQ, Lau SW, Chin NL, Talib RA, Basha RK (2023) Fermented beverage benefits: a comprehensive review and comparison of Kombucha and Kefir microbiome. Microorganisms 11(5):1344

Dajanta K, Chukeatirote E, Apichartsrangkoon A, Frazier RA (2009) Enhanced aglycone production of fermented soybean products by *Bacillus* species. Acta Biol Szeged 53:93–98

Dajanta K, Janpum P, Leksing W (2013) Antioxidant capacities, total phenolics and flavonoids in black and yellow soybeans fermented by *Bacillus subtilis*: a comparative study of Thai fermented soybeans (*Thua nao*). Int Food Res J 20:3125–3132

Davani-Davari D, Negahdaripour M, Karimzadeh I, Seifan M, Mohkam M, Masoumi SJ, Berenjian A, Ghasemi Y (2019) Prebiotics: definition, types, sources, mechanisms, and clinical applications. Foods 8(3):92. https://doi.org/10.3390/foods8030092

De Di Cagno R, Filannino P, Gobbetti M (2016) Fermented foods: fermented vegetables and other products. In: Benjamin C, Paul MF, Fidel T (eds) Encyclopedia of food and health. Academic Press, pp 668–674. ISBN 9780123849533. https://doi.org/10.1016/B978-0-12-384947-2.00284-1

De Mejia EG, Dia VP (2009) Lunasin and lunasin-like peptides inhibit inflammation through suppression of NF-κβ pathway in the macrophage. Peptides. 30:2388–2398. https://doi.org/10.1016/j.peptides.2009.08.005

De Vuyst L (1994) Bacteriocins and bacteriocin-like substances from *Lactobacillus*. In: De Vuyst L, Vandamme EJ (eds) Bacteriocins of lactic acid bacteria. Springer, Boston. https://doi.org/10.1007/978-1-4615-2668-1_11

Di Biase M, Le Marc Y, Bavaro AR, Lonigro SL, Verni M, Postollec F, Valerio F (2022) Modeling of growth and organic acid kinetics and evolution of the protein profile and amino acid content during *Lactiplantibacillus plantarum* itm21b fermentation in liquid sourdough. Food Secur 11(23):3942. https://doi.org/10.3390/foods11233942

Di Cagno R, De Angelis M, Lavermicocca P, De Vincenzi M, Giovannini C, Faccia M, Gobbetti M (2002) Proteolysis by sourdough lactic acid bacteria: effects on wheat flour protein fractions and gliadin peptides involved in human cereal intolerance. Appl Environ Microbiol 68(2):623–633. https://doi.org/10.1128/AEM.68.2.623-633.2002

Dimidi E, Cox SR, Rossi M, Whelan K (2019) Fermented foods: definitions and characteristics, impact on the gut microbiota and effects on gastrointestinal health and disease. Nutrients 11(8):1806. https://doi.org/10.3390/nu11081806

Dinesh K, Renu A (2008) Preparation of probiotic fermented milk using a native isolate of Pediococcus pentosaceous MTCC 5151. Res J Biotechnol 3(4): 28–31

Dwiatmaka YL, Endang Y, Nunung W, Subagus W (2021) Fermentation of soybean seeds using *Rhizopus oligosporus* for tempeh production and standardization based on isoflavones. Int J Appl Pharma. https://doi.org/10.22159/ijap.2022v14i6.43785

Ellington AA, Berhow MA, Singletary KW (2006) Inhibition of Akt singaling and enhanced ERK1/2 activity are involved in induction of macroautophagy by triterpenoid B-group soya saponins in colon cancer cells. Carcinogenesis 27:298–306

Emmons DB, Modler HW (2010) Invited review: a commentary on predictive cheese yield formulas. J Dairy Sci 93(12):5517–5537

FAO/WHO (2002) Report of a Joint FAO/WHO working group on drafting guidelines for the evaluation of probiotics in food. In: Guidelines for the evaluation of probiotics in food. London: Food and Agriculture Organization/World Health Organization. ftp://ftp.fao.org/es/esn/food/wgreport2.pdf

Feher J, Lengyel G, Lugasi A (2007) The cultural history of wine- theoretical background to wine therapy. Cent Eur J Med 2:379–391

Femia AP, Luceri C, Dolara P, Giannini A, Biggeri A et al (2002) Anti tumorigenic activity of the prebiotic inulin enriched with oligofructose in combination with the probiotics *Lactobacillus rhamnosus* and *Bifidobacterium lactis* on azoxymethane-induced colon carcinogenesis in rats. Carcinogenesis 23:1953–1960

Gabriele M, Arouna N, Árvay J, Longo V, Pucci L (2023) Sourdough fermentation improves the antioxidant, antihypertensive, and antiinflammatory properties of *Triticum dicoccum*. Int J Mol Sci 24(7):6283. https://doi.org/10.3390/ijms24076283

Gallaher DD, Khil J (1999) The effect of synbiotics on colon carcinogenesis in rats. J Nutr 129(7 Suppl):1483S–1487S. https://doi.org/10.1093/jn/129.7.1483S

Gayathri B, Agrawal R (2012) Antioxidant activity and fatty acid profile of fermented milk prepared by *Pediococcus pentosaceus*. J Food Sci Technol. https://doi.org/10.1007/s13197-012-0891-9

Ge J, Ping W, Song G, Du C, Ling H, Sun X, Gao Y (2009) Paracin 1.7, a bacteriocin produced by *Lactobacillus paracasei* HD1.7 isolated from Chinese cabbage sauerkraut, a traditional Chinese fermented vegetable food. Wei Sheng Wu Xue Bao 49(5):609–616

Ghosh D, Chattopadhyay P (2011) Preparation of idli batter, its properties and nutritional improvement during fermentation. J Food Sci Technol 48:610–615

Goldin BR, Gorbach SL (1980) Effect of *Lactobacillus acidophilus* dietary supplements on 1, 2-dimethylhydrazine dihydrochloride induced intestinal cancer in rats. J Natl Cancer Inst 64(2):263–265. https://doi.org/10.1093/jnci/64.2.263

Granier A, Goulet O, Hoarau C (2013) Fermentation products: immunological effects on human and animal models. Pediatr Res 74:238–244. https://doi.org/10.1038/pr.2013.76

Grosu-Tudor SS, Zamfir M (2013) Functional properties of LAB isolated from Romanian fermented vegetables. Food Biotechnol 27:235–248

Gueimonde M, Frias R, Ouwehand A (2006) Assuring the continued safety of lactic acid bacteria used as probiotics. Biologia 61(6):755–760. https://doi.org/10.2478/s11756-006-0153-2

Gupta RK, Gangoliya SS, Singh NK (2015) Reduction of phytic acid and enhancement of bioavailable micronutrients in food grains. J Food Sci Technol 52(2):676–684. https://doi.org/10.1007/s13197-013-0978-y

Hameed A, Anwar MJ, Perveen S, Amir M, Naeem I, Imran M, … Awuchi CG (2023) Functional, industrial and therapeutic applications of dairy waste materials. International Journal of Food Properties, 26(1), 1470–1496. https://doi.org/10.1080/10942912.2023.2213854

Han BZ, Rombouts FM, Nout MJ (2001) A Chinese fermented soybean food. Int J Food Microbiol 65(1–2):1–10. https://doi.org/10.1016/s0168-1605(00)00523-7

Han SC, Kang GJ, Ko YJ, Kang HK, Moon SW, Ann YS, Yoo ES (2012) Fermented fish oil suppresses T helper 1/2 cell response in a mouse model of atopic dermatitis via generation of CD4+CD25+Foxp3+T cells. BMC Immunol 9(13):44. https://doi.org/10.1186/1471-2172-13-44

Han ES, Kim HJ, Choi HK (2015) Health benefits of Kimchi. In: Tamang JP (ed) Health benefits of fermented foods. CRC Press, New York, pp 343–370

Haskard C, Binnion C, Ahokas J (2000) Factors affecting the sequestration of aflatoxin by *Lactobacillus rhamnosus* strain GG. Chem Biol Interact 128(1):39–49. https://doi.org/10.1016/s0009-2797(00)00186-1

Hertzler S, Clancy S (2003) Kefir improves lactose digestion and tolerance in adults with lactose maldigestion. J Am Diet Assoc 103:582–587. https://doi.org/10.1053/jada.2003.50111

Hirayama K, Rafter J (2000) The role of probiotic bacteria in cancer prevention. Microbes Infect 2:681–686

Holzapfel WH, Wood BJB (2014) Lactic acid bacteria. In: Biodiversity and taxonomy. Wiley-Blackwell, New York, p 632

Hong WS, Chen YP, Chen MJ (2010) The antiallergic effect of kefir *Lactobacilli*. J Food Sci 75(8):H244–H253. https://doi.org/10.1111/j.1750-3841.2010.01787.x

Ichimura T, Hu J, Aita DO, Maruyama S (2003) Angiotensin I-converting enzyme inhibitory activity and insulin secretion stimulative activity of fermented fish sauce. J Biosci Bioeng 96:496–499

ICMR-DBT, Indian Council of Medical Research Task Force, Co-ordinating Unit ICMR; Co-ordinating Unit DBT (2011) ICMR-DBT guidelines for evaluation of probiotics in food. Indian J Med Res 134(1):22–25

Ishii Y, Tanizawa H (2006) Effects of soyasaponins on lipid peroxidation through the secretion of thyroid hormones. Biol Pharm Bull 29:1759–1763

Jalili M, Nazari M, Magkos F (2023) Fermented foods in the management of obesity: mechanisms of action and future challenges. Int J Mol Sci 24(3):2665. https://doi.org/10.3390/ijms24032665

Jo SY, Choi EA, Lee JJ, Chang HC (2015) Characterization of starter kimchi fermented with *Leuconostoc kimchii* GJ2 and its cholesterol lowering effects in rats fed a high fat and high cholesterol diet. J Sci Food Agric 95(13):2750–2756. https://doi.org/10.1002/jsfa.7018

Jung JY, Lee SH, Jin HM, Hahn Y, Madsen EL, Jeon CO (2013) Metatranscriptomic analysis of lactic acid bacterial gene expression during kimchi fermentation. Int J Food Microbiol 163:171–179

Jung SJ, Park SH, Choi EK, Cha YS, Cho BH, Kim YG et al (2014) Beneficial effects of Korean traditional diets in hypertensive and Type 2 diabetic patients. J Med Food 17:161–171

Keuth S, Bisping B (1994) Vitamin $B_{12}$ production by *Citrobacter freundii* or *Klebsiella pneumoniae* during tempeh fermentation a proof of enterotoxin absence by PCR. Appl Environ Microbiol 60:1495–1499

Kim KY, Hahm YT (2002) Recent studies about physiological functions of Chungkkokjang and Functional enhancement with genetic engineering. Instit Mol Biol Genet 16:1–18

Koistinen VM, Mattila O, Katina K et al (2018) Metabolic profiling of sourdough fermented wheat and rye bread. Sci Rep 8:5684. https://doi.org/10.1038/s41598-018-24149-w

Kotb E (2014) The biotechnological potential of fibrinolytic enzymes in the dissolution of endogenous blood thrombi. Biotechnol Prog 30:656–672. https://doi.org/10.1002/btpr.1918

Koutnikova H, Genser B, Monteiro-Sepulveda M, Faurie JM, Rizkalla S, Schrezenmeir J, Clément K (2019) Impact of bacterial probiotics on obesity, diabetes and non-alcoholic fatty liver disease related variables: a systematic review and meta-analysis of randomised controlled trials. B M J Open 9(3):e017995

Krajka KV, Szaefer H, Bartoszek A, Baer-Dubowska W (2011) Modulation of rat hepatic and kidney phase II enzymes by cabbage juices: comparison with the effects of indole-3-carbinol and phenethyl isothiocyanate. Br J Nutr 105(6):816–826. https://doi.org/10.1017/S0007114510004526

Kudou S, Fleury Y, Welti D, Magnolato D, Uchida T, Kitamura K et al (1991) Malonyl isoflavone glycosides in soybean seeds (*Glycine max* Merrill). Agric Biol Chem 55:2227–2233

Kumar M, Kumar A, Nagpal R, Mohania D, Behare P, Verma V, Kumar P, Poddar D, Aggarwal PK, Henry CJ, Jain S, Yadav H (2010) Cancer preventing attributes of probiotics: an update. Int J Food Sci Nutr 61(5):473–496. https://doi.org/10.3109/09637480903455971

Kwak CS, Park S, Song KY (2012) Doenjang, a fermented soybean paste, decreased visceral fat accumulation and adipocyte size in rats fed with high fat diet more effectively than nonfermented soybeans. J Med Food 15:1–9

Kwak SH, Cho YM, Noh GM, Om AS (2014) Cancer preventive potential of kimchi Lactic acid bacteria (*Weissella cibaria, Lactobacillus plantarum*). J Cancer Prev 19(4):253–258. https://doi.org/10.15430/JCP.2014.19.4.253s

Ledesma ME, Aguíñiga SI, Weiss SB, Rivera MAR, Santiago OE (2019) Casein and peptides derived from casein as antileukaemic agents. J Oncol 8:8150967. https://doi.org/10.1155/2019/8150967

Lee CH (2004) Creative fermentation technology for the future. J Food Sci 69:33–34

Lee MY, Ahn KS, Kwon OK, Kim MJ, Kim MK, Lee IY, Oh SR, Lee HK (2007a) Anti-inflammatory and anti allergic effects of kefir in a mouse asthma model. Immunobiology 212(8):647–654. https://doi.org/10.1016/j.imbio.2007.05.004

Lee YW, Kim JD, Zheng JZ, Row KH (2007b) Comparisons of isoflavones from Korean and Chinese soybean and processed products. Biochem Eng J 36:49–53

Lee YJ, Pan Y, Kwack KB, Chung JH, Park KY (2023) Increased anticancer activity of organic kimchi with starters demonstrated in HT-29 cancer cells. Appl Sci 13(11):6654. https://doi.org/10.3390/app13116654

Li C, Peng K, Xiao S, Long Y, Yu Q (2023) The role of *Lactobacillus* in inflammatory bowel disease: from actualities to prospects. Cell Death Discov 9(1):361. https://doi.org/10.1038/s41420-023-01666-w

Liang Z, Xu Y, Zhang Y, Zhang X, Song J, Jin J, Qian H (2023) Anticancer applications of phytochemicals in gastric cancer: effects and molecular mechanism. Front Pharmacol 12(13):1078090. https://doi.org/10.3389/fphar.2022.1078090

Lidbeck A, Geltner AU, Orrhage KM, Ottova L, Brismar B et al (1991) Impact of *Lactobacillus acidophilus* supplements on the fecal microflora and soluble fecal bile acids in colon cancer patients. Microb Ecol Health Dis 4:481–488

Lim JH, Jung ES, Choi EK, Jeong DY, Seung WJO, Jin JH et al (2014) Supplementation with *Aspergillus* oryzae fermented kochujang lowers serum cholesterol in subjects with hyperlipidemia. Clin Nutr 34:383–387

Liu CF, Pan TM (2010) In vitro effects of lactic acid bacteria on cancer cell viability and antioxidant activity. J Food Drug Anal 18:77–86

Lu Y, Wang W, Shan Y, Zhiqiang E, Wang L (2009) Study on the inhibition of fermented soybean to cancer cells. J North East Agric Univ 16:25–28

Lv XC, Huang XL, Zhang W, Rao PF, Ni L (2013) Yeast diversity of traditional alcohol fermentation starters for Hong Qu glutinous rice wine brewing, revealed by culture dependent and culture independent methods. Food Control 34:183–190

Macouzet ML, Byong RN (2009) Production of conjugated linoleic acid by probiotic *Lactobacillus acidophilus* La5. J Appl Microbiol 106:1886–1891. https://doi.org/10.1111/j.1365-2672.2009.04164.x

Mann GV (1977) A factor in yogurt which lowers cholesteremia in man. Atherosclerosis 26(3):335–340. https://doi.org/10.1016/0021-9150(77)90086-7

Markowiak KP, Śliżewska K (2020) The effect of probiotics on the production of short-chain fatty acids by human intestinal microbiome. Nutrients 12(4):1107. https://doi.org/10.3390/nu12041107

Mazziotta C, Tognon M, Martini F, Torreggiani E, Rotondo JC (2023) Probiotics mechanism of action on immune cells and beneficial effects on human health. Cells 12(1):184. https://doi.org/10.3390/cells12010184

Meizi C, Hee YK, Kyung HL, Ji KJ, Ji HH, Kyu YY, Byung HR, Jung HC, Kun YP (2015) Antiobesity effects of kimchi in diet induced obese mice. J Ethnic Foods 2(3):137–144. ISSN 2352-6181. https://doi.org/10.1016/j.jef.2015.08.001

Meyer J, Butikofer U, Walther B, Wechsler D, Sieber R (2009) Hot topic: changes in angiotensin converting enzyme inhibition and concentration of the teripeptides Val-Pro-Pro and Ile-Pro-Pro during ripening of different Swiss cheese varieties. J Dairy Sci 92:826–836

Mo H, Zhu Y, Chen Z (2008) Review. Microbial fermented tea—a potential source of natural food preservatives. Trends Food Sci Technol 19:124–130

Moktan B, Saha J, Sarkar PK (2008) Antioxidant activities of soybean as affected by *Bacillus* fermentation to Kinema. Food Res Int 4:586–593

Montriwong AK, Sasitorn R, Sureelak R, Sittiruk Y, Jirawat (2012) Novel fibrinolytic enzymes from *Virgibacillus halodenitrificans* SK1-3-7 isolated from fish sauce fermentation. Process Biochem 47:2379–2387. https://doi.org/10.1016/j.procbio.2012.09.020

Moslehishad M, Ehsani MR, Salami M (2013) The comparative assessment of ACE- inhibitory and antioxidant activities of peptide fractions obtained from fermented camel and bovine milk by *Lactobacillus rhamnosus* PTCC1637. Int Dairy Res 29:82–87

Murata M, Houdai T, Yamamoto H, Matsumori M, Oishi T (2006) Membrane interaction of soya saponins in association with their antioxidation effect –analysis of biomembrane interaction. Soy Protein Res 9:82–86

Myungjin K, Chun J (2005) Bacterial community structure in kimchi, a Korean fermented vegetable food as revealed by 16S rRNA gene analysis. Int J Food Microbiol 103:91–96. https://doi.org/10.1016/j.ijfoodmicro.2004.11.030

Nagai T (2015) Health benefits of Natto. In: Tamang JP (ed) Health benefits of fermented foods. CRC Press, New York, pp 433–453

Nagai T, Tamang JP (2010) Fermented soybeans and non soybeans legume foods. In: Tamang JP, Kailasapathy K (eds) Fermented foods and beverages of the world. CRC Press, New York, pp 191–224

Nagpal RK, Ashwani K, Manoj B, Pradip J, Shalini Y, Hariom Y (2012) Probiotics, their health benefits and applications for developing healthier foods: a review. FEMS Microbiol Lett 334. https://doi.org/10.1111/j.1574-6968.2012.02593.x

Nakajima N, Nozaki N, Ishihara K, Ishikawa A, Tsuji H (2005) Analysis of isoflavone content in tempeh: a fermented soybean product and preparation of a new isoflavone-enriched tempeh. J Biosci Bioeng 100:685–687

Nakamura Y, Masuda O, Takano T (1996) Decrease of tissue angiotensin I-converting enzyme activity upon feeding sour milk in spontaneously hypertensive rats. Biosci Biotechnol Biochem 60:488–489

Nout MJR, Aidoo KE (2002) Asian fungal fermented food. In: Osiewacz HD (ed) Industrial applications. The Mycota. Springer, Berlin. Heidelberg. https://doi.org/10.1007/978-3-662-10378-4_2

Nurrahman AM, Suparmo M, Soesatyo HNE (2013) The role of black soybean tempe in increasing antioxidant enzyme activity and human lymphocyte proliferation in vivo. Int J Curr Microbiol Appl Sci 2:316–327

Ogunremi O, Sanni A, Renu A (2015a) Probiotic potentials of yeasts isolated from some cereal-based Nigerian traditional fermented food products. J Appl Microbiol 119(3):797–808

Ogunremi O, Agrawal R, Sanni A (2015b) Development of cereal-based functional food using cereal-mix substrate fermented with probiotic strain – Pichia kudriavzevii OG32. Food Science & Nutrition. 3. https://doi.org/10.1002/fsn3.239.

Okada N (1989) Role of microorganism in tempeh manufacture. Isolation of vitamin $B_{12}$ producing bacteria. Jpn Agric Res Q 22:310–316

Omizu Y, Tsukamoto C, Chettri R, Tamang JP (2011) Determination of saponin contents in raw soybean and fermented soybean foods of India. J Sci Indus Res 70:533–538

Omotade RO, Abiodun IS, Agrawal R (2015) Hypolipidaemic and antioxidant effects of functional cereal-mix produced with probiotic yeast in rats fed high cholesterol diet. J Funct Foods 17:742–748

Oskar A, Simin NM, Robert MR (2004) Yogurt and gut function. Am J Clin Nutr 80(2):245–256. ISSN 0002-9165. https://doi.org/10.1093/ajcn/80.2.245

Otles S, Çağındı Ö (2003) Kefir: a probiotic dairy composition, nutritional and therapeutic aspects. Pakistan J Nutr 2. https://doi.org/10.3923/pjn.2003.54.59

Panda BP, Javed S, Ali M (2010) Production of Angkak through co-culture of *Monascus Purpureus* and *Monascus ruber*. Braz J Microbiol 41(3):757–764. https://doi.org/10.1590/S1517-83822010000300028. Epub 2010 Sep 1

Papadimitriou CG, Vafopoulou-Mastrojiannaki A, Silva SV, Gomes AM, Malcata FX, Alichanidis E (2007) Identification of peptides in traditional and probiotic sheep milk yoghurt with angiotensin I-converting enzyme (ACE) inhibitory activity. Food Chem 105:647–656

Papagianni M, Anastasiadou S (2009) Pediocins: the bacteriocins of *Pediococci*. Sources, production, properties and applications. Microb Cell Fact 8:3. https://doi.org/10.1186/1475-2859-8-3

Park JE, Moon YJ, Cha YS (2008) Effect of functional materials producing microbial strains isolated from Kimchi on antiobesity and inflammatory cytokines in 3T3-L1 preadipocytes. FASEB J 23:111

Park JM, Shin JH, Gu JG, Yoon SJ, Song JC, Jeon WM et al (2011) Effect of antioxidant activity in kimchi during a short term and over ripening fermentation period. J Biosci Bioeng 112:356–359

Pasquale C (2011) Chapter 4-Sourdough Breads. In: Victor RP, Ronald RW, Vinood BP (eds) Flour and breads and their fortification in health and disease prevention. Academic Press, pp 37–46. ISBN 9780123808868. https://doi.org/10.1016/B978-0-12-380886-8.10004-2

Patel P, Butani K, Kumar A, Singh S, Prajapati BG (2023) Effects of fermented food consumption on non-communicable diseases. Foods 12(4):687. https://doi.org/10.3390/foods12040687

Pattanagul P, Pinthong R, Phianmongkhol A, Tharatha S (2008) Mevinolin, citrinin and pigments of adlay angkak fermented by *Monascus* sp. Int J Food Microbiol 126:20–23

Paucar Menacho LM, Amaya FJ, Berhow MA, Mandarino JMG, de Mejia E, Chang YK (2010) A high-protein soybean cultivar contains lower isoflavones and saponins but higher minerals and bioactive peptides than a low-protein cultivar. Food Chem 120:15–21

Pereira DA, Gibson GR (2002) Cholesterol assimilation by lactic acid bacteria and *Bifidobacteria* isolated from the human gut. Appl Environ Microbiol 68(9):4689–4693. https://doi.org/10.1128/AEM.68.9.4689-4693.2002

Pérez AO, Zepeda HA, LE Garcia A, Requena T, Vinderola G, García CT (2022) Role of lactic acid bacteria and yeasts in sourdough fermentation during breadmaking: Evaluation of post biotic like components and health benefits. Front Microbiol 8(13):969460. https://doi.org/10.3389/fmicb.2022.969460

Phelan M, Kerins D (2011) The potential role of milk derived peptides in cardiovascular diseases. Food Funct 2:153–167

Philip R, Epstein LB (1986) Tumour necrosis factor as immunomodulator and mediator of monocyte cytotoxicity induced by itself, gamma interferon and interleukin 1. Nature 323(6083):86–89. https://doi.org/10.1038/323086a0

Ping SP, Shih SC, Rong CT, King WQ (2012) Effect of isoflavone aglycone content and antioxidation activity in natto by various cultures of *Bacillus subtilis* during the fermentation period. J Nutr Food Sci 2:153

Pircher A, Bauer F, Paulsen P (2007) Formation of cadaverine, histamine, putrescine and tyramine by bacteria isolated from meat, fermented sausages and cheeses. Eur Food Res Technol 226:225–231. https://doi.org/10.1007/s00217-006-0530-7

Prado MR, Blandón LM, Vandenberghe LP, Rodrigues C, Castro GR, Thomaz-Soccol V, Soccol CR (2015) Milk kefir: composition, microbial cultures, biological activities, and related products. Front Microbiol 6:1177. https://doi.org/10.3389/fmicb.2015.01177

Prokopidis K, Giannos P, Kirwan R, Ispoglou T, Galli F, Witard OC, Triantafyllidis KK, Kechagias KS, Morwani MJ, Ticinesi A, Isanejad M (2023) Impact of probiotics on muscle mass, muscle strength and lean mass: a systematic review and meta-analysis of randomized controlled trials. J Cachexia Sarcopenia Muscle 14(1):30–44. https://doi.org/10.1002/jcsm.13132

Qian B, Xing M, Cui L, Deng Y, Xu Y, Huang M et al (2011) Antioxidant, antihypertensive and immunomodulatory activities of peptide fraction from fermented skim milk with *Lactobacillus delbrueckii* ssp bulgaricus LB340. J Dairy Res 78:72–79

Qiu KY, Isabella W, Blaide H, Yirui A, Aaron A (2020) Polymeric films for the encapsulation, storage and tunable release of therapeutic microbes. Adv Healthcare Mat 9. https://doi.org/10.1002/adhm.201901643

Quiros A, Hernandez-Ledesma B, Ramos M, Amigo L, Recio I (2005) Angiotensin-converting enzyme inhibitory activity of peptides derived from caprine kefir. J Dairy Sci 88:3480–3487

Rabie MA, Siliha H, Soherel S, el Ahmed AB, Xavier FM (2011) Reduced biogenic amine contents in sauerkraut via addition of selected lactic acid bacteria. Food Chem 129(4):1778–1782. ISSN 0308-8146. https://doi.org/10.1016/j.foodchem.2011.05.106

Rafter JJ (1995) The role of lactic acid bacteria in colon cancer prevention. Scand J Gastroenterol 30:497–502

Ramrez JF, Sanchez MA, Alvarez MM, Valyasebi R (2004) Industrialization of Mexican pulque. In: Steinkraus K (ed) Industrialization of indigenous fermented foods, 2nd edn. Marcel Deckker, New York, pp 547–586

Rauscher GE, Grossgut R, Bauer F, Paulsen P et al (2009) Assessment of alimentary histamine exposure of consumers in Austria and development of tolerable levels in typical foods. Food Control 20(4):423–429

Ravinder N, Ashwani K, Manoj K, Pradip VB, Shalini J, Hariom Y (2012) Probiotics, their health benefits and applications for developing healthier foods: a review. FEMS Microbiol Lett 334(1):1–15. https://doi.org/10.1111/j.1574-6968.2012.02593.x

Rayes N, Seehofer D, Theruvath T, Schiller RA, Langrehr JM et al (2005) Supply of pre and probiotics reduces bacterial infection rates after liver transplantation- a randomized, double-blind trial. Am J Transplant 5:125–130

Reddy NR, Salunkhe DK (1980) Effect of fermentation on phytate phosphorus, and mineral content in black gram, rice, and black gram and rice blends. J Food Sci 45:1708–1712

Rizzoli R (2014) Dairy products, yogurts, and bone health. Am J Clin Nutr 99(5 Suppl):1256S–1262S. https://doi.org/10.3945/ajcn.113.073056

Rogeaux M (2010) Consumer understanding and reaction to health claims: insights and methodology. In: Sara RJ, MacFie H (eds) Consumer-driven innovation in food and personal care products, Woodhead Publishing Series in Food Science, Technology and Nutrition. Wood head Publishing, pp 277–302. ISBN 9781845695675. https://doi.org/10.1533/9781845699970.3.277

Sabeena FKH, Baron CP, Nielsen NS, Jacobsen C (2010) Antioxidant activity of yoghurt peptides: part 1- in vitro assays and evaluation in ω-3 enriched milk. Food Chem 123:1081–1089

Saha TN, Chung R, McIntyre L (2024) A review of biogenic amines in fermented foods: occurrence and health effects. Heliyon 10(2):e24501. https://doi.org/10.1016/j.heliyon.2024.e24501

Salminen S, Marteau P (1997) Safety of probiotic lactic acid bacteria and other probiotics. Lactic 97 (Proceedings) Adria Normandie Caen 71–72

Sanklecha M, Verma L, Pai U, Mishra S, Maqsood S, Birla A (2022) *Lactobacillus rhamnosus* GG evaluation in acute diarrhea (LEAD): an observational study. Cureus 14(4):e24594. https://doi.org/10.7759/cureus.24594

Satish KL, Pugalenthi LS, Ahmad M, Reddy S, Barkhane Z, Elmadi J (2022) Probiotics in irritable bowel syndrome: a review of their therapeutic role. Cureus 14(4):e24240. https://doi.org/10.7759/cureus.24240

Shah NP (2007) Functional cultures and health benefits. Int Dairy J 17:1262–1277. https://doi.org/10.1016/j.idairyj.2007.01.014

Shah NP (2015) Functional properties of fermented milks. In: Tamang JP (ed) Health benefits of fermented foods. CRC Press, New York, pp 261–274

Shahbazi R, Sharifzad F, Bagheri R, Alsadi N, Yasavoli-Sharahi H, Matar C (2021) Anti-inflammatory and immunomodulatory properties of fermented plant foods. Nutrients 13(5):1516. https://doi.org/10.3390/nu13051516

Sharma R (2013) Chapter 41 - Biochemical mechanisms of fatty liver and bioactive foods: fatty liver, diagnosis, nutrition therapy. In: Ronald RW, Victor RP (eds) Bioactive food as dietary interventions for liver and gastrointestinal disease. ISBN 978-0-12-397154-8. https://doi.org/10.1016/C2011-0-07464-1

Shin DH, Jeong D (2015) Korean traditional fermented soybean products. In: Jang J (ed) Ethnic foods, vol 2, pp 2–7

Shivangi S, Devi PB, Ragul K, Shetty PH (2020) Probiotic potential of *Bacillus* strains isolated from an acidic fermented food idli. Probiotics Antimicrob Proteins 4:1502–1513. https://doi.org/10.1007/s12602-020-09650-x. PMID: 32266650

Shobharani P, Agrawal R (2007) Therapeutic importance of volatile compounds produced by *Leuconostoc paramesenteroides*. Turkish J Biol 31:35–40

Shobharani P, Agrawal R (2008) Effect on cellular membrane fatty acids in the stressed cells of *Leuconostoc mesenteroides* sub sp. Dextranicum. A native probiotic lactic acid bacterium. Food Biotechnol 22:47–63

Shobharani P, Agrawal R (2010) Enhancement of cell stability and viability of probiotic *Leuconostoc mesenteroides* MTCC 5209 on freeze drying to be used in food formulation. Int J Food Sci Nutr 60(56):70–83

Shobharani P, Agrawal R (2011) Isolation and characterization of lactic acid bacteria from cheddar cheese as a potent probiotic strain. Indian J Microbiol 51(3):251–258

Shobharani P, Ramesh B, Agrawal R (2006) Strain improvement by mutagenesis and optimum conditions for culture parameters by response surface methodology for lactose tolerance in a novel native culture isolate *Leuconostoc mesenteroides* sub sp. Res J Biotechnol 1(2):5–11

Shon MY, Lee J, Choi JH, Choi SY, Nam SH, Seo KI et al (2007) Antioxidant and free radical scavenging activity of methanol extract of chungkukjang. J Food Composit Anal 20:113–118

Singh TA, Devi KR, Ahmed G, Jeyaram K (2014) Microbial and endogenous origin of fibrinolytic activity in traditional fermented foods of Northeast India. Food Res Int 55:356–362

Sipola M, Finckenberg P, Korpela R, Vapaatalo H, Nurminen M (2002) Effect of long term intake of milk products on blood pressure in hypertensive rats. J Dairy Res 69:103–111

Spagnuolo C, Russo GL, Orhan IE, Habtemariam S, Daglia M, Sureda A, Nabavi SF, Devi KP, Loizzo MR, Tundis R, Nabavi SM (2015) Genistein and cancer: current status, challenges, and future directions. Adv Nutr 6(4):408–419. https://doi.org/10.3945/an.114.008052

Sreeramulu G, Zhu Y, Knol W (2000) Kombucha fermentation and its antimicrobial activity. J Agric Food Chem 48(6):2589–2594. https://doi.org/10.1021/jf991333m

Srividya D, Chandra Prakash SV, Shylaja D, Agrawal R (2015) Antishigella dysenteriae activity by probiotic lactic acid bacteria (*Pediococcus pentosaceus*); an *in Vitro* study. J Microbiol Biotechnol Food Sci 4(4):317–320. https://doi.org/10.15414/jmbfs.201.jmbfs

Stanley NR, Lazazzera BA (2005) Defining the genetic differences between wild and domestic strains of *Bacillus subtilis* that affect poly-γ-dl-glutamic acid production and biofilm formation. Mol Microbiol 57:1143–1158

Steidler L (2003) Genetically engineered probiotics. Best Pract Res Clin Gastroenterol 17:861–876. https://doi.org/10.1016/S1521-6918(03)00072-6

Su HK, Chen WC, Lu JH, Chao HR, Liang YF, Haruka S, Hsu WL, Wu ML, Tsai MH (2023) The effects of using Tempeh as a supplement for type 2 diabetes. Food Sci Nutr 11(6):3339–3347. https://doi.org/10.1002/fsn3.3319

Sudha N, Shobharani P, Agrawal R (2006) Studies on the stability and viability of a local probiotic isolate *Pediococcus pentosaceous* (MTCC 5151) under induced gastrointestinal tract conditions. J Food Sci Technol 43(6):677–678

Suganuma T, Fujita K, Kitahara K (2007) Some distinguishable properties between acid stable and neutral types of α-amylases from acid producing koji. J Biosci Bioeng 104:353–362

Surya C R, Rachappa S B and Renu A (2007) Role of oxygen scavengers in improving the stability and viability of probiotic lactic acid bacteria. Res J Biotechnol 2(1): 26–32

Tamang JP (2015) Naturally fermented ethnic soybean foods of India. J Ethnic Foods 2:8–17

Tamang JP, Fleet GH (2009) Yeasts diversity in fermented foods and beverages. In: Satyanarayana T, Kunze G (eds) Yeasts biotechnology: diversity and applications. Springer, New York, pp 169–198

Tamang JP, Nikkuni S (1996) Selection of starter culture for production of kinema, fermented soybean food of the Himalaya. World J Microbiol Biotechnol 12:629–635
Tamang JP, Dewan S, Tamang B, Rai A, Schillinger U, Holzapfel WH (2007) Lactic acid bacteria in Hamei and Marcha of North East India. Indian J Microbiol 47:119–125
Tamang JP, Watanabe K, Holzapfel WH (2016) Review: diversity of microorganisms in global fermented foods and beverages. Front Microbiol 7:377
Thapa N, Tamang JP (2015) Functionality and therapeutic values of fermented foods. In: Tamang JP (ed) Health benefits of fermented foods. CRC Press, New York, pp 111–168
Tolhurst G, Heffron H, Lam YS, Parker HE, Habib AM, Diakogiannaki E, Cameron J, Grosse J, Reimann F, Gribble FM (2012) Short chain fatty acids stimulate glucagon like peptide 1 secretion via the G protein coupled receptor FFAR2. Diabetes 61(2):364–371. https://doi.org/10.2337/db11-1019
Tolonen M, Taipale M, Viander B, Pihlava JM, Korhonen H, Ryhänen EL (2002) Plant derived biomolecules in fermented cabbage. J Agric Food Chem 50(23):6798–6803. https://doi.org/10.1021/jf0109017
Tsuyoshi N, Fudou R, Yamanaka S, Kozaki M, Tamang N, Thapa S et al (2005) Identification of yeast strains isolated from marcha in Sikkim, a microbial starter for amylolytic fermentation. Int J Food Microbiol 99:135–146
Turan İ, Dedeli Ö, Bor S, İlter T (2014) Effects of a kefir supplement on symptoms, colonic transit, and bowel satisfaction score in patients with chronic constipation: a pilot study. Turk J Gastroenterol 25(6):650–656. https://doi.org/10.5152/tjg.2014.6990
Vanaja G, Villetchka G, Angel A, Agrawal R (2010) Volatile profile of compounds with therapeutic importance produced by probiotic culture strain of *Lb. plantarum* LB cfr. J Food Sci Technol 69481:110–113
Vanajakshi V, Varadaraj MC, Venkateswaran G, Agrawal R (2015) Optimization of a probiotic beverage based on Moringa leaves and beetroot. LWT Food Sci Technol 63(2):1268–1273
Vanitha R, Agrawal R (2012) Purification of protein from probiotic *Leuconostoc mesenteroides* active against *V. cholerae*. Res J Biotechnol 7(1):38–42
Verna EC, Lucak S (2010) Use of probiotics in gastrointestinal disorders: what to recommend? Ther Adv Gastroenterol 3(5):307–319. https://doi.org/10.1177/1756283X10373814
VidyaLaxme B, Rovetto A, Grau R, Agrawal R (2012) Synergistic effects of probiotic *Leuconostoc mesenteroides* and *Bacillus subtilis* in ragi (*Eleucine corocana*) malt for antagonistic activity against *Vibrio cholerae* and other beneficial properties. J Food Sci Technol. https://doi.org/10.1007/s13197-012-0834-5
Vuyst LD, Falony G, Leroy F (2008) Probiotics in fermented sausages. Meat Sci 80(1):75–78. https://doi.org/10.1016/j.meatsci.2008.05.038
Wang LJ, Li D, Zou L, Chen XD, Cheng YQ, Yamaki K et al (2007a) Antioxidative activity of douchi (a Chinese traditional salt-fermented soybean food) extracts during its processing. Int J Food Propert 10:1–12
Wang LJ, Yin LJ, Li D, Zou L, Saito M, Tatsumi E et al (2007b) Influences of processing and NaCl supplementation on isoflavone contents and composition during douchi manufacturing. Food Chem 101:1247–1253
Willcox BJ, Willcox DC, Suzuki M (2004) The Okinawa diet plan. Three Rivers Press, New York
Won TJ, Kim B, Song DS, Lim YT, Oh ES, Lee DI, Park ES, Min H, Park SY, Hwang KW (2011) Modulation of Th1/Th2 balance by *Lactobacillus* strains isolated from Kimchi via stimulation of macrophage cell line J774A.1 in vitro. J Food Sci 76(2):H55–H61. https://doi.org/10.1111/j.1750-3841.2010.02031.x
Yadav H, Jain S, Sinha PR (2007) Antidiabetic effect of probiotic dahi containing *Lactobacillus acidophilus* and *Lactobacillus casei* in high fructose fed rats. Nutrition 23(1):62–68. https://doi.org/10.1016/j.nut.2006.09.002
Yanagisawa Y, Sumi H (2005) Natto *Bacillus* contains a large amount of water soluble vitamin K (menaquinone-7). J Food Biochem 29:267–277
Yanping W, Nv X, Aodeng X, Zaheer A, Bin Z, Xiaojia B (2009) Effects of *Lactobacillus plantarum* MA2 isolated from Tibet kefir on lipid metabolism and intestinal microflora of rats

fed on high-cholesterol diet. Appl Microbiol Biotechnol 84:341–347. https://doi.org/10.1007/s00253-009-2012-x

Yin LJ, Li D, Zou L, Saito M, Tatsumi E, Li LT (2007) Influences of processing and NaCl supplementation on isoflavone contents and composition during douchi manufacturing. Food Chem 101:1247–1253

Yoon S, Do J, Lee S, Chag H (2000) Production of poly-δ-glutamic acid by fed batch culture of *Bacillus lichenifomis*. Biotechnol Lett 22:585–588. https://doi.org/10.1023/A:1005625026623

Young CC, Hong MJ, Yin C, Dal SK, Jin MJ, Jong Il P, Eun SJ, Eun KC, Soo WC (2014) Fermented milk of *Lactobacillus helveticus* IDCC3801 improves cognitive functioning during cognitive fatigue tests in healthy older adults. J Funct Food 10:465–474. ISSN 1756-4646. https://doi.org/10.1016/j.jff.2014.07.007

Yun YR, Lee JJ, Lee HJ, Choi YJ, Lee JH, Park SJ, Park SH, Seo HY, Min SG (2021) Comparison of quality characteristics of commercial kimchi manufactured in Korea, China, and the United States. Food Secur 10(10):2488. https://doi.org/10.3390/foods10102488

Zabat MA, Sano WH, Wurster JI, Cabral DJ, Belenky P (2018) Microbial community analysis of sauerkraut fermentation reveals a stable and rapidly established community. Food Secur 7(5):77. https://doi.org/10.3390/foods7050077

Zeng W, Li W, Shu L, Yi J, Chen G, Liang Z (2013) Non-sterilized fermentative coproduction of poly (γ-glutamic acid) and fibrinolytic enzyme by a thermophilic *Bacillus subtilis* GXA-28. Bioresour Technol 142:697–700

Zeng Y, Fadaak A, Alomeir N, Wu Y, Wu TT, Qing S, Xiao J (2023) Effect of Probiotic Lactobacillus plantarum on Streptococcus mutans and Candida albicans Clinical Isolates from Children with Early Childhood Caries. *Int J Mol Sci.* Feb 3;24(3):2991. https://doi.org/10.3390/ijms24032991

# Bacteriocins

# 10

**Abstract**

When bacteria get infected, they adopt a defence system by producing antimicrobial compounds known as bacteriocins. The bacteriocins vary in their molecular weight. This chapter includes details on the classification, the most recent scientific works, and the understanding on bacteriocins. The role of bacteriocins in improving human health and animals has been discussed. All over the globe studies are being done to achieve food preservation methods by natural means. It discusses the methods of food preservation using bacteriocins in a natural way as they come under the GRAS category. Various types of bacteriocins are produced by different bacteria. It is the probiotic cultures that produce the bacteriocins and have antimicrobial activity. They are helpful in killing many pathogens and can, therefore, prevent many harmful diseases. All these aspects have been well described in this chapter. The bacteriocins are the natural alternative to kill multidrug-resistant pathogens.

**Keywords**

Antimicrobial · Peptides · Gram positive · Gram negative · Lactic acid bacteria · Bacteriocin · Killing · Peptide · Antibiotics · Bacteriophages

## 10.1 Introduction

These are the peptides that can kill pathogenic microorganisms and are formed by bacterial ribosomes. However, they do not harm themselves as they have specific immunity proteins. Cotter et al. (2005) have found that these bacteriocins develop an innate immunity in foods. These help in the preservation of food, in preventing ulcers, can be used as spermicidal agents, are able to prevent cancer, are useful to animals, prevent many dermatitis problems, improve the care of the mouth, and are



R. Agrawal, *Textbook of Industrial Microbiology*,
https://doi.org/10.1007/978-981-97-9582-6_10

also helpful in the growth of the plants. They are produced by many of the Gram (+) and also Gram-negative (−) bacteria. The quantity of bacteriocin required to kill any pathogenic bacteria population is much less as compared to the antibiotics. Bacteriocins, antibiotics and bacteriophages in combination, form a spectrum of antimicrobial compounds. Bacteriocins and bacteriophages work like antibiotics. Antibiotics have been used commonly in the west as the production is very easy and they have a wide activity spectrum. However, recently, the emergence of antimicrobial resistance (AMR) has been a cause of great concern and a threat to human health. Bacteriophages are viruses that infect bacteria. Phages have many advantages as they are specific to each species. When a phage gets integrated into the bacterial genome, the process is called as lysogeny. A minute level of phage is enough to kill the bacteria. Under these situations, the bacteriocins are very helpful. Bacteriocins are a natural alternative to the widely used antibiotics which are causing a great concern to the health of human beings. They are used as food preservatives and as next generation antibiotics which can help in targeting multiple drug-resistant pathogens. According to Perez et al. (2014), many novel bacteriocins are produced from lactic acid bacteria (LAB). They are useful due to their tolerance to high temperatures and wide range of pH. As they do not have any color, smell, and taste, the usefulness is very high. They can be added into many food products. Even at low concentrations, the action is rapid. They usually act by forming pores in the membrane of bacteria which needs to be killed. However, the proteolytic enzymes can decompose them. Then the bacteriocin fragments are not able to survive for long in the human body. This becomes a limiting factor and reduces the time to interact with the target. This initiates the cause of antibiotic resistance. Bacteriocins are primary metabolites and have simple biosynthetic mechanism as compared to antibiotics. LAB bacteriocins have many advantages and can be supplemented in food as they are food-grade. According to the U.S. Food and Drug Administration (FDA) these are classified as Generally Regarded as Safe (GRAS). They can be used in place of antibiotics and have the specificity against pathogens. Cotter et al. (2005) classified the bacteriocins into two categories: lantibiotics (class I) and nonlanthionine-containing bacteriocins (class II), as opposed to the four classes given earlier (Klaenhammer classification scheme 1993). The class III bacteriocins known as bacteriolysins have been reclassified, as they are lytic enzymes and are not peptides. According to Heng et al. (2006), the circular bacteriocins must be grouped differently. Class I bacteriocins or lantibiotics (lanthionine-containing antibiotics) are peptides with small molecular weight (<5 kDa). They have residues of translation reaction which include lanthionine and/or 3-methyl lanthionine. The bacteriocin nisin A has been studied extensively which comes under lantibiotics. The use of nisin in foods has been approved by global agencies. It is the approval of The Joint Food and Agriculture Organization/World Health Organization Expert Committee on Food Additives, as well as the US Food and Drug Agency (FDA) which is required to use any additive in foods.

This bacteriocin can also be used in pasteurized and processed foods (FDA 1988). A lot of work is underway to produce other bacteriocins that can kill pathogens in food and can be approved by the authorities. Scientists have found that the production of lantibiotic bacteriocins involves a conjugative transposon or plasmid in the beginning. The synthesis is in the form of a prepeptide which is nonactive in the biological form. For the formation, a N-terminal leader peptide gets attached to the C-terminal propeptide. The lantibiotics are biosynthesised after the prepeptides get translated. These are made up of a leader peptide and a propeptide moiety which can be modified. After modification, it moves across the cytoplasmic membrane. The leader peptide gets broken down by specific proteolytic enzymes. Many genes are associated with immunity proteins which regulate the production of lantibiotics. These form a cluster around the bacteriocin structural gene. The nonlantibiotics are Class II bacteriocins, which are small (<10 kDa), heat stable, and nonlanthionine. The peptides cannot be modified after the translation has been made. These have been subdivided into four subclasses. Class II bacteriocins have a double-glycine proteolytic processing site. These are inactive prepeptides in which the posttranslational modification does not take place to a large extent. After translation, the prepeptide uses the specific enzymes that cleave the leader peptide concomitant by using an ABC-transporter. An accessory protein is required for this reaction. Although the bacteriocins must be utilized in the pure form, as the purification system is highly expensive, time consuming, and a difficult procedure, it is usually avoided. Recently, many new and innovative rapid methods have been developed to screen the novel bacteriocins. The analysis of bacteriocin-producing LAB strains is done by molecular mass technique with electrospray ionization liquid chromatography/mass spectrometry (ESI-LC/MS) along with the principal component analysis (PCA) of the antimicrobial activity spectrum. This is an extremely precise method which gives accurate results. For the evaluation of molecular mass of the bacteriocin, the supernatant of overnight culture is injected into the ESI-LC/MS system for analysis. If the quantity is low, then the mass detector is set manually at the mass range of *m/z* 500–3000, which is the range of molecular mass of the bacteriocins. The spot-on-lawn method with a control set of indicator strains is a good method to analyze the samples statistically. PCA (plate count agar) can be used for analysis.

## 10.2 Bacteriocins Produced from Gram-Negative Bacteria

- **Colicins** These proteins kill pathogenic microorganisms. The protein consists of three specific parts. First is the terminal with amino translocation (T). This helps to get transferred across the outer membrane, the second is the central part, which binds to the receptor (R) and is in the outer membrane of the bacteria, and the third part has a carboxy terminal. This is a cytotoxic (C) domain, which is responsible for killing the bacteria. Specific immunity proteins can inactivate the colicins. If in case the translocator protein system is absent in bacteria, then they are known as tolerant strains. As the immunity proteins are produced by them, the classification is made under the immune strains. The protein, which is responsible for lysis, is known as bacteriocin-release protein (BRP). This helps in the release of colicins from the bacteria. The translocation of colicins can take place in two ways across the outer membrane. These are termed as groups A and B. In group A, the bacteriocin is present on a small plasmid, which possesses a lytic gene and gets released from the bacteria. The bacteriocin in group B is present on a large plasmid and they do not have any lytic gene. Once these colicin bacteriocins enter the pathogen cell, the action of killing takes place in three ways:

- Pore-forming type colicins The bacteriocin forms pores in the inner membrane of the pathogen cell. With this, the cytoplasm along with the other compounds moves out from the cell. Due to this leakage, the gradient of electrochemicals gets destroyed. There is a loss of the ions which results in the cell death. These include colicin A, B, E1, Ia, Ib, K, and N.
- Nuclease-type colicins These colicins include from E2 to E9. These specifically contain nucleases like DNase, 16S rRNase, and tRNase. They digest the DNA and RNA of the pathogens in a nonspecific manner.
- Peptidoglycanase-type colicins These colicins are responsible for digesting the precursor of peptidoglycan. Due to this, the cells are not able to synthesize peptidoglycan; therefore, the bacteria dies.
- Microcins These are hydrophobic antimicrobial peptides (<10 kDa) that are synthesized in the ribosome. There is an N-terminal leader peptide along with the core peptide. Usually, they are produced by the bacteria belonging to Enterobacteriaceae family. They are beneficial as they can act with high activity under a vast range of temperature, pH, and proteases.

## 10.3 Bacteriocins from Gram-Positive Bacteria

These are antimicrobial peptides which contain less than 60 amino acids. These bacteriocins vary in length, structures, physical and chemical properties, and in the inhibition of the pathogen microorganisms. The stability is obtained due to the thioether linkages between the two β-carbons (Willey and Van Der Donk 2007).

These are lanthipeptides, sactipeptides, circular peptides, and glycocins produced from the Gram-positive bacteria (Cotter et al. 2013).

These are divided into class I (modified peptides that have less than 28 amino acids known as lantibiotics). These peptides can be straight/linear as in the case of nisin. They can be globular peptides that form the class II (30–60 amino acids) bacteriocins like mersacidin. These come under the category of unmodified peptides (called as nonlanthionine). These are heat tolerant, unmodified non-lanthionine, and positively charged. Arnison et al. (2013) have described the further division of Class I bacteriocins. Scientists (Jack et al. 1995; McAuliffe et al. 2001; Willey and Van Der Donk 2007) have characterized the presence of unusual aminoacids, namely, lanthionine and β-methyllanthionine, which results after the dehydration of serine and threonine. It produces didehydroalanine and didehydrobutyrine residues, respectively.

There are class IIa, which has pediocin-like bacteriocins; class IIb has bacteriocins, which have two components; class IIc has bacteriocins, which are circular in nature; and class IId include the bacteriocins, which are not modified and do not resemble the pediocin-like bacteriocins.

1. **Novel class IIa**: anti-*Listeria* bacteriocins

   Activity of some bacteriocins has been shown against the food pathogen *L. monocytogenes*, the organism which usually survives at low temperatures. These have been classified under class IIa bacteriocins. A probiotic product, the Thai fermented fish Plara, has been found to produce bacteriocin which can kill *Listeria* sp. The bacteriocin, namely, Enterocin NKR-5-3C (Ent53C), produced by *Enterococcus faecium* NKR-5-3, has been isolated which has anti-listeria activity. These have been classified under Class IIa bacteriocins. These possess a conserved sequence (YGNGV). The N terminal region is called as the pediocin box. The bacteriocins under this category have disulfide bridges which decide the level of antimicrobial activity. Most of the resistant pathogens can be killed by these bacteriocins (Saïd et al. 2000). They are very effective when used in the purified form. However, the inoculation of the specific LAB culture that produces this particular bacteriocin is effective. In a traditional Thai fermented pork sausage (Nham), the infection of *L. monocytogenes* could be controlled with the bacteriocin, namely, pediocin PA-1/AcH which is produced by *Pediococcus pentosaceus* BCC 3772 (Yutthana et al. 2012). Recently, it has been shown (Thapar et al. 2023) that the incorporation of bacteriocins into the food packaging film or on the surface of the food product gives positive results by inhibiting the cells of pathogenic bacteria. Biofilm-associated infections on medical devices have been a challenge for medical researchers which can be inhibited by these bacteriocins. Nisin A and lacticin Q have been shown to possess strong bactericidal activities both in their planktonic and biofilm cell forms.
2. **Novel class IIb** The bacteriocins which contain two peptides are classified under this category. Bacteriocins of Class IIb need two different types of peptides that are not modified. They form an active poration complex, namely, lactacin F and ABP-118. To kill the pathogenic organism, each peptide is needed in equal amounts. The best example of a two-peptide bacteriocin is Enterocin X. This is produced from *E. faecium* KU-B5. The activity depends on the component peptides which are enterocins X$\alpha$ and X$\beta$.
3. **Novel class IIc** These bacteriocins are circular in nature. The unique feature of bacteriocin in this category is that they are cyclized from head to tail. As they are circular, the structural stability increases. These include carnocyclin A and enterocin AS-48. This also increases the resistance to high temperatures which makes the bacteriocin more stable. Due to these properties, they are widely used in food industries for food preservation and in killing pathogenic organisms.
4. **Novel leaderless bacteriocins (class IId)** The bacteriocins, which are linear and single, are classified under Class IId. Bacteriocins like epidermicin NI01 and lactococcin A come under this category. These bacteriocins are produced from the leader peptide without breaking them. This makes them very useful at commercial scale. Lacticin Q comes under this category which is very helpful in permeabilizing the pathogen membrane. It does not need any specific receptors. This bacteriocin is known to form a huge toroidal pore (HTP). As it is very big (4.6–6.6 nm) in size, the leakage of intracellular components consisting of ions, ATP, and proteins is very fast from the pathogen cell. The cell death becomes

rapid, and it is easier to get rid of the pathogenic bacteria. However, when the lacticin Q molecules move from the outer membrane to the inner cell membrane, the pores which are formed in the membrane get closed. This happens due to the closure of the HTP pores. Killing by this method using lacticin Q via the formation of HTP is very selective. The killing rate also depends on the physiology of the outer membrane of the pathogen cells. This explains the nontoxicity of lacticin Q against Gram-negative bacteria. The lacticin Q molecules are cationic in nature. They get bound to the phospholipid bilayer membrane which is negatively charged. As the HTP's get bound to the membrane lipids, there is a leakage, which releases the intracellular components, along with ions like ATP and small proteins. However, as the pore closes, the lacticin Q molecules move into the membrane. Many studies have shown that the movement of lacticin Q to the extracellular space is associated with the genes.
5. **Novel class II e** The peptides that get modified after translation and are not produced by the ribosomes come under the category of Class IIe bacteriocins. Siderophores are a good example, in which the modification takes place at the serine-rich carboxy-terminal region. The common example of this type of bacteriocin is microcin E492. *Klebsiella pneumonia* is a Gram-negative bacteria, which produces microcin E492.
6. **Class III bacteriocins** These consist of large proteins, which lack stability to heat. They have been categorized under Class III bacteriocins. These are further divided into two distinct groups. The sensitive strains are killed when the cell wall gets lysed. The best example is Enterolisin A. These fall under group A bacteriocins. There is another category of nonlytic proteins which form group B bacteriocins. These include Caseicin 80 and Helveticin J as bacteriocins.

As most bacteriocins are synthesized by the ribosomes, the biosynthetic mechanisms become simple. As it is associated with genes the activity can be increased easily. Even the latest biotechnological methods are quite helpful in killing pathogens. The infections of the oral cavity could be effectively controlled by using the bacteriocin named as Oralpeace™. The bacteriocin is also effective in controlling bacterial gingivitis (caused by *Porphyromonas gingivalis*). For veterinary purposes, the use of nisin has been found to control bovine mastitis. Many of the food and pharmaceutical industries are using these bacteriocins as these are natural and are classified under Generally Recognized as Safe (GRAS) category. Overuse of antibiotics for a long time has made the infectious organisms adapt to these antibiotics and has reduced their effectiveness causing multidrug resistance (MDR). In order to have safe foods and better quality of foods, work is being undertaken vigorously to find alternative ways in place of chemical additives and antibiotics. According to Hatakka et al. (2001), the peptides with antimicrobial activities are either bacteriostatic or bactericidal. The cells, which produce bacteriocin, protect themselves by producing immunity proteins. Many a times, they produce efflux pumps. However, sometimes they may need both the systems to protect themselves to be in the live state (de Freire et al. 2015; Bountra et al. 2017).

These bacteriocins are of various types (Cotter et al. 2005, 2013; Yang et al. 2014). There is a database named BACTIBASE (http://bactibase.hammamilab.org), which provides structural and functional information of >200 bacteriocins (Hammami et al. 2010). Important bacteriocins have been worked out in detail. Linear azole(ine)-containing peptides (LAP) and lasso peptides have been studied both for Gram-positive and Gram-negative bacteria. The nucleotide peptides and siderophore peptides are produced from Gram-negative bacteria (Duquesne et al. 2007; Gabrielsen et al. 2014; Norris and Patchett 2016). Claesen and Bibb (2010) have worked on newer enzymes, namely, linaridins which have biological properties. The scientists have found the role of bacteriocins in peptide engineering. The thiopeptides have been worked out by Bagley et al. (2005). These enzymes are essentially produced by Actinobacteria. These are anti-bacterial enzymes that are mainly produced by *Micrococin* sp. bacteria. There are cyanobactin bacteriocins, which are produced from diverse cyanobacteria (Amran et al. 2023; Martins and Vasconcelos 2015; Martins et al. 2018) and have been found very useful.

## 10.4 Application of Bacteriocins in Food Industries

Due to the properties of bacteriocins being colorless, odorless, and tasteless (Perez et al. 2014), they are utilized in many food products. They do not change the organoleptic properties of the food. They are stable in a wide range of pH, temperature, and salt concentrations (Wilaipun et al. 2004; Fatima and Mebrouk 2013) and can therefore be supplemented into wide range of foods. Using bacteriocin as a food preservative has many advantages. They can enhance the storage life of the food, can survive the temperature variation, and prevent the product from foodborne pathogens in the food chain. When the food gets spoiled in big food companies, it causes huge losses. These losses can be protected if proper preservation techniques are in place (Thomas et al. 2000; Gálvez et al. 2007). The use of bacteriocins into the food products as antimicrobial compounds can be done in many ways. There are many commercial products available. The product MicroGARD® fermentates (DuPont) and Dura Fresh (Kerry) can kill the cells of yeasts, molds, and Gram-positive and/or Gram-negative bacteria. The use of encapsulation technology is very useful as it can release the bacteriocin in a controlled manner. This also enhances the efficiency of bacteriocins, as the release of bacteriocin can be monitored. The use of bacteriocin Plantaricin BM1 on the surface of ham has been shown to prevent the growth of *Listeria monocytogenes* to grow (Zhou et al. 2015). It has been observed that after encapsulation the activity of nisin could increase by many folds (Benech et al. 2002; Hassan et al. 2019). Use of bacteriocins in packaging has shown great promise as it increases the storage life of the product by preventing the growth of pathogens. However, the demand by the food industries will definitely become much more in case the bioavailability can be increased and the production costs can be reduced. This will increase the market accessibility. A combination of bacteriocins and antibiotics helps in increasing their antimicrobial activity and also reduces antibiotic resistance (Cavera et al. 2015).

Studies have been done toward hospital-acquired infections using bacteriocins. It has been found that they have inhibitory effects against many pathogens especially, the methicillin-resistant *Staphylococcus aureus* (MRSA) (Benatez-Chao et al. 2021), vancomycin-resistant *Enterococcus* (Oman and van der Donk 2009), *Clostridium difficile* (Le Lay et al. 2015), and for Gram-negative pathogenic bacteria which includes *Moraxella catarrhalis, Neisseria* spp., and *Haemophilus influenza* (Soltani et al. 2021). Scientists have found that bacteriocin, microcin J25, can kill the resistant strains of *Salmonella* and *E. coli* (Yu et al. 2019). *Pseudomonas aeruginosa, Klebsiella pneumoniae, Acinetobacter* spp., and *Enterobacteriaceae* which are resistant to many drugs have been found to be killed by bacteriocins (Falagas et al. 2008; Kumarasamy et al. 2010). Treatment of infectious diseases of the oral cavity, the lungs, the digestive tract, and urinary tract has been done successfully by the bacteriocins. Respiratory infections have been well addressed using mutacin 1140 and salivaricin D which have shown high inhibitory activities against the pathogens. It has been found that infections caused by *Streptococcus pneumoniae, S. aureus,* and *P. aeruginosa* have also been prevented when food supplemented with specific bacteriocins was consumed.

It has been worked out by tscientists that typhoid in pigs could be prevented by using a consortium of probiotic strains which helps in the killing of *Salmonella enterica* serovar typhimurium organism cells. *Lactobacillus salivarus* str. DPC 6005 has shown the ability to control *Salmonella* in the ileal portion and the mucosa of the digestive tract. *Escherichia coli* NISSLE 1917 has been known to enhance the immunity in the intestine (Westendorf et al. 2005). For the production of pharma-grade bacteriocins, optimization of parameters is necessary in order to obtain an economical product. Recent updates on the clinical studies of bacteriocins is available on the "Global clinical trials data" (https://globalclinicaltrialdata.com). Anticancer activities of bacteriocins, namely, Microcin E492, colicins (A, D, E1, E2, E3), and nisin are very beneficial as they induce apoptosis or depolarize the cell membrane which changes the permeability (Kaur and Kaur 2015). Cancer causing tumors in mice could be brought down with the use of Nisin (Joo et al. 2012). As per the WHO website, about 8.2 million people died of cancer in 2012 around the world which were mainly due to lung cancer. Bacteriocins have shown promise in the prevention of cancer. They are natural (GRAS category) and are allowed as additives in foods. Colicin A and E1 can inhibit the growth of one human standard fibroblast line MRC5 and 11 human tumor cell lines; however, pore-forming colicin U and RNAase activity colicin E3 did not have the ability to inhibit. Colicin D, E2, E3, and pore-forming colicin A have the ability to inhibit the murine leukemia cells P388, whereas pore-forming colicin E1 and colicin E3 can suppress v-myb-transformed chicken monoblasts cells. The bacteriocin nisin has shown a natural way to prevent cancer. The genes, which are responsible for immune functions in the bacteriocins, are present on the same plasmid. Sometimes, they can be in the adjacent regions on the chromosome. Therefore, by the method of conjugation, they can easily enter into the other bacterial cells. Bacteriocins such as Nisin, subtilosin, fermenticin, and lacticin 3147 have been proven to have birth control activity. Work has been done to show the spermicidal activity of these bacteriocins. Therefore, these are used as natural contraceptives (Dicks et al. 2018). The work of Reddy et al. (2004) has shown that when the bacteriocin nisin was applied inside the vagina of rabbits, fertilization did not take place. Also, the rabbits did not suffer from any side effects.

**Anti-viral activity** *Herpes simplex* virus (HSV-1 and HSV-2) causes many types of infections. The bacteriocins, namely, Enterocins CRL35 and ST4V, have activity against these viruses. It has been observed that these bacteriocins were able to inhibit the multiplication of these viruses. Scientists have worked on the bacteriocin produced by *Bacillus subtilis*, namely, Subtilosin A and on *E. faecium* which produces a bacteriocin, which resembles pediocin. It is named as ST5Ha. Both these bacteriocins have been found to kill virus HSV. However, both these bacteriocins have a selectivity index (CC50/EC50) of 173 (Todorov et al. 2010; Quintana et al. 2014). The microorganism *Actinomadura namibiensis* produces bacteriocin, Labyrinthopeptin A1 (LabyA1), which has carbacyclic post-translationally modified labionin residue and can act against HIV and HSV infection and transmission (Férir et al. 2013). The bacteriocin LabyA1 does not allow the viruses to move from the T cells which are infected with HIV to the uninfected CD4 + T cells. Thus, preventing the infection from one cell to the other. The infection was also found to be inhibited from HIV DC-SIGN+ cells to the CD4 + T cells which did not have the infection (Férir et al. 2013). Bacteriocins, which possess activity against viruses, are hydrophobic in nature. They have the ability to bind to the cell membrane as it is made up of lipids.

## 10.5 Veterinary Use

When allopathic antibiotics are used by animals repeatedly, they cause resistance in animals. This not only harms the animals but also humans eating these animals. The use of bacteriocins is a safe alternative to antibiotics. In a study, the growth of broiler chickens could be increased by supplementing Pediocin PA-1, and the infection of *Clostridium perfringens* was also inhibited (Grilli et al. 2009). Similar studies have been done using gassericin A and Divercin AS7 which have been found to be very effective. Mastitis could be inhibited successfully by the use of nisin and lacticin 3147. This bacteriocin is being produced at large scales and utilized commercially. The market product, namely, Mast Out® is an intramammary product which is marketed by Immucell Corporation. Cow farms are using Teatseal® (Cross Vetpharm Group Ltd) for the beneficial effects. Immucell Corporation is approved by the FDA (Food and Drug Administration), which produces "Nisin incorporated wipes" with the name "Wipe Out." Gram-negative bacteriocins like Microcin J25 can control *Salmonella* in poultry. The sensitivity of class II bacteriocins to the protease enzymes present in the intestine is well known. Due to this, the activity of these bacteriocins against the pathogen bacteria gets reduced. In the stomach, pediocin is found to be stable. However, it is not able to sustain the harsh small intestine conditions and gets decomposed. In comparison to class II bacteriocins, the bacteriocins that come under class I are more resistant to proteases. A study has shown that intestine proteases can degrade nisin A (Heinemann and Williams 1966; Jarvis and Mahoney 1969). With recent technologies of peptide engineering and peptide encapsulation, it has been possible to keep the bacteriocins of class II active in the small intestine. The toxicity studies determine the toxic character of the bioactive compound. The lethal dose or LD50 values of a product help in the determination of the adverse effects. Fermented foods are consumed around the globe, which are rich in lactic acid bacteria. Bacteriocins are produced by these LAB bacteria,which help in combating the infections and killing the pathogenic bacteria (Swain et al. 2014). The bacteriocins are formed naturally and are classified under GRAS category. These are known not to impart any toxic effects. Many of the bacteriocins, such as nisin (Maher and McClean 2006), colicin E1, E3, E7, K (Murinda et al. 2003), enterocin DD14 (Caly et al. 2017) carnobacteriocins Cbn BM1 and Cbn B2 (Jasniewski et al. 2009), plantaricin DM5 (Das and Goyal 2014), enterocin AS-48 (Abengózar et al. 2017) also semi-purified bacteriocins produced by *Lactococcus lactis subsp. lactis* and *E. durans* (Cerqueira et al. 2018), have low levels of cytotoxicity at MIC concentrations against various eukaryotic cell lines. However, nisin, pediocin PA-1, and colicin E1, E3, E7, K may be toxic when higher concentrations are used (Murinda et al. 2003). Even wound healing is possible using specific bacteriocins at lower levels. At higher concentrations, toxicity has been observed in the eukaryotic cells (Chalekson et al. 2003). A two-peptide lytic toxin has been studied in the strains of *E. faecalis,* namely, cytolysin. This can kill the cells and has toxicity against many cells (Cox et al. 2005). It is known to be toxic to human retinal cells, erythrocytes, leucocytes, and also intestinal epithelial cells. It is therefore important to check the bacteriocin before using it. This will prevent any adverse effects on the body of humans or animals (Lohans and Vederas 2011). According to the studies, the bacteriocin, pediocin AcH (Bhunia et al. 1990), and TSU4 (Sahoo et al. 2017) are non-immunogenic. The addition of these peptides as bacteriocins does not initiate any immune response. Work has been done to check the role of bacteriocin on the microbiota in the different regions of the intestine (Kheadr et al. 2010; Le Blay et al. 2012; Le Lay et al. 2015). Padilla et al. (2004) isolated the strains of *E. coli* and *Bacterioides fragilis* from the healthy gut. However, these are extremely sensitive to the peptides produced by *Shigella flexneri*. The bacteriocin effect can be studied using metagenomic/metabolomic studies. The immune system of mice can be modified using nisin which has shown an increase of CD4, CD8, and T-lymphocytes cells.

## 10.6 Mechanisms of Action in Bacteriocins

The bacteriocins have been classified on the basis of their mechanism of action. Bacteriocins classified under class II are cationic in nature. Therefore, they can kill the pathogen by interacting with the cytoplasmic membrane of the pathogen (Héchard and Sahl 2002). It has been found that the first mechanistic step, which initiates the killing activity, is the electrostatic interaction. These are carried out between the cationic bacteriocins and the negatively charged components (phospholipids and teichoic acids) present in the bacterial membrane of Gram-positive bacteria. In case of Gram-negative bacteria, these are lipopolysaccharides. This allows the bacteriocin to get inside the cytoplasmic membrane and form pores. With this, the intracellular metabolites like potassium and magnesium ions, amino acids, and ATP move out of the cell. As the important ions required for the cell growth get dissipated and the cell contents get leaked out of the cell, it results in killing the pathogen cell (Todorov 2009). There are many autolytic enzymes present. These are linked to (teichoic, lipoteichoic, and teichuronic) acids present in the cell membrane. These get active and help in the cell lysis (Karpiński et al. 2016). In many cases, it has been observed that these bacteriocins interfere with the nucleic acid synthesis. This has been found in case of Ruminococcin C. Peptidoglycan is formed through an intermediate, which is dependent on a metallopeptidase. This uses zinc as a metal. In case of Gram-positive bacteriocins, many receptors have been identified. The studies have identified undecaprenyl pyrophosphate phosphatase and mechanosensitive channel MscL for nisin and other lanthipeptides. Gram-positive (pediocin-like bacteriocins as sublancin) and Gram-negative bacteriocins (microcin E492) utilize sugar-phosphotransferase systems (PTS). These include mannose-PTS or glucose-PTS. Many other Gram-negative bacteriocins, like microcin B17, microcin J25, and microcin C, get bound and ultimately do not allow the formation of DNA gyrase, RNA polymerase, and Asp-tRNA synthetase, respectively, targeting the formation directly. Even when the germination of these pathogenic bacteria spores is prevented, it helps in killing them. The resistance by the Gram-negative bacteria for not allowing the bacteriocin to act is due to a barrier it makes for the cationic bacteriocins in the outer cell membrane. However, this barrier can be broken down by using porins. It allows the diffusion into the cell. Another way is the Trojan horse approach. In this method, the receptor requires iron uptake. This helps the bacteriocin to move through the outer membrane and enter the periplasm. It is important to study the resistance to bacteriocins of any pathogen before the work is undertaken. The resistance can be either innate, or it can also be acquired. Innate resistance to bacteriocins may be different among various bacteria which is linked to different genetic loci in bacteria. Some strains of *L. monocytogenes* are resistant to pediocin and other strains have been found to be resistant to pediocin PA-1.

When the bacteriocin is broken down at the protein level, then the resistance is developed in nonbacteriocin-producing bacteria. There are many reasons when the resistance gets acquired. This could be due to spontaneous mutations occurring on any associated gene. There could also be a change in the specific receptors, in the cell wall synthesis, in the transcription or metabolism and transport. Even any change in the fluid movement in the bacterial membrane, lipid composition, electrical load, the thickness of the cell wall, or a combination of these factors can bring in, an acquired resistance in the action of the bacteriocin. Other reasons are the changes in the cytoplasm membrane synthesis or reduction in the permeability of the membrane. Even with the change in the hydrophobic nature of the cell membrane, the fatty acid composition gets affected which gives resistance to the *L. monocytogenes* cells against nisin. When the bacteriocin is not able to reach to their receptors or in case it gets downregulated, the receptors get changed which increases the resistance to bacteriocin. This has been documented in cases of lipid II (vanA), Man-PTS (mPTS), synthesis of peptidoglycans (dal and ddl), proteolytic cleavage, and other proteases. Multiple resistance mechanisms have been developed in some strains of bacteria that may operate simultaneously, and give rise to the overall resistance. Resistance to both bacteriocins as well as antibiotics at the same time is termed as cross-resistance. Some strains of *L. monocytogenes* are resistant to class II as well as class I bacteriocins. The work of Crandall and Montville (1998) has documented that nisin-resistant *L. monocytogenes* had cross-resistance to pediocin PA-1. To overcome this, it has been suggested to utilize a mixture of bacteriocins which are taken from different classes and which undergo different mechanism of action. This approach has shown beneficial results in killing the pathogen cells. The cross resistance of bacteriocin resistant variants have been studied by many antimicrobial agents like nisin, pediocin 34, and enterocin FH99. The regulatory agencies consider bacteriocins as new food additives.

When bacteriocins are supplemented to foods, the nutrition, storage life, and appearance of the product get enhanced. They are more effective in pure forms and when the amino acid sequence is well understood, it helps in the use of the particular bacteriocin. The gold standard biochemical and molecular techniques are acceptable method of analysis that have been established by the regulators. The final concentration of the bacteriocin should be known in the finished food. It is vital to use GMP (Good manufacturing practices). The structure and mode of action of microbisporicin has been determined by Castiglione et al.(2008) which is active against multiresistant pathogens. Soltani et al. (2021) have studied the toxicity aspects and regulations of various bacteriocins.

## 10.7 Use of Nisin and Other Bacteriocins Approved as Food Additive

Depending on the laws of each country, the use of bacteriocins in food products is different. Nisin has been added to the European food additive list in 1983 (Directive 83/463/EEC; Directive 95/2/EC) and is used as a food preservative (E234). The FDA Federal Register (1988; 21CFR) and Health Canada (NOP/ADP-0028; 2017) have approved the use of nisin in foods. In a global survey, it has been found that more than 80 countries are using nisin as a food additive. It is a unique bacteriocin as it has homology with a wide variety of foods. Presently, nisin is being used in meat, processed and pasteurized cheese, fruit, and vegetable industries. The amount to be added should be optimized before using it on a commercial scale. The quantity should be added in a way that the bacteriocin must stop the infection and simultaneously improve the palatability of the food product.

Most of the times the limits of bacteriocin to be added are fixed by each country (Gálvez et al. 2011; López-Cuellar et al. 2016). The safety and efficacy of the bacteriocin should be checked after performing animal experiments and by clinical trials. In Canada, the regulation has been done for veterinary drugs. These are The

Food and Drugs Act and Regulations administered by Health Canada. This helps in the safety and effectiveness of the product (https://www.canada.ca/en/health-canada/services/drugs-health-products/veterinary-drugs/legislation-guidelines/guidance-documents/guidance-document-classification-veterinary-drugs-livestock-feeds/document.html).

Udder in cows, gets infected very easily. This is of great concern. To overcome this problem, a nisin-based udder disinfectant has been prepared which is available commercially and is named as Wipe Out® dairy wipes. It has been approved by the FDA. An act has been formed known as "The Federal Food Drug and Cosmetic Act" (FFDCA) which determines the quantity of bacteriocins to be used. In Canada, therapeutic agents are not allowed in feeds. Use of bacteriocins in humans and animals requires a toxicity test. FDA Regulations have been provided as FDA Ingredients and Additives (2018) and are available at https://www.fda.gov/food/food-ingredients-packaging/determining-regulatory-status-food-ingredient.

Antimicrobial compounds are produced by many living organisms apart from microorganisms. Animals, plants, and insects also have the ability to produce antimicrobial compounds. It is not only bacteriocins that can kill the pathogens. Many other compounds that can kill pathogenic bacteria are hydrogen peroxide, fatty acids, organic acids, ethanol, and antibiotics. However, the antimicrobial peptides (AMPs) or proteins are produced only by bacteria. Therefore, these are named as bacteriocins. The defence system is the antimicrobial peptides or proteins. Most of the bacteria have the ability to produce at least one bacteriocin. The bacteriocins are synthesized by the ribosomes. They have become in common use due to the many beneficial attributes to the consumer and to the environment. Moreover, the use of chemical preservatives can hamper the health as they are harmful. Therefore, most of these chemical compounds have been banned by governments around the world in most of the countries. The assessment studies have been done. Piper et al. (2009) studied the activities of lacticin 3147 and nisin against the drug-resistant *Staphylococcus aureus* and *Enterococcus* species. The cytotoxicity of nisin, pediocin, and colicins has been studied (Murinda et al. 2003) on mammalian cells. Nisin has been termed as the new natural food antimicrobial systems (Thomas et al. 2000).

Colicin is a potent bacteriocin which is known to be produced by 20 different strains of *E. coli* (Jordi et al. 2001). Colicin can prevent the production of different types of Shiga toxin which are produced by pathogenic strains of *E. coli* (O26, O111, O128, O145, and O157:H7). Nisin has been known to kill the dangerous human gastroenteritis pathogen *Campylobacter jejuni*. The potential of bacteriocins to be used in place of antibiotics in poultry and other animal feeds has been very effective and this indirectly improves the human health. A mixture of bacteriocins has shown good promise in killing the pathogens. Acuña et al. (2012) have integrated enterocin CRL35 and microcin V genes and have obtained a bacteriocin called Ent35-MccV. This bacteriocin has been obtained by using recombinant PCR technique which has bactericidal activity against clinically isolated enterohemorrhagic *E. coli* and *Listeria monocytogenes*. This novel bacteriocin has the ability to kill Gram-positive as well as the Gram-negative bacteria.

## FAQs

1. Which microorganisms produce bacteriocins?
2. Which media is recommended for bacteriocin production?
3. What are the bacteriocins used for?
4. What are the important properties of a bacteriocin?
5. Can bacteriocin work as best alternative for antibiotic?
6. What is the impact of bacteriocins on antibiotic resistance profile of drug-resistant bacteria?
7. Is it possible to control food-borne bacteria with bacteriocins?
8. What type of bacteriocins are produced by Gram-negative bacteria?
9. What are the nutritional properties of a bacteriocin?
10. What are lantibiotics?

## References

Abengózar MÁ, Cebrián R, Saugar JM, Gárate T, Valdivia E, Martínez-Bueno M, Maqueda M, Rivas L (2017) Enterocin AS-48 as evidence for the use of bacteriocins as new leishmanicidal agents. Antimicrob Agents Chemother 61(4):e02288–e02216. https://doi.org/10.1128/AAC.02288-16

Acuña L, Picariello G, Sesma F, Morero RD, Bellomio A (2012) A new hybrid bacteriocin, Ent35-MccV, displays antimicrobial activity against pathogenic Gram-positive and Gram-negative bacteria. FEBS Open Biol 2:12–19. https://doi.org/10.1016/j.fob.2012.01.002

Amran RJ, Mamdoh S, Fotoon B, Saba S, Satheesh S (2023) A mini review on biotechnological potentials of bioactive compounds and bioproducts isolated from cyanobacteria. Acta Biologica Marisiensis 6:62–86. https://doi.org/10.2478/abmj-2023-0012

Arnison PG, Bibb MJ, Bierbaum G et al (2013) Ribosomally synthesized and post-translationally modified peptide natural products: overview and recommendations for a universal nomenclature. Nat Prod Rep 30:108–160

Bagley MC, Dale JW, Merritt EA et al (2005) Thiopeptide antibiotics. Chem Rev 105:685–714

Benatez C, Diego F, Le ÃB, Angel LE, Jordy A, Morones RÃ (2021) Bacteriocins: an overview of antimicrobial, toxicity and biosafety assessment by in vivo models. Front Microbiol 12:630–695. https://www.frontiersin.org/journals/microbiology/articles/10.3389/fmicb.2021.630695, https://doi.org/10.3389/fmicb.2021.630695, ISSN 1664-302X

Benech RO, Kheadr E, Laridi R, Lacroix C, Fliss I (2002) Inhibition of *Listeria innocua* in cheddar cheese by addition of nisin Z in liposomes or by in situ production in mixed culture. Appl Environ Microbiol 68:3683–3690

Bhunia AK, Johnson M, Ray B et al (1990) Antigenic property of pediocin AcH produced by *Pediococcus acidilactici* H. J Appl Bacteriol 69:211–215

Bountra K, Hagelueken G, Choudhury HG, Corradi V, El Omari K, Wagner A et al (2017) Structural basis for antibacterial peptide self immunity by the bacterial ABC transporter McjD. EMBO J 36:3062–3079

Caly DL, Chevalier M, Flahaut C, Benoit C, Ahmed Al AK, Chataigné G, D'Inca R, Auclair E, Drider D (2017) The safe enterocin DD14 is a leaderless two-peptide bacteriocin with anti *Clostridium perfringens* activity. Int J Antimicrob Agents 49(3):282–289. ISSN 0924-8579. https://doi.org/10.1016/j.ijantimicag.2016.11.016

Castiglione F, Lazzarini A, Carrano L, Corti E, Ciciliato I, Gastaldo L, Candiani P et al (2008) Determining the structure and mode of action of microbisporicin, a potent lantibiotic active against multiresistant pathogens. Chem Biol 15(1):22–31

Cavera VL, Arthur TD, Kashtanov D, Chikindas ML (2015) Bacteriocins and their position in the next wave of conventional antibiotics. Int J Antimicrob Agents 46:494–501

Cerqueira J, Dimitrov S, Silva A et al (2018) Inhibition of herpes simplex virus 1 and poliovirus (PV-1) by bacteriocins from *Lactococcus lactis* subsp. lactis and *Enterococcus durans* strains isolated from goat milk. Int J Antimicrob Agents 51:33–37

Chalekson CP, Neumeister MW, Jaynes J (2003) Treatment of infected wounds with the antimicrobial peptide D2A21. J Trauma Acute Care Surg 54:770–774

Claesen J, Bibb M (2010) Genome mining and genetic analysis of cypemycin biosynthesis reveal an unusual class of post translationally modified peptides. Proc Natl Acad Sci USA 107:16297–16302

Cotter P, Hill C, Ross R (2005) Bacteriocins: developing innate immunity for food. Nat Rev Microbiol 3(10):777–788. https://doi.org/10.1038/nrmicro1273

Cotter PD, Ross RP, Hill C (2013) Bacteriocins - a viable alternative to antibiotics? Nat Rev Microbiol 11(2):95–105

Cox CR, Coburn PS, Gilmore MS (2005) Enterococcal cytolysin: a novel two component peptide system that serves as a bacterial defense against eukaryotic and prokaryotic cells. Curr Protein Pept Sci 6:77–84

Crandall AD, Montville TJ (1998) Nisin resistance in *Listeria monocytogenes* ATCC 700302 is a complex phenotype. Appl Environ Microbiol 64:231–237. https://doi.org/10.1128/AEM.64.1.231-237

Das D, Goyal A (2014) Characterization of a noncytotoxic bacteriocin from probiotic *Lactobacillus plantarum* DM5 with potential as a food preservative. Food Funct 5(10):2453–2462. https://doi.org/10.1039/c4fo00481g

Dicks LMT et al (2018) A review: the fate of bacteriocins in the human gastrointestinal tract: do they cross the gut–blood barrier? Front Microbiol 9:2297

Duquesne S, Destoumieux GD, Peduzzi J, Rebuffat S (2007) Microcins, gene encoded antibacterial peptides from enterobacteria. Nat Prod Rep 24:708–734

Falagas ME, Rafailidis PI, Makris GC (2008) Bacterial interference for the prevention and treatment of infections. Int J Antimicrob Agents 31(6):518–522. ISSN 0924-8579

Fatima D, Mebrouk K (2013) Characterization and determination of the factors affecting anti listerial bacteriocins from *Lactobacillus plantarum* and *Pediococcus pentosaceus* isolated from dairy milk products. Afr J Food Sci 7(2):35–44

FDA Federal Register (1988) Nisin preparation: affirmation of GRAS status as a direct human food ingredient. 21 CFR Part 184, Fed Reg 53:11247–11251

FDA Ingredients and additives (2018). Available at https://www.fda.gov/food/food-ingredients-packaging/determining-regulatory-status-food-ingredient (10 March 2020, date last accessed)

Férir G, Petrova MI, Andrei G, Huskens D, Hoorelbeke B, Snoeck R et al (2013) The lantibiotic peptide labyrinthopeptin A1 demonstrates broad anti-HIV and anti-HSV activity with potential for microbicidal applications. PLoS One 8(5):e64010. https://doi.org/10.1371/journal.pone.0064010

de Freire Bastos Md C, Coelho MLV, da Silva Santos OC (2015) Resistance to bacteriocins produced by Gram-positive bacteria. Microbiol 161:683–700

Gabrielsen C, Brede DA, Nes IF (2014) Circular bacteriocins: biosynthesis and mode of action. Appl Environ Microbiol 80:6854–6862

Gálvez A, Abriouel H, López RL, Ben Omar N (2007) Bacteriocin based strategies for food biopreservation. Int J Food Microbiol 120(1–2):51–70

Gálvez A, Abriouel H, Omar NB, Lucas R (2011) Food applications and regulation. In: Drider D, Rebuffat S (eds) Prokaryotic antimicrobial peptides. Springer, New York. https://doi.org/10.1007/978-1-4419-7692-5_18

Grilli E, Messina MR, Catelli E, Morlacchini M, Piva A (2009) Pediocin A improves growth performance of broilers challenged with *Clostridium perfringens*. Poult Sci 88:2152–2158

Hammami R, Zouhir A, Le Lay C et al (2010) BACTIBASE second release: a database and tool platform for bacteriocin characterization. BMC Microbiol 10:22

Hassan H, Gomaa A, Subirade M et al (2019) Novel design for alginate/resistant starch microcapsules controlling nisin release. Int J Biol Macromol 153:1186–1192

Hatakka K, Savilahti E, Ponka A, Meurman JH, Poussa T, Nase L, Saxelin M, Korpela R (2001) Effect of long term consumption of probiotic milk on infections in children attending day care centres: double blind, randomised trial. BMJ 322:1327

Héchard Y, Sahl HG (2002) Mode of action of modified and unmodified bacteriocins from Gram-positive bacteria. Biochimie 84:545–557

Heinemann B, Williams R (1966) Inactivation of nisin by pancreatin. J Dairy Sci 49:312–314

Heng NW, Philip B, Jeremy J, Ralph T, John RT (2006) The diversity of bacteriocins in Gram-positive bacteria. In: Bacteriocins: ecology and evolution. Springer Berlin Heidelberg. https://doi.org/10.1007/978-3-540-36604-1_4

Jack RW, Tagg JR, Ray B (1995) Bacteriocins of gram-positive bacteria. Microbiol Rev 59(2):171–200. https://doi.org/10.1128/mr.59.2.171-200

Jarvis B, Mahoney RR (1969) Inactivation of nisin by alpha-chymotrypsin. J Dairy Sci 52:1448–1450

Jasniewski J, Cailliez-Grimal C, Chevalot I, Millière JB, Revol-Junelles AM (2009) Interactions between two carnobacteriocins Cbn BM1 and Cbn B2 from *Carnobacterium maltaromaticum* CP5 on target bacteria and Caco-2 cells. Food Chem Toxicol 47(4):893–897. https://doi.org/10.1016/j.fct.2009.01.025

Joo NE, Ritchie K, Kamarajan P, Miao D, Kapila YL (2012) Nisin, an apoptogenic bacteriocin and food preservative, attenuates HNSCC tumorigenesis via CHAC1. Cancer Med 1(3):295–305

Jordi BJ, Boutaga K, van Heeswijk CM, van Knapen F, Lipman LJ (2001) Sensitivity of Shiga toxin-producing *Escherichia coli* (STEC) strains for colicins under different experimental conditions. FEMS Microbiol Lett 204:329–334

Karpiński TM, Szkaradkiewicz AK, Caballero B et al (2016) Bacteriocins. In: Encyclopedia of food and health. Elsevier Sci, Amsterdam, pp 312–319

Kaur S, Kaur S (2015) Bacteriocins as potential anticancer agents. Front Pharmacol 6:272

Kheadr E, Zihler A, Dabour N et al (2010) Study of the physicochemical and biological stability of pediocin PA-1 in the upper gastrointestinal tract conditions using a dynamic in vitro model. J Appl Microbiol 109:54–64

Klaenhammer TR (1993) Genetics of bacteriocins produced by lactic acid bacteria. FEMS Microbiol Rev 12:39–85. https://doi.org/10.1111/j.1574-6976.1993.tb00012.x

Kumarasamy KK, Toleman MA, Walsh TR, Bagaria J, Butt F, Balakrishnan R, Chaudhary U et al (2010) Emergence of a new antibiotic resistance mechanism in India, Pakistan, and the UK: a molecular, biological, and epidemiological study. Lancet Infect Dis 10(9):597–602

Le Blay G, Hammami R, Lacroix C et al (2012) Stability and inhibitory activity of pediocin PA-1 against *Listeria* sp. in simulated physiological conditions of the human terminal ileum. Probiotics Antimicrob Proteins 4:250–258

Le Lay C, Fernandez B, Hammami R et al (2015) On *Lactococcus lactis* ul719 competitivity and nisin (Nisaplin®) capacity to inhibit *Clostridium difficile* in a model of human colon. Front Microbiol 6:1020

Lohans CT, Vederas JC (2011) Development of class IIa bacteriocins as therapeutic agents. Int J Microbiol 386410

López-Cuellar MR, Rodríguez-HernándezA I, Chavarría-Hernández N (2016) LAB bacteriocin applications in the last decade. Biotechnol Biotechnol Equip 30:1039–1050

Maher S, McClean S (2006) Investigation of the cytotoxicity of eukaryotic and prokaryotic antimicrobial peptides in intestinal epithelial cells in vitro. Biochem Pharmacol 71:1289–1298

Martins J, Vasconcelos V (2015) Cyanobactins from cyanobacteria: current genetic and chemical state of knowledge. Mar Drugs 13:6910–6946

Martins J, Leikoski N, Wahlsten M et al (2018) Sphaerocyclamide, a prenylated cyanobactin from the *Cyanobacterium Sphaerospermopsis* sp. LEGE 00249. Sci Rep 8:1–9

McAuliffe O, Ross RP, Hill C (2001) Lantibiotics: structure, biosynthesis and mode of action. FEMS Microbiol Rev 25:285–308

Murinda SE, Rashid KA, Roberts RF (2003) In Vitro assessment of the cytotoxicity of nisin, pediocin, and selected colicins on simian virus 40–transfected human colon and vero monkey kidney cells with trypan blue staining viability assays. J Food Prot 66(5):847–853

Norris GE, Patchett ML (2016) The glycocins: in a class of their own. Curr Opin Struct Biol 40:112–119

Oman TJ, van der Donk WA (2009) Insights into the mode of action of the two-peptide lantibiotic haloduracin. ACS Chem Biol 4:865–874

Padilla C, Lobos O, Hubert E (2004) *Shigella flexneri* strains produce bacteriocins active against members of the human microbial intestinal flora. Rev Latinom Microbiol 46:85–88

Perez RH, Zendo T, Sonomoto K (2014) Novel bacteriocins from lactic acid bacteria (LAB): various structures and applications. Microb Cell Fact 13 Suppl 1(Suppl 1):S3. https://doi.org/10.1186/1475-2859-13-S1-S3

Piper C, Draper LA, Cotter PD et al (2009) A comparison of the activities of lacticin 3147 and nisin against drug-resistant *Staphylococcus aureus* and *Enterococcus* species. J Antimicrob Chemother 64:546–551

Quintana VM, Torres NI, Wachsman MB et al (2014) Antiherpes simplex virus type 2 activity of the antimicrobial peptide subtilosin. J Appl Microbiol 117:1253–1259

Reddy K, Aranha C, Gupta S et al (2004) Evaluation of antimicrobial peptide nisin as a safe vaginal contraceptive agent in rabbits: invitro and in vivo studies. Reproduction 128:117–126

Sahoo TK, Jena PK, Prajapati B et al (2017) In vivo assessment of immunogenicity and toxicity of the bacteriocin TSU4 in BALB/c mice. Probiotics Antimicrob Proteins 9:345–354

Saïd E, Toshihiro S, Kenji S, Ayaaki I (2000) Class IIa bacteriocins: biosynthesis, structure and activity. FEMS Microbiol Rev 24(1):85–106. https://doi.org/10.1111/j.1574-6976.2000.tb00534.x

Soltani S, Hammami R, Cotter PD, Rebuffat S, Said LB, Gaudreau H, Bédard F, Biron E, Drider D, Fliss I (2021) Bacteriocins as a new generation of antimicrobials: toxicity aspects and regulations. FEMS Microbiol Rev 45(1):fuaa039. https://doi.org/10.1093/femsre/fuaa039

Swain MR, Anandharaj M, Ray RC et al (2014) Fermented fruits and vegetables of Asia: a potential source of probiotics. Biotechnol Res Int:250424

Thapar P, Mohinder KS (2023) Bacteriocins: applications in food preservation and therapeutics. *Lactobacillus*-A multifunctional genus. Intech Open. https://doi.org/10.5772/intechopen.106871

Thomas LV, Clarkson MR, Delves-Broughton J (2000) Nisin. In: Naidu AS (ed) Natural food antimicrobial systems. CRC Press, Boca Raton, pp 463–524

Todorov SD (2009) Bacteriocins from *Lactobacillus plantarum* production, genetic organization and mode of action: produção, organização genética e modo de ação. Braz J Microbiol 40:209–221

Todorov SD, Wachsman M, Tomé E et al (2010) Characterisation of an antiviral pediocin-like bacteriocin produced by *Enterococcus faecium*. Food Microbiol 27:869–879

Westendorf AG, Florian E, Deppenmeier T, Damini H, Schmidt J, Alexander MB, Jan B, Dunja B (2005) Intestinal immunity of *Escherichia coli* NISSLE 1917: a safe carrier for therapeutic molecules. FEMS Immun Med Microbiol 43:373–384. https://doi.org/10.1016/j.femsim.2004.10.023

Wilaipun P, Zendo T, Sangjindavong M et al (2004) The two-synergistic peptide bacteriocin produced by *Enterococcus faecium* NKR-5-3 isolated from Thai fermented fish (Pla-ra). Sci Asia 30:115–122

Willey JM, Van Der Donk WA (2007) Lantibiotics: peptides of diverse structure and function. Ann Rev Microbiol 61:477–501

Yang SC, Lin CH, Sung CT et al (2014) Antibacterial activities of bacteriocins: application in foods and pharmaceuticals. Front Microbiol 5:241

Yu H, Li N, Zeng X et al (2019) A comprehensive antimicrobial activity evaluation of the recombinant microcin J25 against the foodborne pathogens *Salmonella* and *E. coli* O157: H7 by using a matrix of conditions. Front Microbiol 10:1954

Yutthana K, Amonlaya T, Takeshi Z, Sittiruk R, Plearnpis L et al (2012) Anti-listeria activity of *Pediococcus pentosaceus* BCC 3772 and application as starter culture for Nham, a traditional fermented pork sausage. Food Control 25(1):190–196. ISSN0956-7135. https://doi.org/10.1016/j.foodcont.2011.10.005

Zhou H, Xie Y, Liu H et al (2015) Effects of two application methods of plantaricin BM-1 on control of *Listeria monocytogenes* and background spoilage bacteria in sliced vacuum-packaged cooked ham stored at 4°C. J Food Prot 78:1835–1841

# Production of Microbial Enzymes

# 11

**Abstract**

Microorganisms have been used for the production of bread, wine, vinegar, and many other products since time immemorial. The use of microbial enzymes in industries and medicines has been very useful and is very cost-effective. Enzymes can be produced easily from microbes and their stability is very high. These enzymes have been permitted by the FDA to be used in food, agriculture, and industries. The microbial enzymes have catalytic properties. The microbial industrial processes that utilize these enzymes do not produce much pollution. This makes them environment friendly. Microbial enzymes have been used in food, feed, pharmaceutical, and biotechnological industries. With recent developments, genetically recombinant enzymes from bacteria and fungi have also been produced. These have shown success in many bioprocesses. This chapter discusses the production of microbial enzymes and also their production by genetic manipulation. With this, the yields of bio product can be increased many folds. This chapter gives an overview on the production of industrially important microbial enzymes, their applications, and means to increase their yields.

**Keywords**

Microbes · Enzymes · Amylase · Lipase · Protease · Pectinase · Fermentation · Genetic manipulation

## 11.1 Introduction

Most of the time when bacteria consume waste, they convert them into safe by-products like carbon dioxide and water. However, when complex wastes like pond sludge are consumed, many enzymes are produced in bacteria that break down the complex waste into simpler compounds. Hydrolytic enzymes (proteases, amylases,



R. Agrawal, *Textbook of Industrial Microbiology*,
https://doi.org/10.1007/978-981-97-9582-6_11

and pectinases) have the ability to degrade natural polymers (proteins, starches, or pectin). These are mostly extracellular. By using suitable nutrients, under optimized conditions, the microbial enzymes can be produced in maximum yields. These can be either intracellular or extracellular. The enzymes either come into the media broth on their own or the cells are lysed to get them. The enzymes can be produced by two methods, namely, submerged fermentation (SmF) and solid-state fermentation (SSF). Submerged fermentation involves growing the microorganisms in liquid nutrient medium with high oxygen.

## 11.2 History

Takamine (1894) was the scientist who discovered the first enzyme and named it as takadiastase. This was a form of diastase, which was obtained from *Aspergillus oryzae*. Later, it was Buchner (1886) who showed that glucose could be converted to ethanol from the cell-free extract of yeast.

Buchner (1907) taught many students on the topic of cell-free fermentation by giving various lectures. Rohm (1908) worked on the application of pancreatic enzymes in tanneries for bating of hides. It was Nelson and Griffin (1916) who showed the adsorption of invertase on charcoal and alumina. A technique was developed to immobilize the enzymes so that they can be reused. There are many advantages of using immobilized enzymes as they are able to produce higher yields of enzymes and the product separation is much easier. It was the work of Sumner (1926) who obtained the first purified enzyme. In 1947, he isolated and crystallized the enzyme urease from Jack beans. It was Novo industry (1960) that produced protease enzyme for the first time on a commercial scale by using *Bacillus lichniformis* (Razzaq et al. 2019). Later, applications of genetic engineering techniques were utilized, which improved the production of enzymes. Traditionally, enzymes have been widely used in the manufacture of linen, leather, and indigo. These enzymes have been produced by microorganisms or from calves' rumen or the papaya fruit. For any commercial process, it is important to optimize the fermentation conditions. This produces enzymes in much higher yields. The microbial enzymes are commonly utilized in the detergent, textile, and starch industries. Singh et al. (2016) have written a good review on the industrial progress of microbial enzymes in the twenty-first century. They have also covered the different aspects on the genetic and protein engineering techniques, which have helped in improving the production processes.

## 11.3 Principles of Industrial Enzymology

Biological catalysts are known as enzymes and they boost up many biochemical reactions. These are utilized in many commercial processes. Usage in the production of sweetening agents, antibiotics, washing powders, and various cleaning products is common. Enzymes are used in analytical devices and in assays that have clinical, forensic, and environmental applications. The word was initially coined by a German physiologist (Kühne 1877). The enzymes are required in very low quantities and are not used during the reaction. They have high specificity and help in lowering of the activation energy (Aldridge 2013).

## 11.4 Major Microbial Enzymes, Classification, and Nomenclature

Mainly, industrial enzymes are hydrolytic in action and are used in the degradation of various natural substances. Proteases are used in the detergent and dairy industries. Various carbohydrates, like amylases and cellulases, are used in starch, textile, detergent, and baking industries. There is a huge global market for industrial enzymes, and it is increasing every year. Many enzymes are used in the production of foods and beverages, cleaning supplies, clothing, paper products, transportation fuels, pharmaceuticals, and monitoring devices. Chiral specificity of enzymes helps in the preparation of enantiomeric pure pharmaceuticals, agrochemicals, chemical feedstock, and food additives. The major microbial enzymes are discussed in detail later in this chapter (11.8).

### 11.4.1 Classification of Enzymes and Their Uses

The rate and the specificity of the metabolic reactions can be increased with enzymes. Therefore, they are also known as catalysts. A biological cell needs the enzymes that are required for all the activities required to live. The enzymes, which are intracellular, are mainly responsible for a particular metabolic pathway of that cell. As per the literature, these catalyze more than 4000 biochemical reactions inside the cell. Both the enzyme and the substrate possess a specific complementary geometric shape that fits exactly into one another. This is known as "the lock and key" model which is very specific (Fischer 1894). It has been found that most of the times the enzymes are much larger than the substrates. Therefore, only a small portion of the enzyme (2–4 amino acids) gets involved in the process of catalysis. This particular region which binds to the substrate, is known as the active site. Enzymes also have sites that bind with the cofactors which are required for the catalytic reaction. Many enzymes can bind with small molecules, which can either increase or decrease the activity of the enzyme and can also provide the feedback regulation. The enzymes have a linear chain of amino acids which gets folded and then produces a three-dimensional product. The enzymes can be denatured and can get deactivated on heating or by using chemical denaturants which could be reversible or irreversible. They are classified depending on the catalyzing reaction. The enzymes are classified on the basis of a numerical classification scheme and an Enzyme Commission number (EC number) is given to them. This number is based on the chemical reactions they catalyze. Usually, these enzymes end as "ase." The enzyme commission was set up in Brussels during the International Congress of

**Table 11.1** Main classes of enzymes in EC system

| Enzyme class | Reaction type |
|---|---|
| Oxidoreductases | Oxidation/reduction |
| Transferases | Atom/group transfer |
| Hydrolases | Hydrolysis |
| Lyases | Group removal |
| Isomerases | Isomerization |
| Ligases | Joining of molecules linked to the breakage of a pyrophosphate bond. |

Biochemistry (1955). Later (1992), it was edited by the International Union of Biochemistry and Molecular Biology. This body worked on the standards of enzyme nomenclature. However, later, an Enzyme Commission was set up which took over the work. Within this system, all enzymes have been described by a four-part Enzyme Commission (EC) number. The classification of these enzymes is based on the specificity of the reaction. Thus they are classified into six groups: Oxidoreductases, transferases, hydrolases, lyases, isomerases, and ligases. Oxidoreductases are the enzymes that catalyze oxidation–reduction reactions. These are responsible for the production of heat and energy. Transferases are the enzymes that catalyze reactions where the transfer of functional groups takes place between the two substrates. Hydrolases are the hydrolytic enzymes that catalyze the hydrolysis of carbohydrates, proteins, and esters.

Lyases are the enzymes that form double bonds and help in removing the groups from the substrates without undergoing hydrolysis. There are two isomers known as cis and trans. The interconversion between the two can be done with the help of enzyme isomerases. The coupling of two compounds can be done by using the enzyme Ligases. Therefore, they are also known as synthases. The reactions are usually done by the hydrolysis of the pyrophosphate bonds. The information for any individual enzyme can be obtained from the Enzyme Nomenclature Database (available at http://enzyme.expasy.org) ENZYME—Enzyme nomenclature database (Table 11.1).

### 11.4.2 Nomenclature of Enzymes

The nomenclature is based on the name of the substrate/the type of the reaction involved which ends with -ase. For example; for the enzyme termed as maltase, the substrate is maltose.

### 11.4.3 Functions of Enzymes

During the processing of different food items, utilization of enzyme (s) is very common. This can be either a single enzyme or a combination of many enzymes. These enzymes are obtained from different microbial sources. When these enzymes are added to the food, then, some preservatives/stabilizers are also required to be added.

However, care has to be taken to remove all the residues from the fermentation medium in which the microorganism has been growing. This should be done before using the enzyme. In the food processing units, the enzymes that are used for hydrolyzing starch can be removed during purification, and the enzymes that are used in bakerie, can be inactivated at high temperatures during baking.

1. Regulation of enzyme preparations has been done (Code of Federal Regulations - Title 21 - Food and Drugs (database), Part 173 (21 CFR 173). As per section 409(b)(1) of the Federal Food, Drug, and Cosmetic Act (the Act) [21 U.S.C. 348(b)(1)], a petition can be filed for the issuance of a regulation. Section 409(b)(2) of the Act [21 U.S.C. 348(b)(2)] explains the statutory requirements for food additive petitions. The details can be obtained in 21 CFR 171.1. Section 201(s) of the Act [21 U.S.C. 321(s)] gives an exemption on the use of a substance that comes under the generally recognized as safe (GRAS) category which is given if the substance is safe. FDA has given Title 21 of the CFR to the food ingredients that fall under GRAS category.
2. A review of all the enzymes, which come under GRAS category, has been done by FDA and is given in the regulation in 21 CFR, Part 184. Enzyme preparations that are approved as food additives are listed in 21 CFR 173 and are affirmed as GRAS in 21 CFR 184 Section. Description of Enzyme Preparation 173.110 Amyloglucosidase is for use in degrading gelatinized starch into constituent sugars. 173.120 Carbohydrase and cellulose are used for clam and shrimp processing. 173.130 Carbohydrases are used for the production of dextrose from starch. 173.135 Catalase is used in the manufacture of cheese. 173.140 Esterase-lipase is used as a flavor enhancer in cheese, fats and oils, and milk products. 173.145 α-galactosidase for use in the production of sucrose from sugar beets. 173.150 aspartic proteinase as a milk-clotting enzyme, and in the production of cheese. 137.105 α-amylase in flour. 184.1012 Alpha-amylase is used for hydrolyzing the edible starch to produce maltodextrin and nutritive carbohydrate sweeteners. 184.1024 Bromelain is used for hydrolyzing proteins and polypeptides. 184.1027 Mixed carbohydrase and protease enzymes are used in the hydrolysis of proteins and carbohydrates. Especially in the preparation of alcoholic beverages, candy, nutritive sweeteners, and protein hydrolysates. 184.1034 Catalase is used to decompose hydrogen peroxide. 184.1316 Ficin (peptide hydrolase) to hydrolyze proteins and polypeptides. 184.1372 is an insoluble glucose isomerase enzyme. 184.1387 Lactase is used in hydrolyzing lactose to glucose and galactose. 184.1388 Lactase is used for hydrolyzing lactose in milk. 184.1415 Animal lipase (triacylglycerol hydrolase) is used to hydrolyze fatty acid glycerides. 184.1420 Lipase is used in the interesterification of fats and oils. 184.1443 Malt (α-amylase and β-amylase) is used for the hydrolysis of starch. 184.1583 Pancreatin (peptide hydrolase) is used for the hydrolysis of proteins or polypeptides. 184.1595 Pepsin (peptide hydrolase) is used for hydrolyzing proteins. 184.1685 Rennet (animal derived) is used for coagulating milk in cheeses and other dairy products. 184.1914 Trypsin is used to hydrolyze proteins. 184.1924 urease, enzyme is used in the production of wine. 184.1985 aminopeptidase, is

used as an optional ingredient for flavor development in the manufacture of cheddar cheese. Barett (1995) has explained the role of aspartic and metallopeptidases as proteolytic enzymes. Bhosale et al. (1995) have worked on the thermostability and high activity of alkaline protease from *Conidiobolus coronatus* (NCL 86.8.20).

3. The activity of an enzyme is expressed by a constant, $k_{cat}$. This constant represents the number of substrate molecules that can be converted to a product by a single enzyme molecule per unit time (usually per minute or per second). It is known as the *turnover rate, turnover frequency,* or *turnover number* (Cleland 2013).

### 11.4.4 Structure of Enzyme

Enzymes, which are based on amino acids, are also known as globular proteins. These can have amino acids from 100 to >2000. Usually, the active site has a small number (less than 10) of the amino acids. The binding of the enzyme to the substrate depends on the correct shape and size of the substrate which would fit into the active site of the enzyme. This is highly specific. As this study was done by Fischer (1894), it has been known as Fischer's "lock and key hypothesis." X-ray crystallography has shown that the shape of the enzymes is very flexible. The active site cannot be separated out from the rest of the protein without the loss of catalytic activity. Most of the enzymes consist only of a protein moiety. However, there are many enzymes that contain a nonprotein component also, which is known as a cofactor. The cofactor is important for the catalytic activity of the enzyme. However, when the nonprotein portion is an organic molecule, then it is called as a coenzyme. When the protein is tightly bound to a coenzyme, then it is known as the prosthetic group of the enzyme. However, when the inorganic form is in the form of a metal ion (iron, manganese, cobalt, copper, or zinc), it is called as a cofactor. When an enzyme requires a cofactor for its activity, the inactive protein component is termed as an apoenzyme, and the apoenzyme plus the cofactor are together termed as a holoenzyme. In the human body, when the cell metabolism takes place, it requires many cofactors and coenzymes. For example, Heme group of cytochromes and bitotin group of acetyl-CoA carboxylase. There are activators which are cations, the positively charged metal ions like Fe-cytochrome oxidase, CU- catalase, Zn- alcohol dehydrogenase, and Mg-glucose-6-phosphate. Coenzymes are organic molecules, usually vitamins which combine with the enzyme-substrate complex temporarily. These are FAD-Flavin Adenine Dinucleotide, FMN-Flavin Mono Nucleotide, NAD-Nicotinamide Adenine Dinucleotide, and NADP-Nicotinamide Adenine Dinucleotide (Tathagata et al. 2010).

**Enzymes and Reaction Equilibrium** These enzymes keep the structure of the products and the reagents intact. They only help in increasing the speed of the reaction. Important parameters are pH and temperature, which affect the reaction. They are required in very less quantities and do not get utilized.

There are three stages in an enzyme-catalyzed reaction: E + S → ES complex → E + P. E represents the enzyme, S represents the substrate, and P represents the product. After the reaction is completed, the product (P) is formed. However, the enzyme dissociates and becomes free which comes out as such. This enzyme molecule can then get bound to another (S) substrate molecule. For some time, when the enzyme gets bound to the substrate it forms an intermediate as the transition phase and then it finally forms the product. The rate of the reaction varies from system to system. With the addition of the enzyme, the amount of energy required (activation energy) for the reaction becomes much less. The reactions can either produce energy (exergonic) or can utilize energy (endergonic). The sequence of aminoacids in an enzyme determines the particular characteristic of that enzyme. When the bonds between the amino acids in a protein are weak, they can be broken down at high temperature or with high acid, but this can deactivate the enzymes. The enzymes remain unchanged as they do not undergo permanent change. The major bacterial enzymes are produced from *Bacillus* which include amylase, glucose isomerase, penicillin amidase, and protease. Asparaginase and lipase are produced from *Escherichia coli* and pullulanase from *Klebsiella.*

There are many fungi that produce enzymes. *Aspergillus* is the most common fungi that produces amylase, aminoacylase, glucoamylase, and catalase. Other common fungal enzymes are cellulase from *Trichoderma*, dextranase from *Penicillium*, glucose oxidase and lactase from *Aspergillus*, lipase from *Rhizopus*, rennet from *Mucor miehei*, pectinase, pectin lyase and protease from *Aspergillus,* and raffinase from *Mortierella*. El-Gendi et al. (2021) have written a very comprehensive review on fungal enzymes, their structure, classification, and their role in humans. There are many enzymes from yeast which include invertase from *Saccharomyces,* lactase from *Kluyveromyces,* lipase from *Candida,* and raffinase from *Saccharomyces.* Hanane et al. (2022) have made a qualitative screening of yeast biodiversity for hydrolytic enzymes isolated from the gastrointestinal tract of a coprophage *Gymnopleurus sturmi* and dung of ruminants. Microbial enzymes can be easily produced in laboratories and are cost-effective. They produce homogeneous products batch after batch. Therefore, these are a preferred source as compared to plants or animals.

## 11.5 Role of Enzymes in Various Industries

**Role of Enzymes in Food Processing Industries** Amylase enzymes from fungi and plants are used in the production of sugars from starch in the manufacture of corn syrup. Catalase helps in breaking the starch into sugar. Yeasts help in the baking industry for raising the dough. Proteases, help in making biscuits. Trypsin is used for the predigestion of baby foods. Barley enzymes are used in brewing industries. Amylases, glucanases, proteases, betaglucanases, arabinoxylases, amyloglucosidases, and acetolactate decarboxylases are used in the production of beer. Enzymes such as cellulases and pectinases are used in the clarification of fruit juices. Renin is used in the manufacture of cheese. Lipases are used in the ripening of blue mold cheese. Lactases break down lactose to glucose and galactose. Softening of the meat is done by using papain enzyme. High amounts of sweet (glucose) can be obtained by using the enzyme glucose isomerase in the process. As the world population is increasing every day, the United Nations Department of Economic and Social Affairs (UNDESA) has estimated that the food demand will also increase by 70% till 2050 (http://www.un.org/waterforlifedecade/food_security.shtml). The biomolecules produced by microbes can help in aspects like flavor, texture, and keeping quality of the food and food safety. This is in turn bringing a better market for quality food. Enzymes also help in the modification of fats, improving the beer quality and also the properties of proteins. They also improve the color and aroma of the products. In bakeries, the enzymes help in improving the stability of the dough texture and shelf life of the products. Even in dairy industries, the microbial enzymes play a vital role especially in making various cheese products. Fernandes (2010) has written a review on the use of enzymes in food processing and the strategies for better biocatalysis.

### 11.5.1 Role of Enzymes in Pharmaceutical Industries

Enzymes are used as therapeutic drugs for betterment of health (Mane et al. 2015). Proteolytic enzymes are helpful in the removal of dead skin and burns. Nattokinase (EC 3.4.21.62) is a potent fibrinolytic enzyme, which is being used for thrombosis therapy (Cho et al. 2010). Enzymes such as dextranase (EC 2.4.1.2) and rhodanase (EC 2.8.1.1) are helpful in the treatment of alimentary dyspepsia, tooth decay, and cyanide poisoning (Okafor 2007). Lipases (triacylglycerol acylhydrolases, EC 3.1.1.3) are known to catalyze the hydrolysis, and they help in the synthesis of esters and long-chain fatty acids. Sharma et al. (2001) worked on the enzyme lipases and their qualities, which are helpful in detergent industries. They also worked on obtaining a recombinant lipase and the immobilization of the enzyme. Tyrosinase is an enzyme (EC 1.14.18.1), which helps in melanogenesis. Hydrolysis of chitosan can be done by using chitosanase (EC 3.2.1.132) (Nidheesh and Velappan 2014). In many cases, the use of these microbial enzymes has helped in reducing the use of radioactive compounds, which are very harmful to human health (Palmer 2001). More so, the quantification of diabetes has been possible with the use of these enzymes. The uses of glucose oxidase (EC 1.1.3.4) for glucose; urease (EC 3.5.1.5) and glutamate dehydrogenase (EC 1.4.1.2) for urea; lipase, carboxyl esterase, and glycerol kinase for triglycerides; urate oxidase (EC 1.7.3.3) for uric acid; creatinase (EC 3.5.3.3) and sarcosine oxidases (EC 1.5.3.1) for creatinine (Le Roes-Hill and Prins 2016) decomposition have been found to be very successful.

### 11.5.2 Role of Enzymes in Baking Industries

The microbial enzymes have been found to be very helpful in improving the quality and efficiency of bakery products. Moisture can be retained in bakery products with the help of amylase, either alone or along with other enzymes to keep them soft and fresh for a longer shelf storage. Stability of the dough can be obtained by using lipase and xylanase (EC 3.2.1.8) enzymes. Collar et al. (2000) studied the effect of enzymes on the making of bread dough by using response surface studies.

### 11.5.3 Role of Enzymes in Dairy Industries

The microbial enzymes such as proteases, lipases, esterases, lactases, aminopeptidases, lysozymes, lactoperoxidases, transglutaminases, and catalases help in improving the organoleptic characters like aroma, flavor, and color of the milk products (Qureshi et al. 2015). Rennet is an enzyme that helps in the coagulation of milk. There are many enzymes that are helpful in preventing the allergenic properties of milk products (Qureshi et al. 2015). The functional properties of dairy products can be enhanced with the help of the enzyme transglutaminase (Kieliszek and Misiewicz 2014). With the deficiency of lactase enzyme, people can develop lactose intolerance and are not able to consume milk or any milk products. In such cases, the use of enzyme lactase (β-galactosidase, EC 3.2.1.23) has been found very helpful. The enzyme has the ability to hydrolyze lactose into glucose and galactose.

### 11.5.4 Role of Enzymes in Beverage Industries

Among the beverages, soft drinks and syrups, packaged water, fruit juices, tea, and coffee fall under nonalcoholic group. The distilled spirits, wine, and beer come under alcoholic group (Encyclopedia of occupational health and safety). Microbial enzymes are used in the brewing industries, which digest the cell wall of any plant material used. This results in better yields, color, aroma, and clear products (Kårlund et al. 2014). In fruit juice industries, the enzymes help in increasing the efficiency in many steps, which involve peeling, juicing, clarification, extraction, and the product quality (Law 2002). The enzymes help in the digestion of pectin, starch, proteins, and cellulose of fruits and vegetables (Mojsov 2012). Usually, the clarification of juices is done with the help of amylases (Sivaramakrishnan et al. 2006), cellulase and pectinase enzymes are used for improving the extraction process (Garg et al. 2016). There are many debittering enzymes like Naringinase (EC 3.2.1.40) and limoninase, which hydrolyze the bitter components (Li et al. 2012). These are used in the citrus juice industry. The haziness in fruit juices is caused due to the presence of pectin, which can be removed with the help of polygalacturonase (EC 3.2.1.15), pectin esterases (EC 3.1.1.11), and pectin lyase (EC 4.2.2.10).

### 11.5.5 Role of Enzymes in the Feed Industries

These enzymes help the animals to utilize the feed in higher quantities and also increase the quantity of the available nutrients to the body at a faster rate (Choct

2006). According to a survey, the global market for enzymes was $ 1.4 billion in 2024. This is expected to reach $1.76 billion by 2029 (Mordor intelligence 2024). The dietary value of feeds and the availability of proteins can be enhanced with these feed enzymes (Collection of information on enzymes 2002). Majorly, phytases, proteases, α-galactosidases, glucanases, xylanases, α-amylases, and polygalacturonases are presently being used by poultry industries (Adrio and Demain 2014). Cereal-based feed contains a large amount of phytic acid, which can be decomposed with phytase enzyme (Frias et al. 2003). Plant-based feeds contain very high amount of celluloses and hemicelluloses, which cannot be digested by the monogastric animals. For better digestion of the plant feeds, enzymes like xylanase and β-glucanase are supplemented in the feeds, which help in the degradation and improve the digestion (Kirk et al. 2002).

### 11.5.6 Role of Microbial Enzymes in Polymer Industries

Biodegradable polymers are produced with the help of microbial enzymes and are termed as biopolymers. These have applications in orthopedic devices and tissue replacement (Ulery et al. 2011). The market for biopolymers is growing rapidly. The global market for biodegradable polymers was $ 19.37 billion, in 2024 which is expected to reach $ 38.69 billion by 2030. Lipases help in the catalysis of lactone polymerization, cyclic diesters, and cyclic carbonates (Kobayashi 2010). Laccase (EC 1.10.3.2), peroxidase, and transglutaminase are the enzymes which help in forming the cross-links in biopolymers (Gurung et al. 2013). Hiraishi and Taguchi (2009) worked on the enzyme-catalyzed synthesis and degradation of the biopolymers. Kadokawa and Kobayashi (2010) worked on the polymer synthesis with the help of enzymatic catalysis.

### 11.5.7 Role of Microbial Enzymes in Paper and Pulp Industries

The use of enzymes in paper and pulp industries is very helpful in reducing the total time taken in the process and in energy consumption. They are also used in bleaching and monitoring effluent waste (Srivastava and Singh 2015). Xylanases and ligninases are the enzymes commonly used by these industries (Maijala et al. 2008). Paper cleaning and improving drainage systems are done with the help of amylase enzyme (Kuhad et al. 2011). In chemical pulping processes, the enzyme laccase can be used in place of chlorine (Fu et al. 2005).

### 11.5.8 Role of Microbial Enzymes in Leather Industries

The wastes from the leather industries are very harmful to human health (Choudhary et al. 2004). Kamini et al. (1999) worked on the microbial enzyme technology as an alternative to conventional chemicals in leather industries. Dehairing of leather is a very important step. When it is done with the help of enzymes, the organic load into the effluent becomes very less and also the leather is not subjected to harsh chemicals. Most of the steps in processing like curing, soaking, liming, dehairing, bating, picking, degreasing, and tanning use these enzymes (Mojsov 2011). Nonfibrillar proteins can be removed with alkaline proteases to make the leather soft, and lipases help in removing the fats (Choudhary et al. 2004; Sharma et al. 2001). De-Souza et al. (2012) studied the use of enzymes in leather processing as compared to the use of chemicals or the use of coenzyme processes.

### 11.5.9 Role of Microbial Enzymes in Textile Industries

The textile industries produce very high amount of environmental pollution (Ahuja et al. 2004). The use of enzymes seems to be a promising. Enzymes like amylase, cellulase, cutinase, protease, pectinase, lipase, and esterase come under hydrolases. These are used for biopolishing of the fabric, anti-felting of wool, in making cotton soft, in finishing process of denim and wool formation (Vaishnav and Demain 2009). Enzymes like catalase, laccase, peroxidase, and ligninase (which come under oxidoreductases) help in bleaching and finishing (Mojsov 2011). Filipsen et al. (1998) have studied the mechanism to remove immobilized triacylglycerol by using lipolytic enzymes in a sequential laundry wash process.

### 11.5.10 Role of Microbial Enzymes in Cosmetics

Usage of enzymes as free radical scavengers has been used in sunscreen creams, toothpastes, mouthwashes, hair waving, and dyeing (Li et al. 2012). The superoxide dismutase (SOD, EC 1.15.1.1) helps in quenching the free radicals and controls the skin from damage. The use of SOD and peroxidase enzymes in sunscreen cream reduces erythema (Singh et al. 2016). For whitening the teeth, enzymes like endoglycosidase and papain are used in toothpastes. Enzymes are used in removing the teeth plaque and bad odor (Paqué et al. 2021). Many enzymes are used in hair dyes (Li et al. 2012). Protein films formed on the contact lens can be removed with the help of enzymes (Alfa and Jackson 2001).

### 11.5.11 Role of Microbial Enzymes in Detergents

The enzymes are used in removing the stains of proteins, starches, oils, and fats (Hasan et al. 2010). The global market was $5710 million in 2023, which is anticipated to increase upto $11,659 million in 2032. The enzymes help in removing the stains, bringing more whiteness to the clothes, and are soft on color and fabric. They are mainly of hydrolase group (amylase and protease) (Li et al. 2012). In dishwashing and laundry, the detergents are incorporated with the enzyme cutinase (EC 3.1.1.74). These enzymes are not harsh and are environmental friendly.

### 11.5.12 Role of Microbial Enzymes for Organic Synthesis in Industries

Many times, specificity of an enzyme is very essential during the usage. The cis- or the trans-form can give entirely a different result. Therefore, chemical synthesis is taking over. These synthetic enzymes are highly specific in chirality and the functional group (Schmid et al. 2001). This also makes the separation of the product much easier. Presently, lipases are used commonly to obtain optically active alcohols, acids, esters, and lactones (Hasan et al. 2010). Polyphenol oxidase is used in the synthesis of 3,4-dihydroxylphenyl alanine (DOPA), which is a chemical used in the treatment of Parkinson's disease (Poewe et al. 2010). The nitrile hydratase enzyme is used for the production of acrylamide. This is commercially produced by the Nitto Chemical Company, Japan. The company produces the compound acrylamide, on a large scale (40,000 tons per year) (Zaks 2001). Another product is high fructose corn syrup (HFCS), which is an alternative sweetener in the food and beverage industries. It is produced in millions of tons every year and uses the enzyme glucose isomerase during the production (Gerhartz 1990).

### 11.5.13 Role of Microbial Enzymes in Waste Management

Enzymes have a major role in the management of waste. They have the ability to degrade toxic pollutants like phenols, aromatic amines, and many more (Pandey et al. 2011). Commonly used microbial enzymes are amylases, amyloglucosidases, cellulases, glucoamylases, lipases, nitrile hydratases, pectinases, and proteases (Karigar and Rao 2011). Oxidoreductases, which include laccase, manganese peroxidase, lignin peroxidase, and tyrosinase, help in the removal of chlorinated phenolic compounds from industrial effluents (Le Roes-Hill and Prins 2016). Microbial oxygenases can degrade halogenated organic compounds (Fetzner and Lingens 1994). Clarke et al. (2000) studied the comparison of enzyme-aided bleaching of softwood pulp by using a combination of xylanase, mannanase, and α-galactosidase and found it very effective. Kalia et al. (2001) studied the use of enzymes in the recovery of oil from edible seeds.

**Growing Enzyme Market** The industrial enzyme market is growing every day at a rapid pace (http://www.sebi.gov.in/cms/sebi_data/attachdocs/1453372309087.pdf). This includes the usage by multinationals. In India, "Advanced Enzymes Technologies Ltd." is the largest manufacturer and exporter of enzyme products, with a share of 30% in the enzyme industry (in 2012). Novoenzyme company is at the top on the global level.

## 11.6 Properties of Enzymes

### 11.6.1 Enzyme Kinetics

The various factors involved for the rate of the reaction of the enzyme are studied in enzyme kinetics. These are of two types, namely, discontinuous and continuous. In discontinuous method, the substrate and the enzyme are mixed and kept for a particular time which was also optimized earlier. *Usually, the rate of reaction is measured based on the Michaelis constant ($K_m$),* which is defined as the amount of substrate concentration to be taken which will give half of the maximum velocity. In continuous enzyme assays, the enzyme is mixed with the substrate and the appearance of product is studied by measuring it over a time continuously. Many times, an artificial substrate is used which is called as a chromogen. This produces a colored product and can be measured easily on a colorimeter or a spectrophotometer. The activity of the enzyme depends on the pH of the buffer solution which should be optimised. The optimized pH needs to be maintained throughout the experiment till it is completed. When the level of the substrate becomes limited, the speed of the reaction goes down. Sometimes, the enzyme can become unstable and can get denatured, if there is a change in the pH of the mixture. The gradient can be measured (by dividing the concentration change to the time interval) in order to evaluate the rate of the reaction over a period. In case, the enzyme is added in higher quantities, the reaction rate becomes faster. It has been found that the amount of enzyme and the reaction rate have a linear relationship. It happens with the enzymes in vivo, which are used in biotechnological applications, and the reaction rate may be controlled by regulating the amount of enzyme present. When a range of different substrate concentrations are used, it has been found that initially the rate of reaction increases, which later goes down with further increase in the substrate concentration and if the substrate is still further increased it has no effect on the reaction rate. At this point, the enzyme becomes close to saturation with the substrate and demonstrates its maximal velocity ($V_{max}$). In enzymology, the Michaelis constant ($K_m$) is defined as the substrate concentration that gives half-maximal velocity. The reaction takes place via the formation of an ES complex which, once formed, can either dissociate (productively) to release the product, or could dissociate in the reverse direction without any formation of the product. In a steady state, although the concentration of the substrate and the product changes, the concentration of the ES complex itself remains constant. The rate of formation of the ES complex and the rate of its breakdown must be balanced, where Rate of ES complex formation = $k_1$[E][S] and Rate of ES complex breakdown = ($k_{-1}$ + $k_2$) [ES]. Hence, at steady state: $k_1$[E][S] = $k_{-1}$ + $k_2$[ES]. On the Lineweaver–Burk plot, there is a significant impact on the line of best fit estimated from the data, and therefore it is based on the extrapolated values both for $V_{max}$ and $K_m$. Enzymes are affected by pH and temperature. Environmental factors affect the rate of enzyme-catalyzed reactions through reversible or irreversible changes in the protein structure. Most enzymes have a characteristic optimum pH at which the velocity of the catalyzed reaction is maximum, and above and below which the velocity goes down. The pH inside a cell is usually different than the optimum pH of an enzyme. Due to this, the enzyme activity can get hampered. Temperature is another major factor that influences the enzyme action. With an increase in the reaction temperature, the movement of the molecules starts at a rapid speed. This influences the reaction rate to be fast. However, at high temperature, the denaturation of the enzyme proteins takes place which reduces the progress of the reaction. Therefore, it is vital to optimize the temperature conditions for optimum enzyme activity known as the temperature optimum ($T_{opt}$). It has been found that for most of the mammalian enzymes, 37 °C and pH of 7.4 are the optimum temperature and pH, respectively. The enzyme activity can suddenly change whenever there is a change in these parameters. When the reaction temperature goes above (>40 °C) or the pH (7.4) goes above, then the enzyme can get denatured. Johannes et al. (2006) found that the enzymes are usually susceptible to the substrates. The work also observed that product inhibition can cause allergic reactions.

### 11.6.2 The Effect of Temperature on Enzyme Activity

To study the effect of the temperature, the enzyme is exposed to a range of temperatures for a fixed period of time. The maximum enzyme activity is obtained at a particular temperature, and this is the optimum temperature for that particular enzyme. It is specific for each enzyme. Usually, the microbial enzymes are highly thermostable as they are obtained from extremely thermophilic microorganisms, such as thermolysin (a protease from *Bacillus thermoproteolyticus*) and Taq polymerase (a DNA polymerase from *Thermus aquaticus*).

### 11.6.3 Catalytic Properties of the Active Site

**Reversible inhibition** A molecule which has a similar structure as that of the substrate, it could bind reversibly to the active site of the enzyme as a competitive inhibitor.

**Competitive Inhibitors** When high concentrations of substrate are used, then they can be displaced from the active site and can restore the enzyme activity. They enhance the $K_m$ of a reaction because they increase the concentration of the substrate required to saturate the enzyme. However, they do not change the $V_{max}$ of the reaction. Sometimes, the higher concentration of the substrate or the product can become inhibitory. Fersht (1985) has explained the structure of enzymes and the mechanism of action.

**Noncompetitive Inhibitors** They react with the enzyme at a site that is different from the active site. The binding of the inhibitor does not physically block the binding site of the substrate, but it prevents any further reaction. They are not related to the substrate, and their inhibition cannot be overcome by increasing the substrate concentration. As they reduce the concentration of the active enzyme the $V_{max}$ of the reaction gets reduced, but it does not change the value of $K_m$.

**Uncompetitive Inhibitors** When a substrate molecule gets bound to the inhibitor, it binds to the enzyme resulting in the reduction of $V_{max}$ and $K_m$ of the reaction. Allosteric enzymes are the key regulatory enzymes that control the activities of the metabolic pathways by responding to the inhibitors and the activators.

### 11.6.4 High Substrate Concentration

**End Product Inhibitors** These are also known as irreversible inhibitors as they bind permanently to an enzyme and are potent toxins. Organophosphorus compounds such as diisopropyl fluorophosphate (DFP) are used as an organophosphate pesticide (parathione, malathione), which inhibits the acetylcholinesterase activity by reacting covalently with serine residues, which are found at the active site of the enzyme

**Enzyme Assay** This measures the enzyme activity within a sample. This is to identify the presence of an enzyme and investigate any specific enzyme kinetics or the activity of inhibition within a sample

### 11.6.5 Common Industrial Microbial Enzymes

Most of the times, these are batch fermentations, which take processing time from 30 to 150 h. Usually, fermentors with a volume of 10–100 $m^3$ are used. The substrates are fed into the reactor in a gradual mode. Carbohydrases and amylases are the enzymes that are used commonly for the hydrolysis of polysaccharides. Therefore, they are important for food and pharmaceutical industries. When amyloglucosidase acts on the starch, the glucose oligosaccharides bearing 1,4-linkage breaks down into glucose molecules. Amylases are used in laundry starches and are also used for removing the wall paper. Fungal-derived amylases have low thermal stability, but at lower temperatures, they act rapidly and help in rapid saccharification. The product yields become much higher due to the addition of the enzyme. It can also hydrolyze starch in sweet syrups. Amylases are widely used in baking industries which help in higher sugar production. It also increases the gas formation and improves the color of the crust. The amylases are utilized in the preparation of chocolate and licorice syrups. Fungal amylases are used in the preparation of clear, starch-free pectin and for the modification of starch in vegetable purees and canned vegetables. For sucrose hydrolysis, both yeasts and fungal invertases are used. However, the yeast invertase is a fructosidase that acts on the fructose, whereas fungal invertase is a glucosidase that acts on the glucose of the sucrose molecule.

Fungal invertase, namely, Raffinose, is utilized for making artificial honey and for making the invert sugar which is much more soluble than sucrose. The use of crude invertase prevents the crystallization in high test molasses, in the manufacturing of confectioneries, liqueurs, and ice creams where high sucrose concentrations can lead to crystallization. Lactose crystallization in ice creams can be prevented by using lactase enzyme, both in whole milk and whey concentrates. Industrially available proteolytic enzymes, namely, proteases, are produced by microorganisms. They are usually mixtures of endopeptidases and exopeptidases. Plant proteases like bromelin, papain, and ficin and the animal proteases like pepsin and trypsin are also widely used in bioprocesses. These proteases are very specific in splitting the peptide linkages. When a proper amount of protease is used, it reduces the mixing time and increases the extensibility of doughs, improving grain texture and loaf volume. To avoid stickiness in the dough, the amount of enzyme should be optimized earlier. Cereal foods are also treated with proteolytic enzymes to improve the product. In beer and ale, these enzymes are added to remove the haziness of the beverages. These contain pepsin, papain, bromelin, fungal, and bacterial proteases in different combinations that decompose the protein. They are also used to tenderize the meats. For meat tenderizing, a combination of plant, fungal, and bacterial proteases has shown to give better results as compared to any single enzyme used. Enzymatic hydrolysis of plant, meat, fish, and milk proteins provides protein hydrolyzates for condiments and animal feeds. Fungal proteases are usually used for digestion, whereas bacterial proteases are used for the treatment of wounds, inflammation, bruises, and blood clots. Many bacterial enzymes are used in the dry cleaning industries. These are used in leather manufacturing for bating and unhairing of hides. They are also used in the recovery of silver from photographic films by enzyme digestion and by solubilization of the gelatin coating. Pectinases are another important group of enzymes of microbial origin that includes pectinesterase and polygalacturonase. Most of the commercial pectin enzymes are a mixture of both. Pectins are colloidal in nature, making solutions viscous and holding other materials in suspension. Pectinesterase removes the methyl groups from the pectin molecules, which exposes the carboxyl groups. In combination with bi or multivalent cations like calcium, these carboxyl groups form insoluble salts that can be removed easily. Polygalacturonase degrades macromolecular pectin, which reduces the viscosity and destroys the protective colloidal action to settle the suspended materials. They increase the fruit juice yields both during crushing and grinding. Grape wines can be clarified faster with better color. The haziness in fresh cider is difficult to remove. However, the use of pectin enzyme makes it very easy. Usually, before concentration, the juices are depectinized and filtered. They are also used in removing the gelatinous coating from coffee beans. Other common hydrolytic enzymes are cellulase, hemicellulase, and pentosanase. Cellulases are used in tenderizing cellulosic food products and in recovering the cellulosic wastes. Hemicellulases can hydrolyze many gums. Use of pentosanase enzymes has shown to increase the yields of starch from wheat. Among the nonhydrolytic enzymes, two enzymes, namely, glucose oxidase and catalase are commonly utilized. Glucose oxidase, which is of fungal origin, helps in the conversion of glucose to gluconic acid and hydrogen peroxide. Commercially, it is used for diabetic patients as paper strip test, which indicates the presence of glucose in the urine by a color change when the strip is dipped into the sample. Ghosh et al. (1996) studied the role of microbial lipases, their production, and applications.

### 11.6.6 Role of Enzymes in Packaging for Enhancing the Product Shelf Life

Enzymes help in the removal of sugar from the eggs which enhances their shelf life and prevents browning. In canned foods, oxygen is a big problem which is removed with the help of the enzyme glucose oxidase. Canned soft drinks have major problems of color loss, alteration in flavor, and corrosion of the can. However, addition of small amounts of glucose oxidase to canned soft drinks has shown an increase in the keeping quality and reduction in corrosion of the cans. In cheese products, the enzyme, glucose oxidase is coated on the inside of the wrapper where it is in direct contact with the cheese. Deoxygenation of food packets is used widely where they are made of a polyethylene film, which is impermeable to water, but could diffuse oxygen. Work has been done to protect the dried and dehydrated products containing fats and other oxygen-sensitive materials. These enzyme packets are a practical solution for packaging dried foods and other items which have a short shelf life and usually get spoilt due to the rancidity of fat or other oxidative changes. Bacterial catalase helps in the cold sterilization of milk especially in cheese processing for removing the residual hydrogen peroxide before it is processed into cheese. Microbial enzymes are natural in origin, nontoxic, have action specificity, and can work under moderate temperatures and near neutral pH. They are preferred in commercial applications and are needed at low concentrations. They can be controlled by optimizing the parameters. Microbial enzymes are commonly used as they are economically cheaper, can be produced in high yields, and have good consistency. They can easily be modified and optimized. The use of microbial enzymes has added advantages, as the supply is regular, they are not dependent on seasonal fluctuations and can be grown easily on inexpensive media. They are stable and have greater catalytic activity. All these properties make microbial enzymes very promising in the fields of diagnosis, treatment, biochemical investigation, and monitoring of many diseases.

### 11.6.7 Selection of Microorganisms for Maximum Enzyme Production

Most classical genetic methods are used for the selection of high-yielding microbial variants. This leads to an economically feasible industrial process. Screening of a large number of microorganisms is required to produce novel products. Advances in genetic and protein engineering have an impact on the screening of microorganisms.

## 11.7 Mechanisms of Enzyme Biosynthesis

The substrate gets converted into product by a multi-step, enzyme-catalyzed process in which simple compounds form macromolecules.

### 11.7.1 Inducible Enzymes

These are utilized for breaking down various compounds inside the cell.

### 11.7.2 Repression Mechanism

It is a regulatory mechanism in which the enzyme for expression of genes and for repression are the same. In an inhibitory feedback system, a substance produced in a metabolic pathway accumulates and represses the synthesis of the enzyme required; therefore, it is called as a repression mechanism.

### 11.7.3 Enhancing Enzyme Biosynthesis

Even when the genes involved in the synthesis of an enzyme are not known, still the yields can be increased by mutations on the chromosomes using ultraviolet (UV) irradiation or mutagens. The conditions for culturing can be optimized to obtain higher yields. Metabolic engineering can be helpful for increasing the microbial production of primary metabolites. A gene from the genomic data can be searched, on the basis of the N-terminal amino acid sequence and molecular weight. This helps in the identification of an enzyme-coding gene. The yield can be increased by knockout of a reaction which can either degrade or convert the target metabolites. If the production of the coenzymes such as ATP, NADH, and NADPH are increased they can help in enhancing the yields of primary metabolites. Production of ricinoleic acid and a hydroxylated fatty acid (an alternative raw material for various petrochemical industrial products) could be enhanced with the over expression of phospholipase A gene in *Schizosaccharomyces pombe* (Tamano 2014). Promoters can overexpress the genes that encode with the enzyme, ultimately producing high yields of the product. Sometimes, heterologous overexpression is also used. Many Gram-negative (*E.coli*) and Gram-positive bacteria (*Bacillus subtilis, Lactobacillus lactis*), yeasts (*Saccharomyces cerevisiae, Pichia pastoris, Hansenula polymorpha,* and *Yarrowia lipolytica*), and filamentous fungi (*A. oryzae, Aspergillus niger,* and *Trichoderma reesei*) are used for heterologous overexpression (Frenzel et al. 2013). Genome sequencers are commercially available. The SOLID system by Life Technologies and the Genome Sequencer FLX system by Roche are available. Genomic DNA of many microbial species and strains has been sequenced and registered in databases. It is important to predict the genes involved in the synthesis, from the data that already exists. This could be by the comparative genomic approach, prediction by the web application for sequence homology (BLAST), or by genome-wide profiling using "Omics" information which is obtained from the transcriptome, proteome, and metabolome. The COBRA toolbox is equipped with some programs like the flux balance analysis (FBA), which is involved in this technology.

### 11.7.4 Cultivation Techniques

The microbial enzymes can be obtained by growing the cultures either by submerged fermentation (SmF) or solid-state fermentation (SSF). When the microbes are in (SmF) in the liquid medium, the culture is constantly in contact with the nutrients. In this condition, the cultures receive high amount of oxygen. However, in anaerobic cultivation methods, the culture is devoid of oxygen supply. Anaerobic fermentation is done under specific laboratory conditions. It involves growing of the culture in an anaerobic hood and anaerobic fermentors. All these conditions make the anaerobic processes very challenging.

## 11.8 Development of Microbial Enzyme Production at Large Scale and Some of the Existing Industrial Processes Using Microbial Enzymes

Enzymes are essential components of animals, plants, and microorganisms, and they catalyze and coordinate the complex reactions of cellular metabolism. Till 1970s, the commercial production of enzymes was only through animal and plant sources. However, with a lot of scientific works on microbial cultures, it was found that enzymes could be produced from microbes at a large scale and were found to be very economical. The production of enzymes by microorganisms in a liquid nutrient media is known as submerged fermentation. The starch hydrolytic amylases (α-amylase, β-amylase, and glucoamylase) are very common. Bacteria and fungi tend to secrete amylases outside the cells to perform extracellular digestion of starch into sugars. The first enzyme was produced in the United States (1896) known as takadiastase which helps in many digestive disorders (Britannica 2024). Yeasts contain a natural enzyme known as zymase that can catalyze the fermentation of sugars into ethanol and carbon dioxide. These enzymes are commercially produced mainly by submerged fermentation in fermentors. Fermentation conditions can be optimized, for the higher yields of enzymes to be produced. Enzymes are biocatalysts which form parts of the metabolic processes of the cells. A concerted action of many enzymes helps in the performance of the metabolic reactions in any living cell. All purified enzymes are proteins. It was the work of Takamine (1894, 1914) to show that the enzymes can be cultivated and can be utilized by the industries. These cultivated enzymes have replaced animal and plant enzymes at a very rapid pace. In textiles, bacterial amylases have been replaced with malt or pancreatin. Each organism produces various types of enzymes like; hydrolyzing, oxidizing/reducing, and others related to metabolism. Microbial enzymes are produced from different strains of fungi, bacteria, and yeasts using both surface and submerged culture methods. For fungal enzymes, the fungi are usually cultivated on the surface of a solid substrate as wheat bran. The bran is steamed for sterilization, then it is cooled, inoculated with the fungal spores, and then spread out on the trays (Underkofler et al. 1947; Forbath 1957). Incubation takes place in chambers where the temperature and humidity can be controlled within the limits by circulating the air. The submerged method was originally developed for the production of penicillin and other antibiotics. The submerged cultures are grown in shake flasks/aerated tubes/flasks. The tanks have provision for sterile air and agitation. The amount of air, degree of dispersion of air, and the amount of agitation are all dependent variables that need to be optimized. In the bran process, the enzyme is extracted from koji into an aqueous solution by percolation. In the liquid process, the microbial cells are filtered. Precipitation of the enzyme is done with the addition of the solvents, like acetone or aliphatic alcohols, to the aqueous enzyme solution, directly or after concentration by vacuum evaporation at low temperature. Then, the enzyme is filtered and dried at low temperature in a vacuum shelf dryer. For dilution, there are diluents like salt, sugar, starch, and wheat flour which are used. It has been observed that commercial enzymes are very stable in the dry form. Sometimes, the addition of stabilizers and activators is required for the maximum stability and efficiency. It is vital for the microbial enzyme manufacturers to have a selected culture, the composition of media, and the culture conditions. Losses in fermentation batches are usually because of contamination, culture variation, and failure of cultural control

The most widely used microbial enzymes by industries are as follows:

**Amylases** Broadly these are classified as $\alpha$, $\beta$, and $\gamma$ subtypes. $\alpha$-Amylase acts faster than $\beta$-amylase. They act on $\alpha$-1-4 glycosidic bonds and are also called as glycoside hydrolases. There are two types of amylases, namely, endoamylases and exoamylases. In endoamylases, the hydrolysis takes place within the starch molecule forming linear and branched oligosaccharides. In exoamylase, the substrate is hydrolyzed at the nonreducing end. According to Saboury et al. (1998, 2002), $\alpha$-amylases are metalloenzymes and need metal ions (calcium) for their stability and activity. Optimization of pH, temperature, carbon, and nitrogen sources is important for obtaining the maximum yield. Using genetic engineering, it has been possible to enhance the enzyme yields. Many bacterial species also produce amylase (*B. subtilis*, *B. stearothermophilus*, *B. amyloliquefaciens*, *B. licheniformis*, *B. coagulans*, *B. polymyxa*, *B. mesentericus*, *B. vulgaris*, *B. megaterium*, *B. cereus*, *B. halodurans*, and *Bacillus* sp. Ferdowsicous). Many halophilic strains produce amylases (*Haloarcula hispanica*, *Halobacillus* sp., *Chromohalobacter* sp., *Bacillus dipsosauri*, and *Halomonas meridiana*). Three-dimensional structural analysis has helped in improving the efficiency and in the alteration or mutation of particular amino acids to improve the efficiency and functions of the enzyme or protein. Amylases from fungi are extracellular which makes the process economically feasible. Genus of *Aspergillus* (*A. oryzae*, *A. niger*, *A. awamori*, *A. fumigatus*, *A. kawachii*, and *A. flavus*), as well as *Penicillium* species (*P. brunneum*, *P. felluta-*

*num*, *P. expansum*, *P. chrysogenum*, *P. roqueforti*, *P. janthinellum*, *P. camemberti*,and *P. olsonii*), *Streptomyces rimosus*, *Thermomyces lanuginosus*, *Pycnoporus sanguineus*, *Cryptococcus flavus*, *Thermomonospora curvata*, and *Mucor* sp. are used for the commercial production of this enzyme.

The enzyme α-amylase is helpful when added to the dough. During baking, it converts starch molecules into smaller dextrin molecules. These are then fermented by yeast during the baking process. The overall bread qualities have been found to be better (Aravindan et al. 2007).

The enzyme produced from *Bacillus amyloliquefaciencs* and other species of *Bacillus* spp. are excellent for the saccharification of starch (van der Maarel et al. 2002).

In ethanol production, initially starch is used as the sugar which gets converted into fermentable sugars with the help of α-amylases. This enzyme is also used in the clarification of fruit juices.

**Glucoamylases (EC 3.2.1.3)** The polysaccharide starch gets hydrolyzed with this enzyme and releases β-glucose. *Aspergillus niger* and *Aspergillus awamori* are the main sources of these saccharifying enzymes. The activity remains good at lower temperatures, however, they lose activity at higher temperatures. They are widely used by food industries especially in the production of high-glucose syrups and high-fructose syrups. When added to the flour in bread making, it helps in converting the starch into maltose and fermentable sugars. Glucoamylases are important in the production of sake, soya sauce, and light beer.

**Recombinant Amylases** An efficient amylase gene is inserted after selection into an appropriate vector system for transformation in an efficient bacterial system. This produces recombinant protein in higher amounts. In *P.furiosus*, the gene expression in *B. amyloliquefaciens* has helped in enhanced amounts of enzyme with stability at higher temperature and lower pH (optimum pH 5.0). Around the globe, amylase enzyme makes around 25% of the market as it is widely used in foods, detergents, pharmaceuticals, paper and textile industries. In the food industries, the enzyme is used as corn syrups, maltose syrups, glucose syrups and in the preparation of juices, alcohol fermentation, and baking. It is commonly used in detergent industry, beer, and liquor brewing. Both submerged fermentation and solid-state fermentation can take place. Most of the times the crude amylase extract is good, but for specific industrial applications, purification of the enzyme is needed. Automation and the use of improved genetically modified strains, in the processes, have made them very cost-effective and to be used often.

**Lipases** Microbial lipase market has been estimated to be USD 590.2 million in 2023. The enzyme is used in the production of biodiesel, foods, drinks, leather, textiles, detergents, pharmaceuticals, and medicals. They act on carboxylic ester bonds and do not require any cofactor and have molecular weight in the range of 19–60 kDa. Metal ions like Co, $Ni^{2+}$, $Hg^{2+}$, and $Sn^{2+}$ inhibit the lipase activities drastically and $Zn^{2+}$, $Mg^{2+}$, EDTA, and SDS inhibit it to a lesser extent.

Lipstatin is a β-lactone molecule that controls the digestive activity of pancreatic lipases. It is a FDA-approved antiobesity medication. Panclicins produced by *Streptomyces* sp. NR 0619 inhibits N-formylalanyloxy or N-formylglycyloxy substituent. Valilactone isolated from *Streptomyces albolongus* MG147-CF2 strain is known to inhibit esterase activity. Ebelactone A and B are isolated from the fermentation broth of Actinomycetes strain G7-Gl. Esterastin, caulerpenyne, vibralactone, and percyquinin are other lipase inhibitors. Yeast and fungal phospholipases are used in the degumming of vegetable oils.

Among the bacterial genera, organisms used for the production of lipases and phospholipases are *Pseudomonas*, *Bacillus*, *Streptomyces*, *Burkholderia*, *Chromobacterium*, *Achromobacter*, *Alcaligenes*, and *Arthrobacter*.

*Achromobacter* sp., *Alcaligenes* sp., *Arthrobacter* sp., *Staphylococcus* sp., and *Chromobacterium* sp.

Major filamentous genera of fungi which have low-cost extraction methods are *Rhizopus, Aspergillus, Penicillium, Mucor, Ashbya, Geotrichum, Beauveria, Humicola, Rhizomucor, Fusarium, Acremonium, Alternaria, Eurotrium* and *Ophiostoma*, *Candida rugosa*, *Candida antarctica*, *T. lanuginosus*, *Rhizomucor miehei, Pseudomonas, Mucor*, and *Geotrichum*. The gel filtration and affinity chromatography are commonly used for the purification, which is followed by precipitating with 60% ammonium sulfate and 35% ethanol. The natural substrate for lipases is a chiral alcohol molecule.

Many flavors like acetoacetate, β-keto acids, flavor esters, methyl ketones, and lactones are synthesized due to the free fatty acids. When there is a change in the location of fatty acid chains the enzyme lipase becomes helpful in changing the lipids into the glyceride form. Specialty fats can be obtained by esterification and interesterification. Interesterification of fat and oil which is done chemically is nonspecific and requires a lot of energy. However, in cases of incorporating nutritionally important polyunsaturated fatty acids (PUFAs) in oils and fats, procedure of enzymatic interesterification is required. However, the interesterification of triglycerides using immobilized lipase has not been found to be commercially viable as the processing costs are very high. The vegetable oils such as palm, rapeseed, canola, and sunflower are being used by food producers to regulate the physical properties of fats and oils since every oil has several types of fatty acids in the 1, 2, 3 positions of triacylglycerides. Microbial lipases, with (1, 3) regio-specificity, are used in the production of cocoa butter-type triacylglycerols. In the cosmetic industries, isopropyl myristate, isopropyl palmitate, and 2-ethylhexylpalmitate are being used for dermal and in sun-tan ointments, bath oils, etc. using lipases. Wax esters (esters of fatty acids and fatty alcohols) are also being enzymatically manufactured (Croda Universal Ltd.) in a batch bioreactor. The company uses *C. cylindracea* lipase. Production of retinoids (Vitamin A and their derivatives) is done with the catalytic reaction of immobilized lipase with the water-soluble retinol derivatives. Lipases are also used for topical antiobese emulsions/oral administration. Tea, made from the *Camellia sinensis* L., can be either unfermented as green tea, semi-fermented as oolong tea, and fermented tea as black tea. In black tea, the enzymes break the membrane lipids to develop the volatiles.

Lipases are used in biosensors. There are two important parts in a biosensor, namely, one is a measurable signal and the second is a biological one. Most of the times, there is a common substrate used, which is tributyrin in many bacteria (*Bacillus subtilis, Chromobacterium viscosum*) and fungus (*Rhizomucor miehei, R. oryzae,* and *Fusarium solani*) which gets hydrolyzed to dibutyrin and butyrate. This is measured as lipases. Immobilized lipases are also used for quantification. Lipases play an important role in fats and oils, soft drinks, drug industries, beverages, and also in medical diagnosis. A surface acoustic wave impedance biosensor has been developed to determine organophosphorous pesticide, using the lipase hydrolysis method which can also analyze dichlorvos insecticide residues in vegetables. Analysis of methylparathion and tributyrin is done using potentiometric biosensor, which is based on *C. rugosa* lipase. Work has been done to purify lipase from *C. rugosa* which has also been immobilized on a glass electrode and has been cross-linked with glutaraldehyde. Recently, enzymes are being incorporated with nanomaterials synergistically with an apparatus that has high sensitivity and stability. This is based on nanotechnology biosensors. It enhances the electrochemical reaction, giving a nanobarcode for biomaterials, and signal amplifying of biorecognition event. Carbon nanotubes (CNT) and gold are used as nanomaterials in these biosensors as they have high catalytic ability for many organic reactions.

On industrial scale, lipases are used for developing a characteristic flavor in dairy products. In cheese, lipase produces a unique flavor by acting on the milk fats. This produces free fatty acids after hydrolysis. When lipases are used from different sources, they produce different types of cheese. When a kid or lamb pre-gastric lipase is used it produces romano cheese. Lipase from *Penicillium camemberti* produces camembert cheese, and when *Aspergillus niger* or *A. oryzae* are used for cheese making, it produces cheddar cheese. When the enzyme lipase is used for catalytic reaction, it has shown to improve the texture and softness of cheese. Flavor development has also been observed in butter and margarine along with enhancing the shelf life of the food products (Aravindan et al. 2007). Even in alcoholic beverages like wine, this enzyme lipase has been found to improve the flavor. Unilever has patented the use of immobilized *Rhizopus miehei* lipase. This has the ability to replace palmitic acid with stearic acid which is a healthful preposition (Colman and Macrae 1980).

*Microorganisms like Helvina lanuginosa, Pseudomonas* sp., and *Geotrichum candidum*. have been utilized to better the sunflower oil. Detection of triglycerides and blood cholesterol has been found to be easier when immobilized lipase is impregnated onto pH/oxygen electrodes along with glucose oxidase (Hasan et al. 2006). Microbial lipases are also used in the production of ice creams, single-cell proteins, and carbohydrate esters.

**Phospholipases** Phospholipases are enzymes that break the phospholipids into fatty acids. This enzyme breaks down lipids by different methods. The enzyme is commonly used in food industries. It is used in oil production, bakeries, and dairy industries. Phospholipase is commercially produced using *Fusarium oxysporum.* It is marketed by company Novozymes A/S (Denmark) especially for baking applications under the name LipopanF®. Another phospholipase is commercialized by DSM Food Specialties (The Netherlands) under the trade name Maxapal® A2 (Zhao et al. 2010). During dough making the enzyme helps between starch and gluten. LysoMax® (DSM Food Specialties), is another commercial phospholipase that is used by baking industries and acts on lecithin (Sirbu and Paslaru 2006). Many studies have shown that when phospholipases are used in the making of cheese, milk, butter, and ice cream, it enhances the yield of the product.

The texture of baked foods can be improved by using lipase enzyme, which is usually provided by the emulsifying properties of the egg lipids (Stemler and Scherf 2022). There are three major companies, namely, Nestlé, Kraft, and Unilever, which have a huge global market for lipase enzyme. Lipase enhances the flavor in cheeses, fastens the ripening of cheese, and is used in cream lipolysis.

Wine aroma-containing esters like ethyl decanoate, ethyl 2-methyl-propionate, ethyl 3-methyl-propionate, ethyl 3-methylbutanoate, ethyl cinnamate, methyl-butyl acetate, 2-phenyl–ethyl acetate and hexyl acetate, 2-ethyl hidroxpropionato, ethyl butanoate, ethyl hexanoate, and octanoate are obtained during aging of the beverage formed by chemical esterification with the help of lipase (Esteban-Torres et al. 2014). In wine making, using a genetically improved strain of *Escherichia coli* BL21, with a gene of lipase/esterase enzyme, has shown to have higher yields of alcohol. The lipid fraction in infant formulations is a source of nutrients and has antiviral and antibacterial activities. Lipases are used in the meat and fish industries for the removal of excess fats. Lipases help in the formation of biopolymers. The enzyme is used in agriculture applications, environment cleaning, and effluent treatment. Effluent can be removed both by aerobic and anaerobic methods using lipases (Rigo et al. 2008). The genetically improved microorganisms have given much better results. Lipases are utilized for bioremediation in oil spills, oil-wet soils, industrial wastes, and lipids from wastewater. Commonly, *Staphylococcus pasteurii* COM-4A, *Bacillus subtilis* COM-B6, and *Arthrobacter* sp. produce lipase. *Geotrichum candidum* is used for the treatment of wastewater from olive mills. The microorganism, namely, *Y. lipolytica* for wastewater treatment and *Mrakia blollopis* SK-4 for remediation of milk fat, have been found successful. Symbiotic cultures of Yeast and bacteria have been found good for the biodegradation of oil and gas. *P. putida* GPo1 alkB; *Acinetobacter* spp. alkM; *Rhodococcus* spp. alkB1, and *Rhodococcus* spp. alkB2 can degrade n-alkanes. Lipases help in increasing the pulping rate in the paper industry. Lipases are used in the cleaning of sheep skin in place of chemical solvents, which are harmful. Degreasing is an important step during the processing of fatty raw ingredients. Lipases are used for the biodegradation of plastics.The major organisms used are *Pseudomonas protegens* BC2-12, *P. protegens* CHA0, *P. protegens* Pf-5, *P. fluorescens* A506 and Pf0-1, *P. chlororaphis.*

Lipases have the ability to degrade polymers. Enzymes obtained from *Penicillium citrinum, Thermobifida fusca, Fusarium solani pisi*, and TLL, when, treated with the linear aromatic polyester poly (trimethylene terephthalate) have been found effective. Lipases can be used through stereospecific hydrolysis for the resolution of racemic mixtures to synthesize the chiral building blocks for pharmaceuticals, agro-chemicals, and pesticides which are in a lot of demand. A thermostable lipase can catalyze bioenergy and pharmaceuticals and esterify palm oil to Fames. They can also prevent hair loss and scalp diseases. For the synthesis of enantiopure compounds *Bacillus* spp., lipases are used in the pharma industries. Lipase helps in knowing the level of TB infection with specificity and sensitivity. The level of lipase in blood serum helps in the detection of acute pancreatitis. They are used in anti-obese creams and in the cure of malignant tumors as lipases initiate the tumor necrosis factor (TNF). Lovastatin, obtained from *Candida rugosa* lipase, helps to reduce the serum cholesterol level. Microbial lipases (EC 1.10.3.2) are used in the manufacture of (2R, 3S)-3-(4-methoxyphenyl) methyl glycidate (a key intermediate for diltiazem) and 3,4-dihydroxylphenyl alanine (DOPA) to cure Parkinson disease. Lipase (EC 3.1.1.3) isolated from *Candida antarctica* (CalB) is used as catalyst in the preparation of cosmetics, pharmacological, and food products as they possess regio-, chemo-, and enantioselective properties. Immobilized CalB is being used in the preparation of Odanacatib, which is a potent cathepsin K inhibitor and can cure osteoporosis in women after menopause. Hepatitis C is an infectious disease affected by Hepatitis C virus (HCV), which is a chronic liver disease and leads to cirrhosis. Lipases are useful as marker enzymes in the medical sector. In *Propionibacterium acnes*, lipase is the pathogenic factor. Lipases from pathogenic bacteria like *P. acnes*, *Corynebacterium acnes*, and *Staphylococcus aureus* produce skin rash in acne patients. Increase in triglycerides (TG) in the serum may give rise to colorectal and pancreatic cancers or precancerous lesions. Malabsorption of fat can decrease the caloric intake. This can cause cystic fibrosis due to the deficiency of fatty acids along with insufficiency in the exocrine pancreatic juice.

The use of free lipases in biodiesel production has been successful at large scales. It is produced from *P. fluorescens* (Lipase AsK, Amano), *B.cepacia* (Lipase PS, Amano) and *T. lanuginosus* (Lipase LA201 and Lipopan 50BG, Novozymes), and immobilized lipases from *T. lanuginosus* (Lipozyme TL IM, Novozymes) and *R. miehei* (Lipozyme RM IM, Novozymes). On large scales, anhydrous alkaline lipase is used to remove the fatty acids from the surface of wool fibers. Even the dewaxing of silk fibers is done with the help of lipase enzymes. The use of lipase enzymes has been a boon which reduces the burden of pollution in the wastewater.

**Lipase inhibitors from microbial sources** Many microbial products have potent pancreatic lipase (PL) inhibitory activity.

**Lipoxygenases (LOX)** This enzyme helps the polyunsaturated fatty acids in lipids to get deoxygenated. They produce precursors for volatile and aroma compounds. The bacterial LOXs vary in their specificity for fatty acids as the oxygenation takes place at a specific site.

In bread dough, this enzyme helps in lightening the color due to the oxidation of the pigment molecules. The main applications of LOXs in dough are based on their ability to bleach the flour pigment carotenoid, by the co-oxidation of pigment along with fatty acids (McDonald 1979). The enzyme improves the handling properties of the dough and the gluten gets strengthened and improves the rheology.

**Cellulases** This enzyme hydrolyzes β-1,4 linkages of cellulose molecule and breaking them into simple glucose molecules. These are of three types. Endo-(1,4)-β-D-glucanase (EC 3.2.1.4), exo-(1,4)-β-D-glucanase (EC 3.2.1.91) and β-glucosidases (EC 3.3.1.21) (Schülein 1988). Based on the sequence of amino acid arrangements, the enzyme comes under glycoside hydrolase (GH) family under which they are put into different groups.

Endoglucanases cleave β-1,4-bonds and expose the non-reducing and reducing ends of cellulosic polymer. Endoglucanases can be obtained from different microorganisms. They belong to different glycoside hydrolase families. It has been found that most of the fungal endoglucanases are catalytic along with a carbohydrate-binding site. However, in bacteria, many such sites are present. Regulation of the cellulolytic property of microbes is done by the feedback inhibition of product glucose. On industrial scale, cellulases are produced from bacteria and fungi. The bacteria used can be either aerobic or anaerobic in nature. However, in anaerobic bacteria, the pathway is different as the cellulosomes are located on the cell surface. These enzymes are majorly used in the food industries (Sukumaran et al. 2005). During juice extraction, this enzyme is added along with other enzymes which reduces the viscosity of various fruit purees. Better flavor and aroma can be obtained in fruits and vegetables when they are treated with the enzyme β-Glucosidase along with pectinase (Humpf and Schreier 1991). The enzyme has the ability to bring down the bitterness in citrus fruits besides improving their aroma and taste (Sajith et al. 2016). Basically, this enzyme is helpful in maceration, color development, clarification, stability, and quality of the product. The aroma of wines can be improved with β-glucosidases through modifications of glycosylated precursors (Raveendran et al. 2018).

### 11.8.1 Xylanases

This microbial enzyme helps in cleaving the xylans which is a major constituent of the hemicellulose. Xylan in hemicellulose can be broken down by using three major enzymes, namely, endoxylanases, exoxylanases, and β-xylosidases. The β-1,4 bonds of xylan get broken down with endoxylanases (EC 3.2.1.8). Hydrolysis of β-1,4 bonds of xylan which is performed from the non-reducing ends is done by exoxylanases (EC 3.2.1.37). However, the breakdown of xylobiose and xylooligosaccharides into xylose is done using β-xylosidases (Sukumaran et al. 2009). These enzymes are usually produced by actinomycetes, bacteria and fungi. Among the actinomycetes and bacterial species majorly they are *Streptomyces* sp., *Bacillus* sp., and *Pseudomonas* sp.. The activity is effective over a large range of pH (5.0–9.0) and temperature (35 and 60 °C). Among fungi, xylanases are produced by *Aspergillus* sp., *Fusarium* sp., and *Penicillium* sp.. The enzyme is produced extracellularly (Nair et al. 2008)

Along with other enzymes, the use of xylanase is common in bakeries. It helps in increasing the volume of the bread, also the hemicelluloses in the wheat flour get broken down with the help of xylanase enzyme. This makes the dough very soft. Xylanase is used in biscuit making to improve the texture, taste, and palatability. Fruit juice extraction, improvement in aroma chemicals, and other micronutrients have been observed when xylanase enzyme is used. It is also beneficial in beer industries, for hydrolyzing the cellular wall of barley

### 11.8.2 Pectinases

This enzyme helps in the hydrolysis of the glycosidic bonds in pectic polymers. The polygalacturonases hydrolyze the glycosidic α-(1-4) bonds). The enzyme pectin esterase removes the acetyl and methoxyl groups from pectin (Saadoun et al. 2013). Pectinase enzymes are utilized by the food and paper industries. The enzyme reduces the turbidity and haziness in fruit juices. It also improves the color and flavor in drinks (Tapre et al. 2014)

### 11.8.3 Glucose Oxidase

Glucose oxidase (EC 1.1.3.4) is classified under oxidoreductases. It is a flavoprotein that was initially discovered by Müller (1928). Under aerobic conditions, this enzyme has the ability to convert glucose into gluconic acid. Initially, β-D-glucose gets oxidized to gluconolactone and then it gets hydrolyzed to gluconic acid. However, the molecular oxygen gets reduced to hydrogen peroxide. The enzyme has two polypeptide chain subunits of similar nature which are bound with a disulfide bond covalently along with one flavin adenine dinucleotide (FAD) molecule which is bound in a non-covalent fashion. The enzyme is produced by *Aspergillus niger* and *Penicillium glaucum* on an industrial scale (Ramachandran et al. 2006). This enzyme is used by many industries such as pharmaceutical, baking, and food. It is also used in the biofuel cells. The enzyme is very helpful in removing glucose and oxygen in the making of diabetic drinks. Many properties in food products can be improved by using this enzyme. These are the color, texture, flavor, and shelf life. In food packaging, the enzyme helps in removing oxygen resulting in increasing the storage life of the product (Hanft et al. 2006)

### 11.8.4 Laccase

Laccase (EC 1.10.3.2) enzyme is classified under oxidases with a subgroup of multicopper enzyme. They oxidize phenolic compounds which are helpful in many industries (Morozova et al. 2022). The major uses of this enzyme are; in the chemical synthesis of compounds, in biobleaching of paper, in bioremediation, in biosensors, wine, and textile. The mode of action is mainly to oxidize phenolics, aromatic amines, and ascorbic acid (Madhavi and Lele 2009). Laccase enzyme is mainly produced by Deuteromycetes, Ascomycetes, and Basidiomycetes (Sadhasivam et al. 2008). Using genetic engineering technique, a recombinant laccase has been produced from *Bacillus licheniformis* which has been taken for industrial applications (Tonin et al. 2016). The enzyme improves the color in food and beverages and for the stabilization of wine. The enzyme is used as an agent for oxygen removal in beer industries which enhances the shelf life of the product. The polyphenols can be oxidized easily by using this enzyme. An enzyme company, namely, Novozyme produces the laccase enzyme under the brand name Flavorstar. Flavorstar is helpful in removing any off-flavor formed in the brewing industry and in bread making it improves the quality of the flour

### 11.8.5 Catalase

Catalase (EC 1.11.1.6) is an enzyme that helps in the breakdown of hydrogen peroxide. It is commonly produced by microorganisms like *Aspergillus niger* and *Micrococcus luteus*. Many potent microorganisms have been isolated from various sources which produce catalase enzyme. The enzyme is used in textile industry to remove hydrogen peroxide. It enhances the food preservation time along with the keeping quality. In wine industry, it helps in removing oxygen when used along with other enzymes. In milk processing, the enzyme removes peroxide (Sîrbu 2011). When this enzyme is incorporated in food wrappers, it helps in preventing oxidation which in turn, increases the shelf storage of food.

### 11.8.6 Peroxidase

This is an oxidoreductase enzyme classified as peroxidase (EC 1.11.1.7). As the prosthetic group iron (III) and protoporphyrin IX are present.

With this enzyme, the peroxide gets reduced. It is also responsible for the oxidation of many inorganic and organic compounds. The molecular mass is in between 30,000 and 150,000 Da. This enzyme is produced by microorganisms, plants and animals. Peroxidase enzyme from microorganism, *Phanerochaete chrysosporium*, is very well characterized (Janusz et al. 2013). This enzyme is found in Firmicutes, Proteobacteria, actinobacteria, and acidobacteria (Ventorino et al. 2015). It was initially isolated from *Streptomyces viridosporus* T7A (Ramachandra et al. 1988). Many food industries use this enzyme for better flavor, color, texture, and nutritional quality. It is also used in biosensors, polymer synthesis and to manage pollutants (Bansal et al. 2013). While using the enzyme, it is important to have caution on the undesirable browning of fruits and off-flavors in vegetables. Therefore, the concentration to be used should be optimized.

### 11.8.7 A-Acetolactate Decarboxylase

This enzyme helps in the beer industry for the rapid maturation of the product (Godtfredsen and Ottesen 1982). It is found in *Brevibacillus brevis. With genetic engineering,* a recombinant, *Saccharomyces cerevisiae,* is designed which has the ability to produce this enzyme (Dulieu et al. 2000). Choi et al. (2015) worked on beer and were successful in bringing down the maturation time. It was found that the usual time of maturation of beer (which is usually 2–12 weeks) could come down to just 24 hours when α-acetolactate decarboxylase enzyme was used in the process. It was also found that the off-note which gets developed due to the presence of diacetyl gets nullified by this enzyme. Immobilized enzymes have an added advantage as they bring stability, can be reused, and enhance the efficiency of the reaction.

### 11.8.8 Asparaginase

This enzyme is widely used by diverse industries. The enzyme breaks down asparagine into aspartic acid and $NH_3$. Although asparagine is a nonessential amino acid for human beings, it inhibits the malignant cells or cancer-growing cells. Therefore, this enzyme acts as an anticancer agent (Krishnapura et al. 2016). During frying/baking in oil (at high temperatures), asparagine gets converted into acrylamide. This is a cancer-causing compound which could be prevented with the addition of asparaginase. The enzyme asparaginase has been found to inhibit the formation of acrylamide to a large extent (Mohan Kumar et al. 2014).

### 11.8.9 Debittering Enzyme, Naringinase

The enzyme naringinase (EC 3.2.1.40) helps in the breakdown of naringin (which is a bitter flavanone glycoside) into a glycon naringenin and rhamnose with the help of two enzymes, namely, α-rhamnosidase and β-glucosidase (Puri and Banerjee 2000). The enzyme nariginase is mainly produced by fungal cultures like *Aspergillus niger, Circinella, Eurotium, Fusarium, Penicillium, Rhizopus*, and *Trichoderma.* Among bacteria, it is produced by *Bacillus* sp., *Burkholderia cenocepacia, Bacteriodes distasonis, Thermomicrobium roseum, Pseudomonas paucimobilis, and many others* (Puri 2012). The enzyme has been immobilized for reusability and better results. This is done by using polyvinyl alcohol cryogels, cellulose triacetate nanofibre, graphene, and other substrates. This enzyme is used in the pulping of tomatoes, peeling of kinnow fruits, and also in waste treatments.

**Proteases** Proteolytic enzymes that hydrolyze proteins are being used commercially. The extracellular proteases have many applications in industries. Genera of *Aspergillus, Penicillium, Rhizopus,* and *Mucor* produce this enzyme protease. Proteases can be of different types such as serine, metallo, carboxyl, acidic, neutral, or alkaline. In detergent industries, the serine proteases are used very commonly. Neutral proteases are used for reducing the dental plaques from teeth. Alkaline proteases are used for the decomposition of the gelatinous coating on the X-ray films, from which silver is recovered. Proteases are used for the cleaning of contact lens and are also used as cleaning agents. As the utility of proteases is enormous, microbial proteases are produced on large scales. Majorly, they are used in leather processing, in dehairing of leather (alkaline protease), and for bating (acid proteases). Proteases are used by the detergent industries to remove stains from clothing. Serine proteases cleave peptide bonds in proteins, which are present both in eukaryotes and prokaryotes. Threonine proteases have a threonine residue within the active site. Aspartase protease uses an activated water molecule that is bound to one or more aspartate residues to catalyze the peptide substrates. Glutamic acid proteases use glutamate carboxylic acid for its activity. Metalloprotease involves a metal which is usually zinc, but may use cobalt also.

The doubling time of microbes is very rapid and therefore, they are preferred for a large-scale production. Using genetic engineering techniques, the strains are being improved to obtain recombinant strains with desired properties and higher yields. They are also used in controlling the clotting of blood. Protease enzymes are used in the diagnosis and treatment of intravascular clots in myocardial infarction, stroke, and deep vein thrombosis. Majorly, proteases like nattokinase, and lumbrokinase, as well as plasminogen activators like urokinase have been found to be very effective in preventing thrombosis (Altaf et al. 2021). They are beneficial in oncology, inflammatory conditions, blood rheology control, and immune regulation. In food industries, they are used in fish and sea food processing as fish meals, for increasing the oil recovery and aqua culture. Functional properties such as emulsification, fat-binding, water-binding, foaming, gel strength, and whipping can be improved with the use of proteases. Digestive proteases are used in bread industry which helps in improving the texture of bread and cheese. Acid proteases are used in beer chill proofing, extraction of fruit juice, fruit pulp, clarification, and cheese production. They are used as therapeutic agents against fatal diseases such as cancer, malaria, and AIDS. Pfizer and Genencor have developed a recombinant rennin (Song et al. 2023).

### 11.8.10 Some of the Existing Industrial Processes Using Microbial Enzymes

**Production of high-fructose syrup** The most significant large-scale application of immobilized enzyme is the production of high-fructose corn syrup (HFCS). Scientists are trying to replace sucrose with cheaper alternatives. HFCS is a soluble sweetener that has been used in many carbonated soft drinks since the 1980s, such as in Coca-Cola and Pepsi-Cola. HFCS is produced by the enzymatic digestion of starch derived from corn (maize). Developments in HFCS production have been most prominent in the USA. Simple corn syrup is manufactured by breaking down starch from corn using the enzyme glucoamylase alone or in combination with α-amylase which are cheap. The added advantage is that they can be used in the soluble form. The starch is extracted from corn at high temperatures and the process utilizes enzymes from the thermophilic organisms, which can sustain very high-temperature optima. To make the syrup more sweeter, the enzyme glucose isomerase is utilized. This produces a mixture of glucose and fructose at equilibrium. As fructose is 150% more sweeter as compared to sucrose, the glucose: fructose mixture has the same level of sweetness as that of the sucrose. It was the work of process engineers in the late 1970s, which led to the production of syrups containing a higher fructose content (55%) which were then used in soft drinks. Production of HFCS (2010, in the USA) was around eight million metric tons, which covered around 37% of the USA sweetener market. The enzyme lactase (β-galactosidase) is used in breaking down lactose into glucose and galactose. These sugars are used as carbon sources in microbial fermentation. Valio Ltd. (Finland) uses lactase enzyme from *Aspergillus* for the treatment of whey. It is immobilized by adsorption and then cross-linked to a support resin. Whey syrups are used in drinks, ice creams, and confectionery products. As it works at low pH (3–5) microbial contamination does not take place. In Finland, plants are able to process 80,000 liters of whey per day. Whey syrups are being produced in England and Norway by Dairy Crest company (Anbu et al. 2015).

**Lactose-free dairy products** These are produced by microbial industrial processes (Dekker et al. 2019). Lactose intolerance has grown in the recent times. Therefore, the consumption of lactose-free dairy products has grown many folds. These products are able to provide the essential nutrients present in dairy products (calcium and vitamins). This has made the lactose-free dairy product market very fast growing. It has been found that fresh dairy products contain a large amount of lactose. These include tmilk, milk drinks, yoghurt, sour cream, kefir, ice cream, dairy cream, and sweetened condensed milk. Products made from whey, such as whey protein concentrates (WPC), whey protein isolates (WPI), and whey permeate. Even the lactose-intolerant people relish to consume the milk products as they have nutritional value and are tasty. The lactose gets hydrolyzed into glucose and galactose with the enzyme lactase. These get adsorbed in the small intestine. Presently, two types of lactases are produced on the large scale. Neutral lactases and acid lactases. The lactose free on the products is one of the top health claims among the milk product and brings in a lot of business.

**Penicillin** The fungus *Penicillium* produces high yields of natural penicillins. Penicillins and cephalosporins represent one of the world's major biotechnology market, with annual sales of about £10,000 million, which accounts for 65% of the entire global antibiotic market. Semi-synthetic derivatives like ampicillin and amoxicillin are used on a regular basis. These are obtained by the cleavage of the natural ones. Immobilized penicillin G acylases are intracellular enzymes present in *E. coli* and a variety of other bacteria. The *E. coli* enzyme has been immobilized on a DEAE ion-exchange support, and it is majorly produced by the Far East (China, Korea, and India) (Tishkov et al. 2010). Penicillin acylase (PA, EC 3.5.1.11) acts as a catalyst for hydrolyzing the amide bond in penicillin antibiotics. This enzyme represents the N-terminal nucleophilic hydrolases. They are present in bacteria, yeast, and fungi. They utilize the heterocyclic compounds as carbon sources. PA is used in the synthesis of various semi-synthetic β-lactam antibiotics. The characteristics of the enzyme, namely, the broad substrate specificity and high regio-, chemo- and stereoselectivity of the enzyme are helpful in the production of chiral compounds and for the protection of hydroxy and amino groups in peptide and in the organic synthesis. PA obtained from *Escherichia coli* (EcPA) is very common. But, the efficiency into β-lactam is very poor. Scientists have been working on obtaining new enzymes that have improved properties. The new methods involve random mutagenesis (directed evolution, gene shuffling) and site-directed mutagenesis which are based on 3D protein structures. Due to antibiotic resistance, new antimicrobial compounds are being obtained, which contain artificial side chains.

## 11.9 Extraction and Separation Methods of Enzymes

Sometimes the enzyme purification procedure separates the enzymatic and non-enzymatic parts from the enzyme mixture. This is a very labor-intensive procedure due to the differences in protein size, physicochemical properties, binding affinity, and biological activity. Initially, the cells are disrupted by repeated freezing and thawing, sonication, homogenization by high pressure (French press), homogenization by grinding, and permeabilization by detergents like Triton X-100 with/without using lysozyme. The cell debris is removed by centrifugation and the protein part comes into the supernatant. During the cell lysis, the proteases get released, which help in the digestion of proteins. In order to protect the protein of interest, the extract is kept under cold conditions and the process is conducted at a very fast speed. In such conditions, protease inhibitors can be added before the cell disruption. Adding DNAse helps in reducing the viscosity of the cell lysate which happens due to the high DNA content.

### 11.9.1 Extraction of Extracellular Enzymes

#### 11.9.1.1 Precipitation and Differential Solubilization

In this method, a precipitant is used, and all the soluble proteins get precipitated. During this reaction, the protein molecules get attracted to other protein molecules and not to the solvent. This results in the formation of an aggregation of protein molecules which then gets precipitated.

The pH of the cell extract is adjusted to the isoelectric point of the protein which needs to be precipitated. At this pH, the protein is not soluble; therefore, it gets precipitated. This is then taken in an appropriate buffer. Salt-like ammonium sulphate helps in removing the proteins. This method is called as "salting out." In this method, when the hydrophobic ends of the proteins are exposed, then they get precipitated easily. It is a very easy method to concentrate proteins from the cell lysates.

#### 11.9.1.2 Ammonium Sulfate Precipitation

Initially, the protein is precipitated with ammonium sulfate $(NH_4)_2SO_4$. The different fractions of the precipitated protein are collected. The ammonium sulfate can be removed by dialysis. The method is economical and can be used for larger volumes.

### 11.9.2 Extraction of Intracellular Enzymes

Where the proteins are integrated with the cell membrane, in that case, the cell membrane needs to be broken down in order to isolate a specific protein. However, if the protein is located in any particular organelle, in that case, the particular organelle should be isolated before purification. For example, if the protein is located in the mitochondrial membrane, then the organelle mitochondria needs to be isolated.

### 11.9.3 Separation Methods for Enzymes

(i) **Sucrose gradient centrifugation** In this method a linear concentration gradient of sugar is made. This could be of sucrose, glycerol, or silica. It is made in such a way that the highest concentration is at the bottom and the lowest is at the top. On the top of the gradient, the protein is layered in such a way as it is usually spun at high speed in an ultracentrifuge. With this, the heavy macromolecules migrate to the bottom of the tube very fast and the lighter ones stay at the top. During centrifugation, the particles move away from the center of rotation due to the centrifugal force.

(ii) **Chromatographic methods** There are different types of chromatographic methods. These include size exclusion chromatography/gel-filtration chromatography: This is a molecular sieve method, where the molecules in the solution get separated on the basis of their size and molecular weight. This method is usually utilized for proteins and industrial polymers. Sample is transported through the column in an aqueous solution. The separation methods have been discussed in detail in Chap. 12 of this book.

(a) **Ion exchange chromatography** Each compound has an ionic charge. Based on their ionic charge, the separation of compounds is done by this technique. The anion exchange resins have a positive charge. Therefore, they have the ability to retain and separate the oppositely charged ions, which are negatively charged compounds (anions). Similarly, the cation exchange resins, which have a negative charge, are able to retain and separate the oppositely charged, positive molecules (cations).

(b) **Affinity chromatography** This method uses specific resins which have specific ligands attached on their surfaces to separate the compounds. This fits between the ligand and its target compound like that of lock and key, makes it very specific, and generates a single peak.

(c) **High-performance liquid chromatography(HPLC)** In this method, high pressure is applied so that the solutes are driven out faster through the column. This makes the diffusion slower which results in better resolution. Reverse phase HPLC method is very common in which the column material is hydrophobic. For the elution of proteins, a gradient with increasing amounts of organic solvent, like acetonitrile is utilized.

## 11.10 Storage of Enzymes

For the stability of the enzymes, they should be stored in their original commercial form (lyophilized or ammonium sulfate suspension) at the appropriate optimized temperature as specified on the label. Usually, the temperatures to store microbial enzymes for a longer period should be in the region of (−) 10 to (−) 25 °C. When stored at higher temperatures, the enzymes gradually get denatured and lose their catalytic activity. When they are used, at that time, they can be diluted with an ice-cold buffer or distilled water depending on case by case basis.

## 11.11 Immobilization of Enzymes

Enzymes are used in batch fermentation for metabolite production. The soluble enzymes are separated after the process is completed. When the product is thermostable in that case, the enzyme is usually removed by thermal denaturation/by ammonium sulfate precipitation/ultrafiltration. However, these methods are not only expensive, they make the enzyme inactive. Therefore, scientists have been working for methods to immobilize the enzymes so that they can be reused from batch to batch. When an enzyme is mixed with warm agar solution and is left to set, then the enzyme gets immobilized inside the solid agar. This can be cut into small pieces and can be reused. After the completion of the process and after downstream processing, the enzyme can be recovered, which is in the active state, and can be reused in the next batch. This makes the process very cost-effective. Immobilization of enzymes is very advantageous as along with reuse, the stability increases at different pH and temperature conditions. The activity between an immobilized and a nonimmobilized system can be calculated as the effectiveness factor ($\eta$). The effectiveness of immobilization can be calculated by the equation below

Effectiveness factor = Activity of immobilized biocatalyst / Activity of non –immobilized biocatalyst.

Therefore, if effectiveness factor is 0.1, then it means 10% activity in a nonimmobilized system as compared to the immobilized under the same conditions.

It also means that the biocatalyst can be reused 10 times. The immobilized enzyme can also be used in continuous flow reactors, namely, continuous stirred tank reactors (CSTRs) and packed-bed reactors (PBRs). On commercial scale, it has been used for the production of high-fructose syrups. Work has shown that the immobilized glucose isomerase enzyme can be used for long durations.

The batch fermentation is much more expensive mainly due to the labor and also many variations take place from one batch to the other as the reaction conditions get changed. Apart from agar, there are many other techniques to immobilize the enzymes like adsorption, covalent bonding, and entrapment. This is done using different support systems which gives thermostability to the enzyme and prevents the loss of enzyme activity. The removal of the enzyme becomes very easy.

(i) **Adsorption for immobilized enzyme** There is physical adsorption of an enzyme on a supporting matrix like cellulose, sephadex, polystyrene, kaolinite, collagen, alumina, silica gel, and glass. The method is easy as the adsorbent and the enzyme need just stirring them together, for a specific optimised time. Forces like van der Waal force, ionic interaction, or hydrophobic interaction bind them together and then immobilization takes place with hydrogen bonding. Due to the weak bonds which have an energy of 20 kJ $mol^{-1}$, the enzyme yields are low and also the adsorption can be easily reversed. This can lead to desorption of the enzyme at a critical time. DEAE–Sephadex is an ion exchange resin that consists of inert dextran particles that remain positively charged at pH 6–8 and are activated by the addition of numerous diethylaminoethyl groups. After long time duration, the enzyme can lose its activity due to denaturation. However, the pH can be adjusted to an acidic value. This allows the support matrix to be used many times, giving the process many economic benefits. Some enzymes can function under acidic conditions for which CM–Sephadex is recommended. These have carboxymethyl groups; therefore, they remain negatively charged at pH 3.5–4.5. The proteins at this low pH are generally positively charged which allows them to bind ionically to the support. Desorption of the enzyme occurs when the pH is alkaline. As the method is simple many industries use this method.

(ii) **Covalent bonding for immobilization of enzyme** In this procedure, immobilized enzyme can be produced by using different polymers and various chemical coupling procedures that firmly bind to the support. Activated cellulose supports are very popular due to their high hydrophobicity and ready availability. They can be derivatized easily. For an effective chemical bridge between the cellulose and the enzyme molecule, it is important that it has a reactive group. An example of such a bridging molecule is glutaraldehyde, which contains two aldehyde groups, one at either end of its $(CH_2)_3$ moiety. At neutral pH, the aldehyde group can react with the free amino group. Thus, one end of the glutaraldehyde molecule may get attached to the support, and the other to the enzyme. Immobilized enzyme, with covalent bonding, gets attached strongly with their support. Therefore, it is very difficult to remove them. However, the operational costs become high.

(iii) **Entrapment for immobilization of enzyme** The enzyme gets included within the matrix of a highly cross-linked polymer. The bulk phase is separated by a semipermeable microcapsule in a nonaqueous phase, and the enzyme is not attached to anything, as it is retained in the agar. The catalyst can be mixed with sodium alginate gel and extruded into a solution of calcium chloride to produce solid calcium alginate particles and does not require high temperatures but it might leak out from the gel.

Immobilized whole yeast cells have been used successfully in the preparation of champagne (Moët and Chandon Industry). The immobilization matrix can act as a barrier to the diffusion of substrates, products, and other molecules. Immobilized systems are useful for the study of membrane-bound proteins in living cells. Industrial enzymes are glucose isomerase for the production of sweetening agents, penicillin acylase, fumarase, β–galactosidase, and amino acid acylase. Since 2000, there has been a consistent growth in the enzyme market (Table 11.2).

**Table 11.2** The major industrial processes that use immobilized enzymes

| Enzyme | Method of immobilization | Support material for binding | Industry of utilization | References |
|---|---|---|---|---|
| Amylase | Covalent binding | Chitosan | Beer | Samia et al. (2020) |
| Lipase | Encapsulation | Lactoalbumin | Cheese flavor | Tong Guan et al. (2021) |
| Pectinase | Adsorption | polymeric matrix | Clarification of apple juice | Kishor Rajdeo et al. (2016) |
| Protease | Entrapment | Zeolite | dairy | Kumari et al. (2015) |
| Cellulase | Adsorption | Kaolin | Textile | de Souza Lima et al. (2022) |
| Laccase | Entrapment | Alginate bead | Dye removal from water | Varga Béla et al. (2019) |
| Tyrosinase | Cross linking | Iron oxide | Phenol removal from water | Abdollahi et al. (2018) |

## 11.12 Biosensors

Glucose biosensor devices have been developed to measure blood glucose levels. Clark and Lyons (1962) have shown that an enzyme electrode is a device in which a traditional electrode is modified to respond to an enzyme layer. Successful commercialization was done (1975) by Yellow Springs Instruments (YSI) as model 23A glucose analyzer.

On a peroxide-sensitive electrode, the enzyme, namely, glucose oxidase was put in. This helps in measuring the hydrogen peroxide ($H_2O_2$). As the metabolite uses glucose along with oxygen, then hydrogen peroxide is produced.

$$\text{Glucose} + O_2 \rightarrow \text{Gluconic acid} + H_2O_2.$$

The reaction rate is dependent on many factors of the reaction taking place as the concentration of initial glucose supplied and the amount of oxygen in the system. These factors decide the amount of $H_2O_2$ formation. The reaction rate follows the principle of Michaelis–Menten equation. The glucose analyzers are commercially available in the market. The YSI model 2900 Series has been found very effective as the time taken to analyze one sample of blood, plasma, or serum does not take even a minute. The sample requirement is also very low (10 μl). The enzyme glucose oxidase can be changed in one month. These factors make the sensor very efficient and cost-effective. The work of Nguyen (2019) has come out with newer inventions of small handy biosensors. The scientists brought the enzyme and the transducer very close to the surface of the sensor. This technology has been used commercially by The ExacTech company. They have brought out a blood glucose meter which appears as a pen. It has disposable electrode strips. Sensors have also been developed which look alike a credit card-style meter. This card-type meter can measure the rate of electron flow from glucose to the electrode surface. Discoveries have been done to make a glucose meter in which the enzyme is immobilized on an osmium-based polyvinyl imidazole hydrogel. The electrons are passed from the enzyme to the electrode by using a series of fixed electroactive osmium centers. This helps in moving the electrons in an onward direction known as electron hopping. Such innovations in sensors have brought light in the diagnosis of diseases at a very rapid rate. Developments are taking place especially in glucose sensors, so that glucose can be measured continuously. This will be linked to a pump, in such a way, so that insulin can be dispensed gradually in order to maintain the level of glucose.

## 11.13 Legal Aspects of Enzyme Regulations

These are different in each/and in between the countries. Any country requires a complete approval process, a notification of enzyme, or no requirement thereof. It is important to know if the enzyme is classified as a processing aid or a food additive before taking the marketing approval. The regulation of enzymes for Australia and

New Zealand depends on the processing procedure. In the USA, the enzymes are classified as food additives or they come under GRAS category. The act permits an individual/company to decide whether an ingredient will come in GRAS category for a particular intended use or not. In the EU, food enzymes are not covered by any food safety regulations.

Worldwide the food enzyme business is worth around 1 billion US$. Looking into the AMFEP (Association of manufacturers and formulators of enzyme products) 247 enzymes have been listed that can be used by the food industries. Due to the expected side effects of genetic modification, a lot of research has been done on the safety concerns and regulatory aspects, especially in Germany, Switzerland, and Austria. Fundamental aspects of the pre-market approval and safety evaluation have come into existence but there is a difference on the basis of food additives and processing aids. Food additives are used for technological purposes wherein the processing aids do not have any technological effect, and there is no risk on the health by using the food product.

In Canada, USA, and Japan, all food enzymes are regulated as food additives, however, according to Australia and EU food legislation they are considered as processing aids. There is an expert committee on Food Additives (JECFA) which is a joint committee constituted by FAO/WHO both. This committee has not differentiated the enzymes either as food additives or processing aids. However, it is FDA that usually needs the maximum information on enzymes in detail. According to FDA, the report must include technical data, structural modifications in case of genetically modified microorganisms (GMMs) and also the purification process during manufacturing of the enzymes. The details on the vectors and the stability of the enzyme in food are other important factors that should be mentioned in the report by the user. The principles for toxicological testing have been laid down by FDA, which should be followed by the manufacturers.

Till now, lysozyme and invertase are considered in the category of additives. The majority of about 160 food enzymes are produced and marketed. However, they come under the processing aids. Their regulation is governed by the national legislation. The EU member states differ in their regulations. In each country like France, Denmark, Poland, and Hungary, the enzymes are subjected to an authorization procedure. In the UK, a voluntary approval is permitted. In EU, the supplementation of food additives is regulated by the council directive 94/34/EEC. These are used in limited quantities in certain foods using GMP (good manufacturing practice), with no health hazards and fulfil the purity criteria of the commission directives. SCF considers enzymes from (edible) plants and animals which should not cause any health problems. It is allowed that industries can submit the information on risk assessment of food enzymes in three and a half years. Recently, it has been decided that the food enzymes must undergo a safety evaluation by EFSA. The EU legislation also includes: Regulation (EC) No 1331/2008 establishes the common authorization procedure for food additives, food enzymes, and food flavorings. Under the regulation (EU) No 234/2011, the common authorization procedure has been implemented from 2011. The regulation was amended by Commission Implementing Regulation (EU) No 562/2012. In certain fermented foods, microbial cultures are utilized which produce enzymes, but these are not considered as food enzymes.

Joint FAO/WHO Expert Committee on Food Additives (JECFA) has been conducting safety evaluations of enzymes to set the guidelines. The guidelines along with the amendments and supplements till 1999 have been published in an updated version of the Compendium of Food Additive Specifications which was published as FAO, Food and Nutrition Papers.

The regulations improve the functioning of the internal market by removing disparities among member states and bringing about legal certainty to the market. It allows only the authorized food enzymes to be commercialized and used in foods that are sold in the EU. This also applies to imported food items. In order to provide safe, nutritious, high quality, and affordable food to the consumers the central objective of EU policy should cover all the stages of the food supply chain. There are many challenges which include the demographic imbalance, climate change, resource and energy scarcity, slowing agricultural productivity, increasing concentration of the supply chain, price volatility, and change in the trends of diet.

Microbial enzymes have enormous potential in various industries.

### FAQs

1. What is the effect of extremely high pH on enzymes?
2. What type of enzymes break down carbohydrates? .
3. How do enzymes affect the microbial growth?
4. What are the factors that affect an enzyme?
5. Which enzyme inhibits the growth of bacteria?
6. Does enzyme rennet fall under vegetarian category?
7. What are the enzymes produced at industrial level?
8. What is an enzyme made up of?
9. In which industries is enzyme amylase used?
10. Which enzyme is used for dehairing of leather and why?

## References


Abdollahi K, Yazdani F, Panahi R, Mokhtarani B (2018) Biotransformation of phenol in synthetic wastewater using the functionalized magnetic nano-biocatalyst particles carrying tyrosinase. 3 Biotech 8(10):419. https://doi.org/10.1007/s13205-018-1445-2

Adrio JL, Demain AL (2014) Microbial enzymes: tools for biotechnological processes. Biomol Ther 4(1):117–139

Ahuja SK, Ferreira GM, Moreira AR (2004) Utilization of enzymes for environmental applications. Crit Rev Biotechnol 24:125–154

Aldridge S (2013) Industry backs biocatalysis for greener manufacturing. Nat Biotechnol 31:95

Alfa MJ, Jackson M (2001) A new hydrogen peroxide based medical device detergent with germicidal properties: comparison with enzymatic cleaners. Am J Infect Control 29(3):168–177

Altaf F, Wu S, Kasim V (2021) Role of fibrinolytic enzymes in anti-thrombosis therapy. Front Mol Biosci 8:680397. https://doi.org/10.3389/fmolb.2021.680397

Anbu P, Gopinath SC, Chaulagain BP et al (2015) Microbial enzymes and their applications in industries and medicine. Biomed Res Int:1–3

Aravindan R, Anbumathi P, Viruthagiri T (2007) Lipase applications in food industry. Indian J Biotechnol 6:141–158

Bansal N, Kanwar SS (2013) Peroxidase(s) in environment protection. Scientific World J 714639. https://doi.org/10.1155/2013/714639

Barett AJ (1995) Proteolytic enzymes: aspartic and metallopeptidases. Methods Enzymol 248:183

Bhosale SH, Rao MB, Deshpande V, Srinivasan MC (1995) Thermostability of high activity alkaline protease from *Conidiobolus coronatus* (NCL 86.8.20). Enzyme Microbiol Technol 17:136–139

Britannica (2024) The editors of encyclopaedia "Jokichi Takamine". Encyclopedia Britannica. https://www.britannica.com/biography/Jokichi-Takamine. Accessed 28 April 2024

Buchner E (1886) Uber den Einfluss des Sauerslolfes auf Garung. Z Physiol Chem 8:380–386

Buchner E (1907) Cell-free fermentation. In: Nobel lectures: chemistry. 1901-1966. Elsevier, Amsterdam, pp 103–120

Chandra P, Enespa, Singh R, Arora PK (2020) Microbial lipases and their industrial applications: a comprehensive review. Microb Cell Factories 19(1):169. https://doi.org/10.1186/s12934-020-01428-8

Cho YH, Song JY, Kim KM et al (2010) Production of nattokinase by batch and fed-batch culture of *Bacillus subtilis*. New Biotechnol 27(4):341–346. https://doi.org/10.1016/j.nbt.2010.06.003

Choct M (2006) Enzymes for the feed industry: past, present and future. Worlds Poult Sci J 62:5–15

Choi EJ, Ahn HW, Kim WJ (2015) Effect of α-acetolactate decarboxylase on diacetyl content of beer. Food Sci Biotechnol 24:1373–1380. https://doi.org/10.1007/s10068-015-0176-y

Choudhary RB, Jana AK, Jha MK (2004) Enzyme technology applications in leather processesing. Ind J Chem Technol 11:659–671

Clark LC, Lyons C (1962) Electrode systems for continuous monitoring in cardiovascular surgery. Ann N Y Acad Sci 102:29–45. https://doi.org/10.1111/j.1749-6632.1962.tb13623.x

Clarke JH, Davidson K, Rixon JE et al (2000) A comparison of enzyme aided bleaching of softwood pulp using a combination of xylanase, mannanase and α-galactosidase. Appl Microbiol Biotechnol 53:661–667

Cleland WW (2013) Substrate binding, catalysis, and product release. In: William J, Lennarz M, Daniel L (eds) Encyclopedia of Biol Chem, 2nd edn. Academic Press, pp 347–349. ISBN 9780123786319. https://doi.org/10.1016/B978-0-12-378630-2.00026-8

Code of Federal Regulations - Title 21 - Food and Drugs (database)

Collar C, Martinez JC, Andreu P et al (2000) Effect of enzyme associations on bread dough performance: a response surface study. Food Sci Technol Int 6:217–226

Collection of information on enzymes (2002) European Communities. http://ec.europa.eu/environment/archives/dansub/pdfs/enzymerepcomplete.pdf. Accessed 02 Mar 2016

Colman MH, Macrae AR (1980) Fat process and composition. UK patent 1577933

De Souza LJ, Immich APS, de Araújo PHH, de Oliveira D (2022) Cellulase immobilized on kaolin as a potential approach to improve the quality of knitted fabric. Bioprocess Biosyst Eng 45(4):679–688. https://doi.org/10.1007/s00449-021-02686-5

Dekker PJT, Koenders D, Bruins MJ (2019) Lactose-free dairy products: market developments, production, nutrition and health benefits. Nutrients 11(3):551. https://doi.org/10.3390/nu11030551

De-Souza FR, Gutterres M (2012) Application of enzymes in leather processing: a comparison between chemical and coenzymatic processes. Braz J Chem Eng 29(3):471–481

Dulieu C, Moll M, Boudrant J, Poncelet D (2000) Improved performances and control of beer fermentation using encapsulated α-acetolactate decarboxylase and modeling. Biotechnol Prog 16:958–965. https://doi.org/10.1021/bp000128k

El-Gendi H, Saleh AK, Badierah R, Redwan EM, El-Maradny YA, El-Fakharany EM (2021) A comprehensive insight into fungal enzymes: structure, classification, and their role in mankind's challenges. J Fungi (Basel) 8(1):23. https://doi.org/10.3390/jof8010023

ENZYME - Enzyme nomenclature database http://enzyme.expasy.org

Esteban-Torres MB, José M, José R, Blanca M, Rosario (2014) Characterization of a versatile arylesterase from *Lactobacillus plantarum* active on wine esters. J Agric Food Chem 62. https://doi.org/10.1021/jf500991m

Fernandes P (2010) Enzymes in food processing: a condensed overview on strategies for better biocatalysis. Enzyme Res. https://doi.org/10.4061/2010/862537

Fersht A (1985) Enzyme structure and mechanism. WH Freeman, New York

Fetzner S, Lingens F (1994) Bacterial dehalogenases: biochemistry, genetics, and biotechnological applications. Microbiol Rev 58(4):641–685

Filipsen JAC, Appel ACM, Van Der Hidjen HTWM et al (1998) Mechanism of removal of immobilized triacylglycerol by lipolytic enzymes in a sequential laundry wash process. Enzyme Microb Technol 23:274–280

Fischer E (1894) Influence of configuration on the action of enzymes. J Am Chem Soc 3:2985–2993

Forbath TP (1957) Flexible processing keys enzymes' future. Chem Eng 64:226–229

Frenzel A, Hust M, Schirrmann T (2013) Expression of recombinant antibodies. Front Immunol Sec B. Cell Biol 4. https://doi.org/10.3389/fimmu.2013.00217

Frias J, Doblado R, Antezana JR et al (2003) Inositol phosphate degradation by the action of phytase enzyme in legume seeds. Food Chem 81(2):233–239

Fu GZ, Chan AW, Minns DE (2005) Preliminary assessment of the environmental benefits of enzyme bleaching for pulp and paper making. Int J Life Cycle Assess 10:136–142

Garg G, Singh A, Kaur A et al (2016) Microbial pectinases: an ecofriendly tool of nature for industries. 3 Biotech 6(1):47–59

Gerhartz W (ed) (1990) Enzymes in industry: production and applications. VCH, New York

Ghosh PK, Saxena RK, Gupta R et al (1996) Microbial lipases: production and applications. Sci Prog 79:119–157

Godtfredsen SE, Ottesen M (1982) Maturation of beer with α-acetolactate decarboxylase. Carlsb Res Commun 47(2):93–102. https://doi.org/10.1007/BF02914029

Gurung N, Ray S, Bose S, Rai V (2013) A broader view: microbial enzymes and their relevance in industries, medicine and beyond. Biomed Res Int:1–18. https://doi.org/10.1155/2013/329121

Hanane T, Najoua B, Salsabil H, Abdellatif JI, Dalila B, Ahmad I, BukharI SAR, Irfan M, Chen L, Hicham B (2022) Qualitative screening of yeast biodiversity for hydrolytic enzymes isolated from the gastrointestinal tract of a coprophage "gymnopleurus sturmi" and dung of ruminants. Fermentation 8(12):692. https://doi.org/10.3390/fermentation8120692

Hanft F, Koehler P (2006) Studies on the effect of glucose oxidase in bread making. J Sci Food Agric 86:1699–1704. https://doi.org/10.1002/jsfa.2455

Hasan F, Shah AA, Hameed A (2006) Industrial applications of microbial lipases. Enzym Microb Technol 39:235–251. https://doi.org/10.1016/j.enzmictec.2005.10.016

Hasan F, Shah AA, Javed S, Hameed A (2010) Enzymes used in detergents: lipases. Afr J Biotechnol 9(31):4836–4844. https://doi.org/10.5897/AJBx09.026

Hiraishi T, Taguchi S (2009) Enzyme-catalyzed synthesis and degradation of biopolymers. Mini Rev Org Chem 6:44–54

Humpf HU, Schreier P (1991) Bound aroma compounds from the fruit and the leaves of blackberry (*Rubus laciniata* L.). J Agric Food Chem 39:1830–1832. https://doi.org/10.1021/jf00010a028

International Congress of Biochemistry (1955) In: Snell EE, Fasella PM, Braunstein A, editors. Chemical and Biological Aspects of Pyridoxal Catalysis: Proceedings of a Symposium of the International Union of Biochemistry, Rome, October 1962. Elsevier; 2013 Oct 22

Janusz G, Kucharzyk KH, Pawlik A, Staszczak M, Paszczynski AJ (2013) Fungal laccase, manganese peroxidase and lignin peroxidase: gene expression and regulation. Enzym Microb Technol 52(1):1–12. https://doi.org/10.1016/j.enzmictec.2012.10.003

Johannes T, Simurdiak MR, Zhao H (2006) Biocatalysis. In: Lee S (ed) Encyclopedia of chemical processing. Taylor and Francis, New York, pp 101–110

Kadokawa JI, Kobayashi S (2010) Polymer synthesis by enzymatic catalysis. Curr Opin Chem Biol 14:145–153

Kalia VC, Rashmi Lal S et al (2001) Using enzymes for oil recovery from edible seeds. J Sci Ind Res 60:298–310

Kamini NR, Hemchander C, Geraldine J et al (1999) Microbial enzyme technology as an alternative to conventional chemical in leather industry. Curr Sci 76:101
Karigar CS, Rao SS (2011) Role of microbial enzymes in the bioremediation of pollutants: a review. Enzyme Res. https://doi.org/10.4061/2011/805187
Kårlund A, Moor U, Sandell M et al (2014) The impact of harvesting, storage and processing factors on health-promoting phytochemicals in berries and fruits. PRO 2(3):596–624
Kieliszek M, Misiewicz A (2014) Microbial transglutaminase and its application in the food industry. A review. Folia Microbiol 59:241–250. https://doi.org/10.1007/s12223-013-0287-x
Kirk O, Borchert TV, Fuglsang CC (2002) Industrial enzyme applications. Curr Opin Biotechnol 13:345–351. https://doi.org/10.1016/S0958-1669(02)00328-2
Kishor R, Tirunagari H, Kuna L, Nitin WF (2016) Immobilization of pectinase on reusable polymer support for clarification of apple juice. Food Bioprod Process 99:12–19. ISSN 0960-3085. https://doi.org/10.1016/j.fbp.2016.03.004
Kobayashi S (2010) Lipase catalyzed polyester synthesis- a green polymer chemistry. Proc Jpn Acad Ser B Phys Biol Sci 86(4):338–365. https://doi.org/10.2183/pjab.86.338
Krishnapura PR, Belur PD, Subramanya S (2016) A critical review on properties and applications of microbial l-asparaginases. Crit Rev Microbiol 42:720–737. https://doi.org/10.3109/1040841X.2015.1022505
Kuhad RC, Gupta R, Singh A (2011) Microbial cellulases and their industrial applications. Enzyme Res 280696. https://doi.org/10.4061/2011/280696
Kühne W (1877) Über das Verhalten verschiedener organisirter und sog. ungeformter Fermente [On the behavior of various organized and so-called unformed ferments]. Verhandlungen des Naturhistorisch-medicinischen Vereins zu Heidelberg. Neue Folge [new series] (in German). Heidelberg vol 1, pp 190–193
Kumari A, Kaur B, Srivastava R, Sangwan RS (2015) Isolation and immobilization of alkaline protease on mesoporous silica and mesoporous ZSM-5 zeolite materials for improved catalytic properties. Biochem Biophys Rep 2:108–114. https://doi.org/10.1016/j.bbrep.2015.05.009
Law BA (2002) The nature of enzymes and their action in foods. In: Whitehurst RJ, Law BA (eds) Enzyme in food technology. Sheffield Academic Press, Sheffileld, pp 1–18
Le Roes-Hill M, Prins A (2016) Biotechnological potential of oxidative enzymes from Actinobacteria. In: Actinobacteria - Basics and biotechnological applications. IntechOpen. https://doi.org/10.5772/61321
Li S, Yang X, Yang S et al (2012) Technology prospecting on enzymes: application, marketing and engineering. Comput Struct Biotechnol J 2:1–11
van der Maarel MJ, van der Veen B, Uitdehaag JCM, Leemhuis H, Dijkhuizen L (2002) Properties and applications of starch converting enzymes of the α-amylase family. J Biotechnol 94:137–155. https://doi.org/10.1016/S0168-1656(01)00407-2
Madhavi, Lele (2009) Laccase: properties and applications. BioResources. 4(4)
Maijala P, Kleen M, Westin C et al (2008) Biomechanical pulping of softwood with enzymes and white-rot fungus *Physisporinus rivulosus*. Enzym Microb Technol 43:169–177. https://doi.org/10.1016/j.enzmictec.2007.11.017
Mane P, Tale V, Gandhi R (2015) Overview of microbial therapeutic enzymes. Biol Med. https://api.semanticscholar.org/CorpusID:42956627
McDonald CE (1979) Lipoxygenase and lutein bleaching activity of durum wheat semolina. Cereal Chem 56:84–89
Mohan Kumar NS, Shimray CA, Indrani D, Manonmani HK (2014) Reduction of acrylamide formation in sweet bread with L-asparaginase treatment. Food Bioprocess Technol 7:741–748. https://doi.org/10.1007/s11947-013-1108-6
Mojsov K (2011) Applications of enzymes in the textile industry: a review. In: 2nd international congress: engineering, ecology and materials in the processing industry: Jahorina, Bosnia and Herzegovina. Tehnoloski Fakultet Zvornik, pp 230–239
Mojsov K (2012) Microbial alpha-amylases and their industrial applications: a review. Int J Manage IT Eng 2(10):583–609. ISSN 2249-0558

Mordor intelligence (2024) Market forecast for enzymes
Morozova OV, Andreeva IS, Zhirakovskiy VY, Pechurkina NI, Puchkova LI, Saranina IV, Emelyanova EK, Kamynina TP (2022) Antibiotic resistance and cold-adaptive enzymes of antarctic culturable bacteria from King George Island. Polar Science 31:100756
Müller D (1928) Studies on the new enzyme glucose oxidase. I Biochem Z 199:136–170
Nair SG, Sindhu R, Shashidhar S (2008) Fungal xylanase production under solid state and submerged fermentation conditions. Afr J Microbiol Res 2:82–86
Nelson JM, Griffin EG (1916) Adsorption of invertase. J Amer Chem Soc 38(5):1109–1115
Nguyen HH, Lee SH, Lee UJ, Fermin CD, Kim M (2019) Immobilized enzymes in biosensor applications. Materials (Basel) 12(1):121. https://doi.org/10.3390/ma12010121
Nidheesh T, Suresh PV (2014) Recent developments in chitosanase research and its biotechnological applications: a review. Food Chem 150:392–399. ISSN 0308-8146. https://doi.org/10.1016/j.foodchem.2013.10.083
Okafor N (2007) Biocatalysis: immobilized enzymes and immobilized cells. In: Okafor N (ed) Modern industrial microbiology and biotechnology. Science pub, Enfield, pp 398–420
Palmer T (2001) Enzymes: biochemistry, biotechnology and clinical chemistry. Horwood Publishing Ltd, Chichester, pp 345–350
Pandey D, Singh R, Chand D (2011) An improved bioprocess for synthesis of acetohydroxamic acids using DTT (dithiothreitol) treated resting cells of *Bacillus* sp. APB-6. Bioresour Technol 102(11):6579–6586. https://doi.org/10.1016/j.biortech.2011.03.071
Paqué PN, Schmidlin PR, Wiedemeier DB, Wegehaupt FJ, Burrer PD, Körner P, Deari S, Sciotti MA, Attin T (2021) Toothpastes with enzymes support gum health and reduce plaque formation. Int J Environ Res Pub Health 18(2):835. https://doi.org/10.3390/ijerph18020835
Poewe W, Antonini A, Zijlmans JC, Burkhard PR, Vingerhoets F (2010) Levodopa in the treatment of Parkinson's disease: an old drug still going strong. Clin Interv Aging 7(5):229–238. https://doi.org/10.2147/cia.s6456
Puri M (2012) Updates on naringinase: structural and biotechnological aspects. Appl Microbiol Biotechnol 93:49–60. https://doi.org/10.1007/s00253-011-3679-3
Puri M, Banerjee UC (2000) Production, purification, and characterization of the debittering enzyme naringinase. Biotechnol Adv 18:207–217. https://doi.org/10.1016/S0734-9750(00)00034-3
Qureshi MA, Khare AK, Pervez A (2015) Enzymes used in dairy industries. Int J Appl Res 1(10):523–527
Ramachandra M, Crawford DL, Hertel G (1988) Characterization of an extracellular lignin peroxidase of the lignocellulolytic actinomycete *Streptomyces viridosporus*. Appl Environ Microbiol 54(12):3057–3063
Ramachandran S, Fontanille P, Pandey A, Larroche C (2006) Gluconic acid: properties, applications and microbial production. Food Technol Biotechnol 44:185–195
Raveendran S, Parameswaran B, Ummalyma SB, Abraham A, Mathew AK, Madhavan A, Rebello S, Pandey A (2018) Applications of microbial enzymes in food industry. Food Technol Biotechnol 56(1):16–30. https://doi.org/10.17113/ftb.56.01.18.5491
Razzaq A, Shamsi S, Ali A, Ali Q, Sajjad M, Malik A, Ashraf M (2019) Microbial proteases applications. Front Bioeng Biotechnol 7:110. https://doi.org/10.3389/fbioe.2019.00110
Rigo ER, Roberta L, Patrícia O, Débora F, Denise L, Marco L (2008) Application of different lipases as pretreatment in anaerobic treatment of wastewater. Environ Eng Sci 25:1243–1248. https://doi.org/10.1089/ees.2007.0197
Rohm O (1908) US Patent 866,411
Saadoun I, Dawagreh A, Jaradat Z, Ababneh Q (2013) Influence of culture conditions on pectinase production by *Streptomyces* sp. (strain J9). Int J Life Sci Med Res 3:148. https://doi.org/10.5963/LSMR0304002
Saboury A (2002) Stability, activity and binding properties study of amylase upon interaction with $Ca^{2+}$ and $CO_2$. Biologia - Section Cell Mol Biol 57
Saboury AA, Saboury M, Umar Dahot S, Ghobadi J, Chamani AA, Moosavi-Movahedi AA (1998) Thermodynamic studies on the interaction of cobalt with alpha-amylase. JCCS 45(5):667–671

Sadhasivam S, Savitha S, Swaminathan K, Lin FH (2008) Production, purification and characterization of mid-redox potential laccase from a newly isolated Trichoderma harzianum WL1. Process Biochemistry 43(7):736–742

Sajith S, Priji P, Sreedevi S, Benjamin S (2016) An overview on fungal cellulases with an industrial perspective. J Nutr Food Sci 6(1):461. https://doi.org/10.4172/2155-9600.1000461

Samia AA, Mohamed AAA, Gamal ME, Atef MI, Aliaa RE, Sherien MMA, Mohamed EH (2020) Catalytic, kinetic and thermal properties of free and immobilized *Bacillus subtilis*-MK1 α-amylase on Chitosan-magnetic nanoparticles. Biotechnol Rep 26:e00443. ISSN 2215-017X. https://doi.org/10.1016/j.btre.2020.e00443

Schmid A, Dordick JS, Hauer B et al (2001) Industrial biocatalysis today and tomorrow. Nature 409:258–268. https://doi.org/10.1038/35051736

Schülein M (1988) Cellulases of *Trichoderma reesei*. Methods Enzymol 160:234–242. https://doi.org/10.1016/0076-6879(88)60125-X

Sharma R, Chisti Y, Banerjee UC (2001) Production, purification, characterization, and applications of lipases. Biotechnol Adv 19(8):627–662. https://doi.org/10.1016/s0734-9750(01)00086-6

Singh R, Kumar M, Mittal A, Mehta PK (2016) Microbial enzymes: industrial progress in 21$^{st}$ century. 3 Biotech 6(2):174. https://doi.org/10.1007/s13205-016-0485-8

Sîrbu T (2011) The searching of active catalase producers among the microscopic fungi. Annals of Oradea University, Biology Fascicle/Analele Universităţii din Oradea, Fascicula Biologie 18(2)

Sirbu A, Paslaru V (2006) Effect of lysomax formulation on rheological behavior of dough. J Agroaliment Process Technol 12:199–208

Sivaramakrishnan S, Gangadharan D, Nampoothiri KM et al (2006) α-Amylases from microbial sources—an overview on recent developments. Food Technol Biotechnol 44(2):173–184

Song P, Zhang X, Wang S, Xu W, Wang F, Fu R, Wei F (2023) Microbial proteases and their applications. Front Microbiol 14(14):1236368. https://doi.org/10.3389/fmicb.2023.1236368

Srivastava N, Singh P (2015) Degradation of toxic pollutants from pulp & paper mill effluent. Discovery 40(183):221–227

Stemler CD, Scherf KA (2022) Improvement of cake baking properties by lipases compared to a traditional emulsifier. Food Chem X 15:100442. https://doi.org/10.1016/j.fochx.2022.100442

Sukumaran RK (2009) Bioethanol from lignocellulosic biomass, Part II. Production of cellulases and hemicellulases. In: Pandey A (ed) Handbook of plant-based biofuels. CRC Press. Taylor & Francis Group, Boca Raton, pp 141–157

Sukumaran RK, Singhania RR, Pandey A (2005) Microbial cellulases - production, applications and challenges. J Sci Ind Res (India) 64:832–844

Sumner JB (1926) The isolation and crystallization of the enzyme urease. Preliminary paper. J Biol Chem 69:435–441

Takamine J (1894) Process of making diastatic enzyme. U.S. Pat. 525,820 and 525,823

Takamine J (1914) Enzymes of *Aspergillus oryzae* and the application of its amyloclastic enzyme to the fermentation industry. Ind Eng Chem 6:824–828

Tamano K (2014) Enhancing microbial metabolite and enzyme production: current strategies and challenges. Front Microbiol 18(5):718. https://doi.org/10.3389/fmicb.2014.00718

Tapre AR, Jain RK (2014) Pectinases: enzymes for fruit processing industry. Int Food Res J 21:447–453

Tathagata M, Kathryn M McCulloch, Steven EE, Tadhg PB (2010) 7.18 - Cofactor catabolism. In: Hung-Wen (Ben) L, Lew M (eds) Comprehensive Nat Prod II. Elsevier, pp 649–674. ISBN 9780080453828. https://doi.org/10.1016/B978-008045382-8.00153-2. https://www.sciencedirect.com/science/article/pii/B9780080453828001532

Tishkov VI, Savin SS, Yasnaya AS (2010) Protein engineering of penicillin acylase. Acta Nat 2(3):47–61

Tonin F, Melis R, Cordes A, Sanchez-Amat A, Pollegioni L, Rosini E (2016) Comparison of different microbial laccases as tools for industrial uses. New biotechnology. 33(3):387–398

Tong G, Bin L, Ran W, Yutao H, Jie L, Yuan L (2021) The enhanced fatty acids flavor release for low-fat cheeses by carrier immobilized lipases on O/W pickering emulsions. Food Hydrocoll 116:106651. ISSN 0268-005X. https://doi.org/10.1016/j.foodhyd.2021.106651

Ulery BD, Nair LS, Laurencin CT (2011) Biomedical applications of biodegradable polymers. J Polym Sci B Polym Phys 49(12):832–864. https://doi.org/10.1002/polb.22259

Underkofler LA, Severson GM, Goering KJ, Christensen LM (1947) Commercial production and use of mold bran. Cereal Chem 24:1–22

Vaishnav P, Demain AL (2009) In: Moselio S (ed) Encyclopedia of microbiology (Third edn) Industrial Biotechnology (overview). Academic Press, pp 335–348. ISBN 9780123739445. https://doi.org/10.1016/B978-012373944-5.00150-4. https://www.sciencedirect.com/science/article/pii/B9780123739445001504

Varga BM, Mónika SV (2019) Immobilization of laccase in alginate beads. Hungarian J Ind Chem 47. https://doi.org/10.33927/hjic-2019-16

Ventorino V, Aliberti A, Faraco V, Robertiello A, Giacobbe S, Ercolini D et al (2015) Exploring the microbiota dynamics related to vegetable biomasses degradation and study of lignocellulose degrading bacteria for industrial biotechnological application. Sci Rep 5:8161. https://doi.org/10.1038/srep08161

Zaks A (2001) Industrial biocatalysis. Curr Opin Chem Biol 5(2):130–136. https://doi.org/10.1016/S1367-5931(00)00181-2

Zhao X, Shi-Jian D, Tao G, Xu R, Wang M, Reuhs B et al (2010) Influence of phospholipase A2 (PLA2)-treated dried egg yolk on wheat dough rheological properties. Lebensm Wiss Technol 43(1):45–51. https://doi.org/10.1016/j.lwt.2009.06.027

# Downstream Processing 12

**Abstract**

With microbial sources, becoming a promising alternative for the production of many useful metabolites, these biotechnological processes have taken an upper hand. Many industries are working using these processes. However, the downstream processing of these processes is very expensive. Various equipments are required for these processes and many parameters need to be optimized. The microbial products can be either intracellular or extracellular. In case the product is intracellular, the cells need to be disrupted which is a very cumbersome procedure. However, with the recent developments, many separation methods have been adopted. The microbial cells vary in size, and this makes the separation process very challenging. For the separation of the cells, they can be either induced/aggregated/enlarged into fibers or into large spheres. With this, the gravity sedimentation or press filtration becomes easier and can separate the cells smoothly. Scientists have worked by genetic manipulation, on the genes which are involved in these processes. During the downstreaming of products, the cells can also be lysed by introducing the lytic genes into the cells. This chapter discusses the different methods of downstreaming of metabolites which are either extracellular or intracellular.

**Keywords**
Separation · Centrifugation · Down streaming · Intracellular · Extracellular · Optimization · Product



R. Agrawal, *Textbook of Industrial Microbiology*,
https://doi.org/10.1007/978-981-97-9582-6_12

## 12.1 Introduction

In downstream processing and analytical bioseparation, the biological products are separated and purified for various purposes. It is important that the purified product be obtained in large quantities, so that it is a profitable preposition during marketing. The maximum product yield can be obtained only after the optimization of various parameters. In analytical bioseparation the product is purified for measuring a component/components in a mixture, starting from a single cell. Downstream processing (DSP) basically involves a series of operation from where a pure and a homogeneous product can be extracted. It could be a single cell, tissue culture fluid, or plant tissue (Ying et al. 2019).

## 12.2 The Various Steps Involved in Downstream Processing

(a) **Removal of Insolubles** Initially the product is separated in the form of whole cells obtained as a solute from the culture broth. It is also termed as harvesting of microbial cells. There are many procedures for this step which include filtration, centrifugation, sedimentation, precipitation, flocculation, electroprecipitation, and gravity settling (Voulgaris et al. 2016). Procedures like grinding, homogenization, or leaching are also done if the product needs to be recovered from solid sources such as plant and animal tissues (Rathore and Shirke 2011; Chaichol and Weeranoppanant 2023).

**Filtration** This is a very common technique to separate the biomass from the filtrate. There are many parameters which affect the efficiency of the process. These are mainly, the size of the microorganism used, if any other organism is present along with, the viscosity of the medium and the temperature at which the process is being carried out. There are many types of filters that can be used in this method of separation. They can be depth filters, absolute filters, rotary drum vacuum filters, and membrane filters. Each type of filter will be discussed as follows:

- **Depth Filters** These filters have a filamentous matrix. Glass wool, asbestos, and filter paper fall in this category. The biomass goes within the matrix; however, the fluid goes out. This is the best method to separate the filamentous fungi.
- **Absolute Filters** These filters have a specific pore size that is smaller than the size of the microbial cells so that these cells can be removed easily. This separation method is excellent in removing bacteria from the culture medium.
- **Rotary Drum Vacuum Filters** When the particle size of the organism is very small (0.5–10 μm), in that case the cell separation is made from the broth by using rotary drum vacuum filter. However, it is important that the broth should contain around 10–40% of solids (by volume). This method is effective in filtering the cells of yeast and filamentous fungi. The operation is easy and it needs very less power. A rotating drum is partially immersed in a tank of broth. Once the power is switched on, then the drum starts rotating.

At this time, the microbial cell biomass gets deposited on the surface of the drum like a cake. Once the process is completed the cake is removed for further use.

- **Membrane Filters** This method of separation for the biomolecules involves the use of membranes which are made up of specific pore sizes and the use can be customized as per the size of the microorganism. In this method, basically, it has a semi-permeable membrane that selectively retains the particles/molecules that are bigger than the pore size. However, the smaller molecules pass through the membrane pores. These membranes are usually made up of polymeric materials like polyethersulfone and polyvinyl difluoride. The method has some drawbacks as it is very difficult to sterilize the membrane filters. However, with recent developments, either ceramic or steel materials are being used for these filters. This has made the cleaning and sterilization procedure easy and smooth (Francy et al. 2013).
- Membrane adsorbers These are porous membranes that have ion exchange groups and/or affinity ligands to which the proteins get bound. Later, the proteins can be eluted by chromatography.
- **Pervaporation** Extraction of volatile compounds can be done successfully by this method. It involves permeation through a membrane along with evaporation. However, with this method large-scale operations have been found to be difficult. Also, the cost of this method is high and therefore, it is not used by many industries (Alharbi et al. 2021).
- **Perstraction** It uses the principle of membrane filtration along with the solvent extraction. Separation of hydrophobic compounds can be done by this method (Almquist et al. 2023).
- The use of membrane has a drawback as the pores get clogged during the separation process. This limits the use of this method. The separation by this method can be of two types. One is static filtration and the other is a cross-flow filtration. In static filtration, the culture broth is passed through the membrane in a vertical manner. However, in cross-flow filtration, the culture broth is passed crosswise through the membrane. This helps in reducing the clogging.
- The particle size and other characters of the filter determine the type of filtration. Based on the membrane size, the technique is differentiated into microfiltration, ultrafiltration, and reverse osmosis (Barman et al. 2021).

(b) **Separation of Biomass**

(i) **Flotation** It is a method of solid-liquid separation of biomass. In this method, when the air is blown into the system, a foam is developed possessing a microorganism-bubble-complex (Zouboulis et al. 2008). The collector substances which include the long chain fatty acids and amines help in the formation of a stable foam. This is stable by collision and adsorption. Due to the difference in the density of the complex with that of the surrounding liquid medium, the complex goes to the surface of the liquid. This microorganism-enriched foam can then be removed from the surface. Optimization of various parameters like the flotation rate, the amount of gas and recycle flow rate, the bubble size, bubble surface area flow rate, saturation pressure, and the energy efficiency of the flotation helps in getting the maximum separation. Gulden et al. (2020) have shown that with the increase in the gas flow rate the flotation rate also increases and the process becomes more energy efficient (Riedele et al. 2020).

(ii) **Centrifugation** This method depends on the difference in the density of the particles to be separated and the liquid medium which determines the separation. The separation under the centrifugal force, happens at a specific rate in a solution (Mortimer et al. 2016). There are various parameters that determine the centrifugation. The optimization of mass and size of the Brownian particles, and also the gravitational field are very important. As the mass and size are different for various particles, it creates a difference in the sedimentation velocity; therefore, the particles get separated differentially. During centrifugation, the centrifugal force ($g$) is applied away from the center, which is outwards. The applied force can be determined quantitatively, by the square of the angular velocity of the rotor ($\omega$ in radians S − 1) and the radians ($r$, in centimeters) of the particle from the axis of rotation. Therefore, separation of solid particles from the liquid phase (fluid/particle separation) can be done using this technique. The method is quite effective at the laboratory scale, but at large scale it has not proved to be effective (Liu et al. 2010). Recently, continuous flow industrial centrifuges, have been developed. These are very effective in separation at large industrial levels. In this method, the separation, is a continuous process in which the sample is fed continuously, and the clarified product is also separated continuously. Garrison and Bochdansky (2015) discovered a simple separation method for downstreaming of aquatic microbes. Heasman et al. (2000) developed an extended shelf-life of microalgae concentrate diets by the method of centrifugation.

There are many types of centrifuges as described below:

1. **Tubular Bowl Centrifuge** It is usually a very small centrifuge and is commonly used in pilot plants. During the operation, it can run at a high centrifugal speed of 15,000 rpm. It has the ability to run both in batch as well as in continuous mode. The solids are removed manually. The centrifugal force by the bowl is the factor that decides the speed of the rotor.
2. **Disc Centrifuge** It has a series of discs which are usually conical in shape. This gives large space for centrifugation. The base of the centrifuge tank is wrapped around a center line. A rapid rotation is given using a centrifugal force/gravitational force (more than 10,000 times greater than the force of gravity). As the centrifugal force is applied, the clarified fluid moves upwards, and the solids come down, and settle at the bottom. The disc centrifuge may have nozzles to allow a continuous exit of the concentrated suspension. The nozzle allows operation in a continuous mode.

3. **Multi-chamber Centrifuge** This type of centrifuge consists of many chambers which are connected in a way that the sample feed passes in a zigzag fashion. Each chamber has a different centrifugal force. The peripheral chambers have higher gravitational force. This results, in the small-sized particles to settle down in the outermost chamber. It is a very efficient method, as the sample runs for a long time in various chambers. However, the cleaning of multi-chamber centrifuges is quite challenging as it takes a long time. Moreover, as the discharge is done manually, it is difficult to separate the samples with a higher content of solids. Usually, samples with less than 4–5% solids by volume can be separated easily by these centrifuges.
4. **Scroll Centrifuge** It is composed of a rotating horizontal bowl that is tapered at one end. This method can be applied to samples that have a high solid concentration and are with a biomass/solid content of 5–80%. The fluids can be easily concentrated by this method. The solids are deposited on the wall of the tank which can be removed after scrapping. This type of decanter centrifuge is also known as a solid bowl centrifuge. The solid particles are separated continuously. The method is commonly used in the wastewater treatment and processing industries which use chemicals, oils, and food.

(iii) **Sedimentation** The higher impurities can be separated by this method. The heavy particles get settled from the suspension with the effect of gravitational force. The particles that settle down from the suspension are known as sediment and therefore the technique is known as sedimentation. Pitt et al. (2020) have studied the factors affecting sedimentation separation of bacteria from blood. The Brownian particles get velocity under a gravitational field. This is also known as sedimentation/settling velocity. The unit of sedimentation coefficient is Svedberg. When water is treated, initially the concentration of solid particles is reduced by sedimentation. It also helps in reducing the requirement of coagulating chemicals. There are many different ways for sedimentation which include horizontal flow, radial flow, inclined plate, ballasted floc, and floc blanket sedimentation.

(iv) **Precipitation** In this method, the separation of compounds is based on the solubility which is dependent on the strength of the ions in the solution, the pH, and the temperature of the solution (Li et al. 2021). When the cations and the anions combine to form an insoluble ionic solid in an aqueous solution, then it is called a *precipitate*. When the separation is made under optimized conditions, the desirable compound becomes an insoluble solid. This comes out of the solution and can then be separated from the remaining liquid. It is a good method to separate a specific compound from a mixture. However, all aqueous reactions do not form precipitates and therefore, all factors and the solubility rules should be studied before using this method. The prediction of these reactions helps in the determination of the ions present in a solution. Industries can produce chemicals by using precipitation techniques. The solids produced in these reactions are crystalline solids, which can either be suspended throughout the liquid or may fall to the bottom of the solution. The remaining fluid is called as the supernatant liquid. The two components of the mixture, namely, the precipitate and the supernatant, can be separated by any of the methods like filtration, centrifugation, or decantation. The precipitation reactions are of two types. These are either single-replacement reactions or double-replacement reactions. In a double replacement reaction, the two ionic reactants dissociate but bind to the respective anion or the cation from the other reactant. The ions replace each other based on their charges as either a cation or an anion. This is termed as switching partners. This type of double replacement reaction is a *precipitation reaction* that happens in aqueous solution and one of the products formed is insoluble. The method is efficient in removing nucleic acids, pigments, neutral salts, organic solvents, and high molecular weight polymers. There are many protein-specific precipitations as affinity precipitation and ligand precipitation.

- Ammonium sulfate precipitation: Ammonium sulfate does not have any toxicity to the proteins and is cost-effective. Therefore, this method is used very commonly for the precipitation of proteins (Wingfield 2001). It increases the hydrophobic interaction between protein molecules which helps in their precipitation. In this method, the precipitation of the protein is dependent on many factors as protein concentration, pH, and temperature. Mirica et al. (2012) worked on the selective precipitation and purification of monovalent proteins using oligovalent ligands and ammonium sulfate.
- Organic solvents for precipitation of proteins The common organic solvents used for protein precipitation are ethanol, acetone, and propanol (Baghalabadi et al. 2021). These solvents reduce the dielectric constant of the medium. This in turn increases the electrostatic interaction between the protein molecules which results in the precipitation of proteins. The process is carried out below 0 °C to protect the proteins from getting denatured.
- Precipitation of proteins using nonionic polymers Nonionic polymers like polyethylene glycol (PEG) help in the precipitation of proteins (Ward et al. 2016). As it has a high molecular weight the proteins get precipitated. This polymer is not toxic and does not denature the proteins.
- Precipitation by using ionic polymers Polyacrylic acid and polyethyleneimine are charged polymers that form complexes with oppositely charged protein molecules, resulting in the precipitation of proteins (Antunes et al. 2023).
- Increase in temperature The heat-sensitive proteins can be precipitated by increasing the temperature (Mackay and Chilkoti 2008).
- Change in pH for precipitation Alterations in pH helps in protein precipitation (Li et al. 2021).

- Affinity precipitation The affinity interaction between antigen and antibody precipitates the proteins (Anaya et al. 2013).
- Precipitation by ligands Ligands have a specific binding site for proteins. This, allows the precipitation to happen in a very selective manner (Abrusán and Marsh 2018).
- Adsorption In earlier times, activated charcoal was used as the adsorbent material. However, recently, the protein is precipitated using cellulose-based adsorbents (Bai et al. 2023). The low molecular weight compounds (vitamins, antibiotics, and peptide matrices) are made up of either polystyrene, methacrylate, or acrylate. During the adsorption process, a bed is made which is prepared with an adsorbent column and the culture broth is passed through this column. The product gets adsorbed into the adsorbent which is then eluted out.

(v) **Flocculation** In this method of separation, the microbial cells form large aggregates and then they settle at the bottom of the fermenting vessel. Later, the cells can be removed easily from the vessel. This method is dependent on the nature of cells and the ionic constituents of the medium. Many times, flocculating agents like inorganic salt, organic polyelectrolyte, and mineral hydrocolloid are added to bring about flocculation effectively (Ojima et al. 2015).

(vi) **Electroprecipitation**

Electrostatic Precipitators It was the work of Hohlfield (1824), who demonstrated the probability of electrical separation for the first time. Later, it was Sir Oliver Lodge (1931) who collected lead fumes using an electrostatic precipitator through a steam engine. This separation method was used by various industries for the collection of solid and liquid particulates. In this method, the dust particles are charged by electrostatic energy. When the particles get charged, they are then deposited on the collection device. When enough charged dust gets inside the collector, then it is shaken with the force of gravity. This makes the dust particles to fall below and then they get collected into a hopper. It is then removed by a conveyor. The collected dust can be either removed or recycled. The electrostatic precipitators vary in sizes and types. The method is not used in high-carbon ferrochrome furnaces as the dust is usually very fine; therefore it limits the dust removal efficiency. Due to high voltage, there is a risk of explosion. This could mainly happen because of carbon monoxide gas. However, the initial cost of this method is very high, which does not make the method effective (Hosansky 2021).

Industrial applications of electroprecipitation are of two types

**The horizontal-flow plate-type precipitator** In this method, a series of parallel plates are arranged in a space of 400 mm. This helps the gas to pass. An insulation is made on the discharge electrodes which are placed in the center of the plates. The plate precipitators are preferred by the industries.

(a) **The tube-type precipitator** In this method, there is a vertical tube and the gas is made to pass upwards through these tubes. The tubes measure about 250 mm in diameter. The discharge electrode is in the form of a coaxial element. They are used for mist collection. This limits the application of the method. The particles need to be removed periodically from the collector. For higher efficiency, the discharge electrodes should be maintained and cleaned periodically. This preserves their sharp edges and also the emission capability. The particle size of the dust decides the diffusion process. If the particles are more than 2 microns in diameter, then the particles get charged due to the ionic collision. If the particles are less than 0.2 microns in diameter, then the charge is acquired by diffusion processes (Sara et al. 2022). There are many advantages of this method of separation as the efficiency is high (99.9%) even for particles of 0.01 micron, the method can work up to 850 °C temperature. The range of pressure survival is also high which could be from minus 0.2 to plus 20 bar. The requirement of energy is low and the process is cost effective with high efficiency.

- **Gravity settling** This method follows Newton's Law for coarse particles. The separation is done on the basis of gravitational force in a liquid that has a mixture of minerals. The method is used in the mineral sand industry like tantalum and tin. The principle is based on the specific gravity which is the ratio of its density to a standard substance. Water is usually taken as the standard which has a unit of one (1). Any particle with less than 1 as specific gravity floats on the surface of water. However, if it is more than 1, the particles sink to the bottom. But, when gravity separators are used, then they utilize air as a standard and not water. This makes the method very precise for accurate separation. The process of stratification of air is used. Depending on the weight of the particles, with respect to air, they either rise or fall and thus get separated (Hogan et al. 2023) (Fig. 12.1).

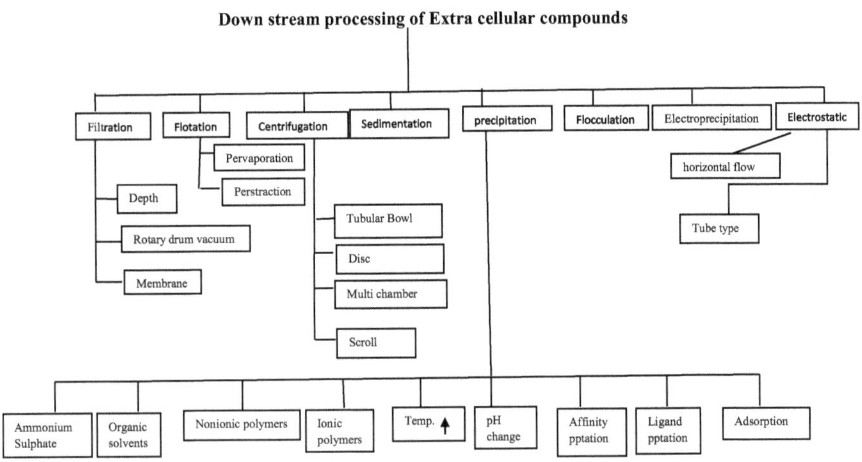


**Fig. 12.1** Downstream processing of extracellular compounds

## 12.3 Downstream Processing of Intracellular Products

Many products as vitamins and enzymes are produced inside the cell (intracellular). Therefore, to obtain these products in an active form, they need to be removed from the cells, for which the cells should be disrupted. The different methods for cell disruption are physical, chemical, or enzymatic methods (details given in Chap. 13 of this book). Low molecular weight compounds are extracted with the help of organic solvents. This is a good method for the lipophilic compounds, but not for hydrophilic compounds. The extraction can be done as follows:

- Physical extraction The sample is distributed between the two liquid phases. The method is good to extract nonionizing compounds.
- Dissociation extraction Ionisable compounds can be extracted by this method of dissociation. It is a good method to extract antibiotics.
- Reactive extraction In this method, a carrier molecule like a phosphorus compound, or aliphatic amine is taken with which the desired product is made to react. Then the product is extracted with an organic solvent.
- Supercritical fluid (SCF) extraction Supercritical fluids (SCFs) which are the intermediates between gases and liquids are used in this method. These are in the form of fluids above their critical temperature and pressure. Supercritical $CO_2$ has a low critical temperature and pressure and therefore, it is effective. However, the high cost of the method makes it useful only for specific extractions. Extraction of caffeine from coffee beans, pigments, and flavour ingredients from biological materials are made by this method.
- Extraction of high molecular weight compounds As proteins get decomposed with organic solvents and lose their biological activities, they are extracted by using an aqueous two-phase system or by reverse micelle formation.
- Aqueous two-phase systems (ATPS) A mixture is prepared by mixing a polymer (polyethylene glycol) and a salt solution (ammonium sulfate). Two different polymers can also be mixed. The two phases are not miscible. The proteins get separated and they come into one phase, leaving the cells and other solids in another phase.
- Reverse miceller systems for separation Aggregates of surfactant molecules are made in organic solvents with water. By forming reverse micelles in the aqueous medium, the proteins are separated out. There is not much loss in biological activity; therefore, the method is used commonly.

Gottschalk (2011) have written an overview on the downstream processing in the biomanufacturing of metabolites in various industries (Fig. 12.2).

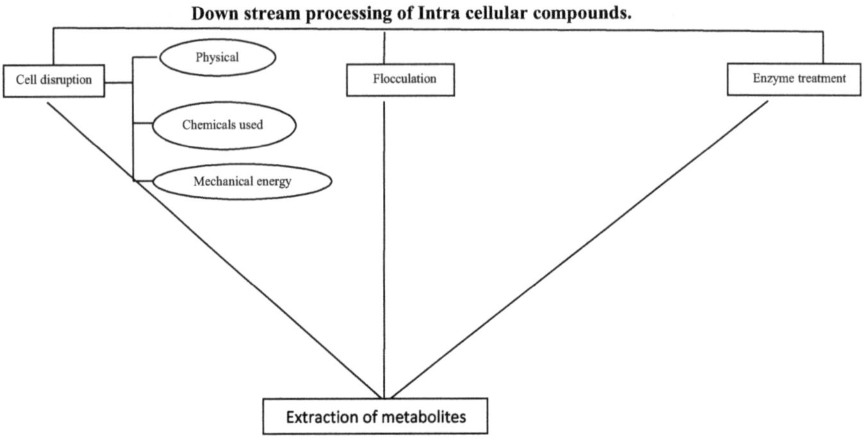


**Fig. 12.2** Extraction methods for Intracellular compounds

## 12.4 Product Isolation and Concentration

The desired product is separated in this step, of the process. Most of the times, the product has water (80–98%) which is the major impurity. This can be removed by concentrating the product. The concentration can be done by using either solvent extraction, adsorption, ultrafiltration, or precipitation methods. The actual procedure to be adopted for concentration depends on the nature of the product, the quality, and the quantity of the material to be utilized. The cost-effectiveness is another major criteria which needs to be looked into. Removal of water from the broth can be done by simple evaporation with a heating device for steam. Zhang et al. (2013) have demonstrated that the apparatus for virus concentration is based on the evaporation and interfacial tension. They have worked on the development of future aerosol detection devices that can control the airborne transmission diseases like COVID-19, influenza, and other airborne transmission viruses.

The evaporator concentrators are of many types as given below.

(i) Plate evaporators For the liquid to get concentrated, the evaporator is placed on the plates, on which the steam is supplied. This results in a viscous consistency.
(ii) Falling film evaporators The sample liquid flows down the long tubes, which then forms a thin film on the heating surface.
(iii) Forced film evaporators The sample liquid film is subjected to move mechanically to get a dry product concentrate.
(iv) Centrifugal forced film evaporators With the help of a centrifugal force the liquid is passed over heated plates to concentrate the sample (Fig. 12.3).

After the extracellular and the intracellular products are extracted and isolated, the steps involved for concentration up to the final product are the same. The

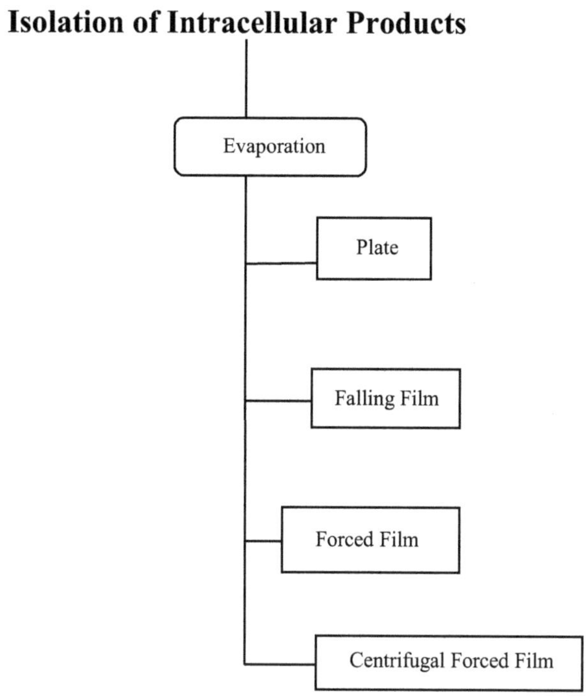


**Fig. 12.3** Centrifugal forced film evaporators

concentrated product can be separated, which is then purified, by various chromatographic methods. These could be either affinity, size exclusion, reversed phase, or ion-exchange chromatographic methods. Then, the product is crystallized and fractionally precipitated. The final processing steps include product polishing followed by packaging of the product in a stable, transportable, and convenient form. Crystallization, desiccation, lyophilization, and spray drying are the unit operations that are often used. Sometimes, sterilization of the product is done to remove or deactivate any trace contaminants which could be harmful for the safety of the product. Such operations might include the removal of viruses or depyrogenation. Affinity chromatography often isolates and purifies the product in a single step. Recovery of the bioproducts is usually done by using, disintegration or separation techniques. The recovery of solids and liquids along with soluble molecules can be done by the finishing operations which include processes such as drying and crystallization. Optimization of the product yield is extremely important. It is an upstream process. If the product concentration is high in the recovery program, then, it helps in reducing the number and size of the downstream processing operations and also the recovery cost. Separation of cells and the extracellular fluid is the first physical step in the recovery process. If the product is a liquid, then the recovery is carried out by unit operations which can concentrate and purify a soluble molecule.

Initially, the unit operation should be done using a high concentration of the raw material, which could be reduced later. During the recovery of a specialized biomolecule, expensive equipments are required, which give a highly purified product.

If the product is a solid and is in the cytoplasm or if it is adhered to the cell surface, in that case, the extraction of the product needs to be done before purification. For this, the cell has to be broken and the cell extract needs to be separated from the cell debris. The cell extract is then concentrated and purified. If products contain *S. cerevisiae* and *S. carlsbergensis* cells (usually common in the food and beverage industries), then drying of the product is required as the finishing operation. It is important to see that the procedure for separating the product is cost-effective. For this, it is important to understand the nature of the product (intra-/extracellular, physicochemical properties), choice of feedstocks, selection of fermentation process (submerged, solid state), free cells/immobilized cell fermentation, and the type of catalysts (types of microorganisms, enzymes) used. The specificity of the separation method depends on the type of microorganism (filamentous fungi, yeast, or bacteria). Other parameters include the nature of the product as intra-/extracellular, the chemical and physical properties, substrates being used in the fermentation process, type of fermentation process which could be submerged/solid state, and the purity specifications. Many byproducts arise from succinic acid fermentation that have similar structures and properties. These have an impact on the cost of separation procedure. It has been observed that in the production of succinic acid, there is a constant release of organic acids. This lowers the pH during the process; therefore, a pH control is needed. The pH is controlled by using calcium, magnesium, sodium or potassium hydroxides, carbonates, ammonium salts or ammonia which produces succinate salt, and through acidification with sulphuric acid it gets converted into succinic acid. Yeasts can tolerate low pH. Many dicarboxylic acids have been found at pH below 3.0 and are present in their undissociated form. Carbon sources like mono or disaccharides are used commonly in the fermentation processes. However, presently, low-cost complex raw materials like agro-industrial wastes are being used by industries. This has replaced the petroleum-based processes with the bio-based industrial fermentation processes. In such processes, a pretreatment is given to the material which can digest and remove the inhibitors. It produces a cleaner broth which improves the fermentation process and reduces the final process cost. Minute amounts of monocarboxylic acids are found which are detrimental to the final product. As they have only one functional group, they act as polymerization terminators. The specificity of any downstream process includes; the physical and chemical characteristics of the material, type of the metabolite to be extracted, polarity, acid dissociation constant, sensitivity, and solubility of the product. Ethanol can be purified by using a combination of distillation and molecular sieving method as it allows the smaller water molecules (0.28 nm) to pass through and retain only the dehydrated ethanol (0.44 nm). The moisture is dehydrated using azeotropic distillation. Molecular sieves can be regenerated by heating or by applying vacuum. Downstream processing equipment includes laboratory systems, pilot-scale equipment, and production scale. Wrongly designed equipment can inhibit the process yield and economy. From a commercial point of view, it is very important that the product recovery should be much more than the cost of production to make the process economically feasible.

Steps have been optimized to obtain PUFAs (polyunsaturated fatty acids). This involves three major steps. Starting from the recovery of the biomass, extraction of the lipids, concentration, and then purification of the PUFAs. In case of microalgae extraction, the processes like centrifugation or filtration are being used for the removal of the biomass from the culture broth. When flocculants are added, then the recovery of biomass yield is much higher. This is attributed to the enhanced cell aggregation. Spray drying and freeze drying are good drying methods for lipid extraction. However, if the drying step can be omitted it can reduce the production cost to a large extent.

An appropriate solvent system should be used for the extraction of lipids from algal cells. The extraction is very efficient, when the solvents are added in the order of increasing polarity (chloroform > methanol > water). After the lipid extraction, PUFAs can be concentrated by various methods which include chromatographic methods, supercritical fluid extraction, or urea fractionation. The HPLC method has been found to be very effective for the recovery of pure products. Methods like microfiltration, ultrafiltration, nanofiltration, and reverse osmosis have been found useful for the purification. Out of these methods, ultrafiltration is a very common method. Li et al. (2004) used 5-, 20-, and 50-kDa spiral wound ultrafiltration membranes for the removal of bioemulsifiers from the fermentation broth which contained, benzylpenicillin antibiotic to improve the efficacy of the solvent extraction process. An antibiotic, surfactin, was produced in a fermentor by using *Bacillus subtilis*. This was purified by using two ultrafiltration membranes (MWCO = 3 and 10 kDa) in a series (Mohd et al. 2007). The market for microbial pharmaceutical proteins (biopharmaceuticals) has grown a lot and is expected to take over the medicine market due to their very low production costs and built-in safety measures. The medicines which are produced using microbial cultures have an added advantage as they can be grown in the laboratory and the supply can be constant without any interruption. It has also shown that the pathogenic cultures do not get replicated. This inhibits any contamination. The final product is lower in cost which makes the process cost-effective. It has also been found that the upstreaming is easy for the production of biopharmaceuticals from microbial cultures. Membrane processes can be scaled up easily, where the size of the target protein is double than that of the size of the host cell protein (Catherine 2012) (Fig. 12.4).

There are many specifications for any product formation. All these specifications are given below in the Fig. 12.5.

## 12.5 Analytical Support for Downstream Processing

During the downstream processing, many impurities must be separated. These should be measured quantitatively. The major ones include the host cell proteins and nucleic acids which may have different molecular weights. For quantitation, it is important to optimize various parameters with the help of immunological assays

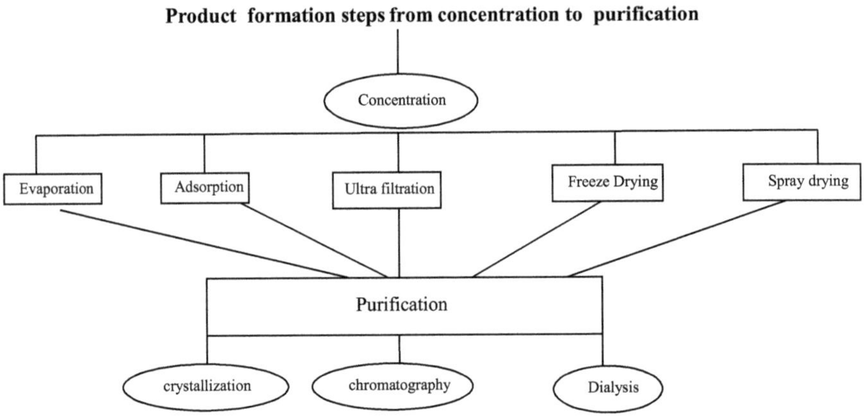


**Fig. 12.4** Product separation steps from concentration to purification

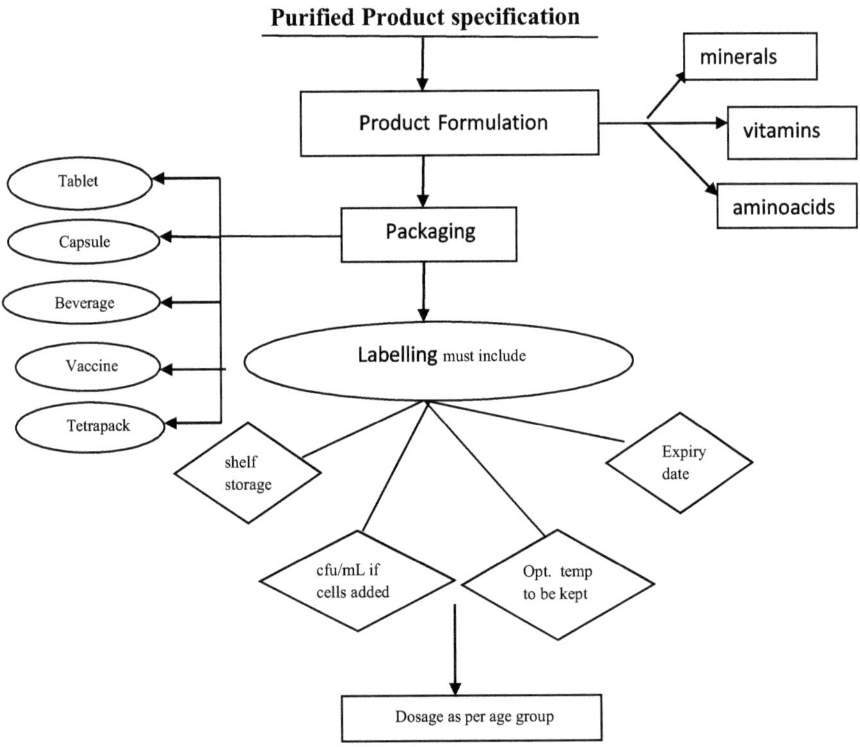


**Fig. 12.5** Product specifications

(qPCR). The use of multiplexed technique has made it very easy as the sample handling and clean-up techniques take less time for large-scale manufacturing. The amount to be loaded on to a column can be determined by a simple titer assay which is based on the absorbance or HPLC (A280/ HPLC based). If a consistent amount of protein is loaded on to a column, the product quality gets ensured. The product elution from the column can also be determined using different assays. Many times, during the downstream processing it has been observed that there is degradation of the protein by either oxidation, deamidation, isomerisation, or clipping. This can also happen if there is a process hold or if light is exposed. These can be well detected by LC-MS technique. UV-based peptide mapping, charge heterogeneity, and reverse phase chromatography are other methods that can identify the degraded protein from the nondegraded protein.

(a) **Reusability of Nanobiocatalysts (NBCs)**

In scaling-up process, the reusability of the nanobiocatalyst plays a major role. The microcarriers can be separated easily from the reaction medium, but it is difficult for nanocatalyst to be separated. Reusability of the NBCs can be improved, when the magnetic field combines with the immobilized enzyme, and it can be easily separated from the reaction media. Nanoparticles initially dissociate and later recluster after the product is formed. The product can be separated using the magnetic field. The results are very promising when polymer-coated nanocarriers are utilized. These nanocarriers can be recycled; therefore, it helps to bring down the bioprocessing cost. Processing steps include harvesting, filtration, primary capture, buffer exchange, concentration, purification, bioconjugation, and formulation.

## FAQs

1. Which is the most important step in downstream processing and why?
2. What are the main steps involved in downstream processing?
3. What are the recent methods used in downstream processing?
4. What is membrane-based bioseparation?
5. What are the different centrifuges used in downstream processing after fermentation?
6. What are the different types of filtration processes?
7. What is the goal of downstream processing?
8. What are the factors that affect downstream processing?
9. What are the steps involved in downstream processing?
10. How is a final product specified. What features are important in product labeling?

## References

Abrusán G, Marsh JA (2018) Ligand binding site structure influences the evolution of protein complex function and topology. Cell Rep 22(12):3265–3276. https://doi.org/10.1016/j.celrep.2018.02.085

Alharbi WN, Saeed WS, Alwarthan AA, Badjah-Hadj-Ahmed AY, Aouak T (2021) Extraction of organic volatile pollutants in over saturated water by pervaporation technique using a poly (dimethylsiloxane) based sealer as a membrane. Water 13(8):1049. https://doi.org/10.3390/w13081049

Almquist CB, Garza L, Flood M, Carroll A, Armstrong R, Chen S, Marcellino C (2023) Perstraction: a membrane-assisted liquid–liquid extraction of pfoa from water. PRO 11(1):217. https://doi.org/10.3390/pr11010217

Anaya JM, Shoenfeld Y, Rojas-Villarraga A et al (eds) (2013) Autoimmunity. From bench to bedside. El Rosario University Press, Bogota (Colombia)

Antunes EC, Oliveira F, Temmink H, Schuur B (2023) Separation of polysaccharide and protein by ionic liquid based extraction techniques. J Ionic Liquids 3(2):Article 100058. https://doi.org/10.1016/j.jil.2023.100058

Baghalabadi V, Razmi H, Doucette A (2021) Salt mediated organic solvent precipitation for enhanced recovery of peptides generated by pepsin digestion. Proteomes 9(4):44. https://doi.org/10.3390/proteomes9040044

Bai X, Zhongxiang L, Pengfei L, Yijun Z, Linfeng H, Tongchao S (2023) An eco-friendly adsorbent based on bacterial cellulose and vermiculite composite for efficient removal of methylene blue and sulfanilamide. Polymers 15(10):2342. https://doi.org/10.3390/polym15102342

Barman SR, Roy U, Das P, Mukhopadhayay A (2021) Chapter 6-Membrane processes for removal of polyaromatic hydrocarbons from wastewater. In: Green chemistry and water remediation, research and application. Elsevier, Pub, pp 189–207. https://doi.org/10.1016/B978-0-12-817742-6.00006-2

Catherine C (2012) Membrane chromatography. In: Catherine C (ed) Membrane processes Biotechnol Pharma. Elsevier, pp 169–212. ISBN 9780444563347. https://doi.org/10.1016/B978-0-444-56334-7.00005-8

Chaichol P, Weeranoppanant N (2023) Advances in insitu and in-line liquidÔÇôliquid extraction for bioprocess intensification. React Chem Eng 8(10):2387–2402. Pub: The Royal Soc Chem. https://doi.org/10.1039/D3RE00266G

Francy DS, Stelzer EA, Brady AM, Huitger C, Bushon RN, Ip HS, Ware MW, Villegas EN, Gallardo V, Lindquist HD (2013) Comparison of filters for concentrating microbial indicators and pathogens in lake water samples. Appl Environ Microbiol 79(4):1342–1352. https://doi.org/10.1128/AEM.03117-12

Garrison CE, Bochdansky AB (2015) A simple separation method for downstream biochemical analysis of aquatic microbes. J Microbiol Meth 111:78–86. https://doi.org/10.1016/j.mimet.2015.01.025

Gottschalk U (2011) Overview of downstream processing in the biomanufacturing industry. In: Murray MY (ed) Comprehensive biotechnology, 2nd edn. Academic Press, pp 669–682. ISBN 9780080885049. https://doi.org/10.1016/B978-0-08-088504-9.00238-5

Gulden S, Riedele C, Kopf MH, Nirschl H (2020) Potential of flotation as alternative separation process in biotechnology with focus on cost and energy efficiency. Chem Eng Sci 218:Pub. Elsevier BV 115117.ISSN0009-2509. https://doi.org/10.1016/j.ces.2019.07

Heasman MJ, Diemar W, O'Connor T, Sushames FL (2000) Development of extended shelf life microalgae concentrate diets harvested by centrifugation for bivalve molluscs—a summary. Aquac Res 31(8–9):637–659

Hogan K, Paul S, Lin G, Fuerte-Stone J, Sokurenko EV, Thomas WE (2023) Effect of gravity on bacterial adhesion to heterogeneous surfaces. Pathogens 15;12(7):941. https://doi.org/10.3390/pathogens12070941

Hohlfeld M (1824) Das Niederschlagen des Rauches durch Elektricitat. Arch Gesammte Nat 2:205–206

Hosansky D (2021) Electrostatic precipitator. Encyclopedia Britannica https://www.britannica.com/technology/electrostatic-precipitator

Li SZ, Li XY, Cui ZF, Wang DZ (2004) Application of ultrafiltration to improve the extraction of antibiotics. Separ Purifn Technol 34:115–123. https://doi.org/10.1016/S1383-5866(03)00185-0

Li LT, Liu J, Guosheng F, Changliang Z, Qu S, Bo Z, Yunfei (2021) Insight into the influence of pH on the precipitation induced by bacteria. Soil Use Manag 38. https://doi.org/10.1111/sum.12779

Liu HF, Ma J, Winter C, Bayer R (2010) Recovery and purification process development for monoclonal antibody production. MAbs 2(5):480–499. https://doi.org/10.4161/mabs.2.5.12645

Mackay JA, Chilkoti A (2008) Temperature sensitive peptides: engineering hyperthermia-directed therapeutics. Int J Hyperth 24(6):483–495. https://doi.org/10.1080/02656730802149570

Mirica KA, Lockett MR, Snyder PW, Shapiro ND, Mack ET, Nam S, Whitesides GM (2012) Selective precipitation and purification of monovalent proteins using oligovalent ligands and ammonium sulfate. Bioconjug Chem 23:293–299

Mohd HMI, Diego EC, Richard AF, Paula J (2007) Recovery and purification of surfactin from fermentation broth by a two-step ultrafiltration process. J Membr Sci 296(1–2):51–57. ISSN 0376-7388. https://doi.org/10.1016/j.memsci.2007.03.023

Mortimer M, Petersen EJ, Buchholz BA, Holden PA (2016) Separation of bacteria, protozoa and carbon nanotubes by density gradient centrifugation. Nanomaterials (Basel) 6(10):181. https://doi.org/10.3390/nano6100181

Ojima Y, Nguyen MH, Yajima R, Taya M (2015) Flocculation of *Escherichia coli* cells in association with enhanced production of outer membrane vesicles. Appl Environ Microbiol 81(17):5900–5906. https://doi.org/10.1128/AEM.01011-15

Pitt WG, Alizadeh M, Blanco R, Hunter AK, Bledsoe CG, McClellan DS, Wood ME, Wood RL, Ravsten TV, Hickey CL et al (2020) Factors affecting sedimentational separation of bacteria from blood. Biotechnol Prog 36(1):e2892. https://doi.org/10.1002/btpr.2892

Rathore A, Shirke A (2011) Recent developments in membrane based separations in biotechnology processes: review. Prepar Biochem Biotechnol 41:398–421. https://doi.org/10.1080/10826068.2011.613976

Riedele SJC, Kopf MH, Nirschl H (2020) Potential of flotation as alternative separation process in biotechnology with focus on cost and energy efficiency. Chem Eng Sci 218:115117, ISSN 0009-2509. https://doi.org/10.1016/j.ces.2019.07.035

Sara BEA, Goran FGA, Johan G (2022) Effect of fluid velocity and particle size on the hydrodynamic diffusion layer thickness. Eur J Pharm Biopharm 180:1–10

Sir Oliver Lodge (1931) Past years. An autobiography. Hodder and Stoughton, London

Voulgaris I, Chatel A, Hoare M, Finka G, Uden M (2016) Evaluation of options for harvest of a recombinant *E. Coli* fermentation producing a domain antibody using ultra scale-down techniques and pilot-scale verification. Biotechnol Prog 32(2):382–392. https://doi.org/10.1002/btpr.2220

Ward KC, Shu S, David (2016) Protein separation using non-ionic and cationic surfactant precipitation. J Chem Technol Biotechnol 91:n/a–n/a. https://doi.org/10.1002/jctb.4942

Wingfield P (2001) Protein precipitation using ammonium sulfate. Curr Protoc Protein Sci Appendix 3:Appendix 3F. https://doi.org/10.1002/0471140864.psa03fs13

Ying W, Chen L, Yong C, Xiaoran J, Guo-Qiang C (2019) Microbial engineering for easy downstream processing. Biotechnol Adv 37(6):107365.ISSN0734-9750. https://doi.org/10.1016/j.biotechadv.2019.03.004

Zhang JY, Mahalanabis M, Liu L, Chang J, Pollock NR, Klapperich CM (2013) A disposable microfluidic virus concentration device based on evaporation and interfacial tension. Diagnostics (Basel) 3(1):155–169. https://doi.org/10.3390/diagnostics3010155

Zouboulis AZ, Lazaridis NK, Matis KA (2008) The process of flotation: an efficient solid/liquid separation technique for biological materials. Int J Environ Pollu 32. https://doi.org/10.1504/IJEP.2008.016896

# 13 Cell Disruption and Extraction Methods for Intracellular Lipids and Proteins


**Abstract**

Most of the microorganisms like bacteria, yeast, and microalgae have the ability to produce metabolites that are useful to human beings. These are important as they are considered natural and come under GRAS category. Important metabolites are alcohol, wine, enzymes, lipids, natural colors, organic acids, and proteins. Microbial metabolites have many advantages as the production on large scale is very cost-effective, and they can be produced easily. However, many of these high-valued biomolecules are produced inside the cells and are, therefore, known as intracellular metabolites. Therefore, to obtain these metabolites, the cells have to be disrupted. The procedure involves many steps for down streaming of the metabolite. The most important is to disrupt the cell wall. There are many different methods for cell disruption, and the selection of the method depends on the cell wall constituents, application feasibility, and the cost of the process. This chapter discusses the various methods of cell disruption/permeabilization in bacteria, yeast, and algae. There are many important factors that decide the cell disruption method in each microorganism. The major being, cell wall composition, nature of biomolecules, and the degree of purity in which the product is required to be extracted. The power consumption in the process and the level at which the product has to be scaled up are vital factors which need to be studied, before taking up the disruption method. This chapter discusses the optimization of parameters for the maximum yield of cell disruption. There are many methods like bead milling and high-pressure homogenization (HPH) that can be utilized on the large scale as they have been found to be very effective. These methods have been explained in this chapter. Energy consumption is another major aspect during downstream processing which involves the heat management. Also, the cooling systems are needed, so that the proteins do not get inactivated. All these aspects of downstream processing have been explained in this chapter.





R. Agrawal, *Textbook of Industrial Microbiology*,
https://doi.org/10.1007/978-981-97-9582-6_13

**Keywords**
Cell disruption · Cell wall · Intracellular · Extracellular · Enzymes · Inactivation · Proteins · Lysis · Cell membrane.

## 13.1 Introduction

Cell disruption, or cell lysis, is a process of breaking the cell wall and/or membrane in order to release intracellular fluids containing molecules or particles of interest, such as proteins or viruses. The cell disruption is usually done with the help of enzymes. These enzymes are used as industrial catalysts, therapeutic agents, analytic reagents, and also as manipulative tools in genetics. Around 95% of the enzymes are available from the suppliers in different quantities ranging from μg to kg. Most of these are essentially used for research purposes. Many enzymes (40–50) are produced on a large scale (from multiple kilograms to tons/annum). Enzymes like hydrolases (proteases, amylases, cellulases, and lipases) have a lot of demand in various industries. These are usually available in the form of liquid concentrates, powders, or granules. After dissolving the cell wall with such concentrates, the soluble enzyme gets released. The major producer of these enzymes is a Danish company Novozyme. This has a market share of 47%. Another U.S. company, DuPont, has a share of 21%. The world market has increased from £110 million in 1960 to £200 millions in 1970, to £270 millions in 1980, £1000 millions in 1990, and more than £2000 millions in 2010. The global enzyme market was about $ 12 billion in 2022 which is expected to grow at 6.8% from 2022 to 2027. Enzymes used in food and beverages are the largest sector in the industrial enzyme market. This was £750 million in 2010. Market for enzymes used in technical applications (for diagnostic applications, research, and biotechnology) accounts for another £700 million. It has been estimated that in future the demand will grow at a rate of 6–7% per year. The developing economies of the Asia-Pacific region, the Middle East, and Africa are now also seen to be emerging as the fastest-growing markets for industrial enzymes (Singh et al. 2016). Microbial enzymes are used widely as compared to plant and animal enzymes as they are easy to produce and the optimization of process parameters is much easier. Also the process can be conducted in a laboratory. They need less energy input during production and are cost-effective. Therefore, their role is increasing in food, agriculture, chemicals, and pharmaceutical industries. But most of the enzymes are intracellular. Therefore, disruption of cells is an important step. In cell disruption, the outer boundary of the cell or the cell membrane is broken down. This allows the cellular materials like DNA, RNA, proteins, or the cell organelles to get released. For best results, it is vital to have a very reliable and efficient cell disruption method. Optimization of different parameters is vital, as the biological macromolecules should not be decomposed. There is a need to get them in the intact form. The temperature, pH, and time of treatment are important factors to obtain the best conditions for cell disruption and also for releasing the product. Most of the times the cell disruption process is done at 4 °C or even at much lower temperatures. Before taking up the cell disruption process, it is important to study the microorganism type, the cell wall constituents, and the physiological state of the cells. The shape of the bacteria affects the disruption of the cell. Cell wall disruption for bacilli bacteria (in the shape of a rod) has been found to be much easier than a circular one, which are known as cocci as they have been found to be resistant to mechanical forces. The Gram-positive bacteria have more rigid cell wall as compared to the Gram-negative bacteria. Therefore, the cell wall of Gram-negative bacteria can be easily disrupted with the help of osmotic shocks. The cell wall surface (S) layer proteins and polysaccharide capsules also affect the cell disruption efficiency. It is known that in the Gram-positive bacteria (*B. subtilis* and *S. aureus*) the cell wall becomes thick in the stationary phase. This reduces the efficiency of disruption (Middelberg 1995). The newer cells can be disrupted very easily as compared to the older cells. Therefore, it is important that while optimization of parameters during disruption of cells; the genera, species, type of cell, and the growth phase of the cells should be taken into consideration. This increases the disruption efficiency to a large extent. It is also vital to check disruption by various methods and then decide the best method for any organism as it may vary in different organisms.

Disruption of cells can be done by three methods, namely, physical, chemical, and biological. In physical methods, the mechanical breakdown is made with the help of contact, pressure, or denaturation (Lakatos et al. 2019). The wastage in these methods is very less but a lot of energy is required. Acid/alkali are used in chemical methods of cell disruption (de Bruin and Birnboim 2016). In the chemical method, the DNA gets released after mild acid and alkali treatment, and the amount of DNA can be calculated from the amount of purines. However, in many other cases, this method produces a lot of wastage and the method has been found to be corrosive. The use of enzymes for cell disruption as a biological means is very expensive but it is not harmful. Parameters like pH and temperature should be optimized before taking up the disruption procedure when enzymes are utilized (Halim et al. 2012). The cell disruption is an important operational step for the diagnosis of pathogens at the molecular level, for vital immunoassays, protein purification and in the study of protein function and structure, cancer diagnostics, drug screening, and mRNA transcriptome determination. Disruption of cells is also very important in the analysis of protein composition, lipids, and nucleic acids either individually or in complexes.

Cell disruption is divided into two categories. It can be complete or partial. Partial technique is used in the testing of drugs and also in the study of intracellular ionic currents. In complete cell lysis, the cell membrane is disintegrated completely. Analysis of DNA, RNA, and subcellular components is usually done by this method. Globally, in 2023, $5.1 billion have been spent on cell disruption techniques to extract metabolites, which is expected to reach $ 10.9 billion by 2032 (rate of 8.6%). Recently, many automated analytical processes are being used. Very small volumes in nano- or pico liters are required in microfluidic technology of cell lysis. This technology is simple, easy, efficient, and cost-effective.

With this latest technique, it is possible to disrupt even one single cell (Shehadul et al. 2017).

## 13.2 Mechanical Disruption Methods

In mechanical cell lysis, cell membrane is physically broken down by using shear force (Grigorov et al. 2021). This method is very popular and is available commercially because of a combination of high throughput and higher lysing efficiency.

Force is required in this method to separate the intracellular proteins without adding any chemical or enzyme. These include:

I. Mortar and pestle grinding
II. Blender
III. Bead beating
IV. Chaotropic Agents
V. Ultrasonication
VI. Homogenization
VII. Microfluidizer
VIII. Sonicator
IX. French press/X-mesh

I. **Mortar and Pestle Grinding**

In this method, the cells are disrupted by grinding. Usually, it is common with plant samples that are frozen in liquid nitrogen. After the material is disrupted, the metabolites are extracted with the addition of solvents.

II. **Blenders**

When blender is used at high speed, the cell wall of the cells gets disrupted. Later the concentrated material can be removed.

III. **Bead Beating**

In order to open/ disrupt the cells, the beads are made up of either glass or ceramic, which are used (0.1–2 mm in diameter). These are mixed with the sample, which is suspended in aqueous media and are then vortexed. A mechanical shear is given which is gentle so that it keeps the organelles intact. This method was developed by Hopkins (1970s). The sample and the bead mix are subjected to high level of agitation by stirring or shaking. When the beads collide with the cellular sample, they open the cells and release the intercellular components. This method can be used for spores and also for animal and plant tissues. As it can break very small sample sizes, it is very advantageous. It inhibits contamination, does not release any harmful aerosols in the process, and is inexpensive. Usually, the bead beating is done in batches. For the process, the material is taken in sealed plastic vials or in centrifuge tubes. Cell disruption takes around 1–3 minutes of shaking. Significantly, faster rates of cell disruption are achieved with a bead beater variation called as Soni Beast. Damage to heat-sensitive proteins is usually prevented by cooling the sample during bead beating. The cooling can be done using ice, keeping the vials in pre-chilled aluminum vial holders, or circulating gaseous coolant through the machine. For high sample volumes, a rotating fluorocarbon rotor is used, inside a 15, 50, or 200 ml chamber which helps in agitating the beads. This is surrounded by a static cooling jacket. Presently, monocellular organisms like yeast, algae, and bacteria are processed by this method. This method is used in a number of industries, like cosmetics, pharmaceuticals, food, and biotechnology.

IV. **Chaotropic Agents**

Many hydrophobic compounds are utilized in aqueous solutions by disrupting the structure of water using chaotropic agents like urea, guanidine, and sodium iodide which are used at high molarities. Guanidine salts and sodium iodide, are used at 6 M whereas urea is used at 9.5 M. These are used in combination with the detergents so that biological systems can be denatured and emulsified. Chaotropes are widely used for nucleic acid isolation which use silica-based resins/gels for purification. Nucleic acids which are lysed with chaotropic agents, like 6 M guanidine, along with Proteinase K get adsorbed to silica gel with the addition of ethanol. Clean nucleic acids are usually eluted with water or TE buffer. Due to the noncovalent interactions, they break and work as surfactants.

V. **Ultrasonication**

In this procedure, vibration is induced inside a titanium probe which is immersed in the cell solution. During the cavitation, tiny bubbles are formed that explode and produce a local shockwave. This disrupts the cell wall due to the change in the pressure. Disruption of plant and fungal cells is usually made by this method.

VI. **Homogenization**

In this method, the samples are forced through a narrow space, and high pressure is applied on them. It is commonly used for soft and solid tissues. With the increase in temperature, the rate of cell disruption also increases. It can be easily adapted to different sample sizes to accommodate the increased demand with a high level of disruption efficiency and can be used for the recovery of recombinant proteins. No harsh chemicals are used in this process, and it can be scaled up for commercial needs. In liquid-based homogenization, the cells are lysed by forcing the cells or tissues into a suspension, which is passed through a narrow space. At this time, a shear force is applied on the cells which helps them to lyse. It can also be done by squeezing the cells, through a tube that is slightly smaller than bead beating. High-pressure homogenizer (HPH) is widely used for lysing the cells on a large scale. The cells in the media are passed through an orifice valve. In the process, the cell membrane gets disrupted as the shear force at the orifice is very high. As the cells enter the orifice, they get compressed. As soon as they come out of the orifice, then they expand.

VII. **Microfluidizer**

In this technique, a z-shaped channel (200–400 μm) is made through which a suspension of cellulose fibres is pumped with a pressure. This is better than the homogenization process, as it produces particles with uniform sizes (Al-aqbi et al. 2021). Due to the simplicity of the design, the maintenance is very low. Results obtained are very consistent and dependable. However, this method is not in common use as very high pressure is required and many safety protocols have to be checked before taking up the procedure. The LM20 Microfluidizer is a digitally controlled high-pressure homogenizer; however, the performance has been optimized for small volumes. It has a high yield of cell disruption and is in common use.

VIII. **Sonicator**

In this method, sound waves with high frequency are used for disrupting the cells. The cells of bacteria and spores/fine tissues can be agitated and lysed easily. The cell membrane is broken down, and the proteins can be easily released into the solution. After obtaining the proteins, the protein can be purified. The cells, which are obtained after centrifugation, can be resuspended in buffer and then disrupted with a sonicator (Kamineni and Huang 2019). These cells can be used in food processing and medical treatment. The sound waves (>20 kHz) are applied in an ultrasonic bath which has an ultrasonic probe. This is called as a sonicator. This technique is also used in surgery. It is very useful in ophthalmology, cataract removal, orthopedics to remove bone cement, and the treatment of fractures. It is also helpful in the removal of small cartilage tissues during orthopedic surgery and in tumor removal. Sonication has been a proven technology to remove bacteria from implants in vitro with High-Intensity Focused Ultrasound (HIFU). Application of this method to disrupt the bacterial cell walls has been found to be useful as the energy requirement is low.

IX. **French Press/X-mesh**

This press can easily break the plasma membrane and the cell walls of various microorganisms. Then, the proteins can be easily isolated. French press has a piston, with which a high pressure can be applied. This forces the sample to pass through a narrow valve. The cell lysis is very fast and requires very little time. As the sample passes through the valve, it undergoes a shear stress and decompression which in turn disrupts the cells. This method has been found to be effective in the disintegration of chloroplasts, animal tissues, and other biological materials. Even after the cell wall is disrupted the cell nucleus is not disturbed in any way. Proteins and other cellular components can easily be obtained after the disruption of cells. This method involves an inside-out membrane vesicle which helps in many biochemical assays.

## 13.3 Nonmechanical Disruption Methods

(a) Decompression, Osmotic shock, Thermolysis, Freeze thaw, Desiccation, and Cell bomb (Harrison 2011).
(b) Chemical methods
(c) Enzymatic methods

(a.1) **Decompression** In nitrogen decompression, large quantities of nitrogen are used for cell lysis. Nitrogen is dissolved in the cells under very high pressure. When the pressure is released, the nitrogen violently escapes the cell, causing disruption and breakage of the cell walls.

(a.2) **Osmotic shock** Through osmosis, water moves into the cell. This increases the volume to a large extent and ultimately the cells get bursted. However, this method is good only for animal cells and protozoans, as they do not have any cell walls.

(a.3) **Thermolysis/microwaves** This method is an alternative to the thermal pretreatment of biomass feedstock like lignocelluloses, microalgae, and macroalgae biomass. The frequency ranges from 0.3 to 300 GHz, of which 2450 MHz waves are used commonly for microalgae cell wall disruption. In this method, the electromagnetic waves interact with the dielectric and polar molecules which produce heat. This, in turn, increases the internal pressure. Pretreatment time and power of the microwaves are important operation parameters. The work of Passos et al. (2014, 2017) has demonstrated that microwaves can increase the soluble fraction of organic matter, proteins, carbohydrates, and lipids. There are many advantages of the microwave technology. These are the effectiveness, robustness, and easy scaling up of the lysed cells. Ultrasound waves induce alternations of high and low-pressure cycles in the liquid. Microbubbles, created during low-pressure cycles ultimately implode in high-pressure cycles and produce local shock waves. This creates acoustic vibrations, extreme temperature peaks, and thermolysis of water around the bubbles. This in turn forms reactive free radicals. Cell disruption can be done at different frequencies (20 kHz, 40 kHz, and 1 MHz). Ultrasonic cavitation is very effective at low (18–40 kHz) frequencies than at higher frequencies (400–800 kHz). However, the efficiency depends on the species of the microalgae (shape, size, intracellular structure), operational parameters (temperature, time, power, number of cycles), and the amount of biomass. This technology has been commonly used during the production of biodiesel, bioethanol, and biogas. The method is effective in breaking the algal cells even at low concentration of biomass. Another novel method is Pulse electric field in which the suspension of microalgae is placed between the two electrodes and a pulsed electric field is applied for electroporation which increases the permeability of the cell wall. This induces a reversible or an irreversible damage based on the strength of electroporation. Microwaves break the bonds within the cell walls and may denature the proteins.

**Freeze and thaw** This method is effective for soft plant material and algae in which the disruption of cell wall is done by using a series of freezing and thawing cycles. During freezing, ice crystals are formed which expand on thawing. This finally results in the rupture of the cell wall (Phothiset and Sanguansri 2014).

**Desiccation** In this method, the cell suspension is dried by air drying or vacuum drying. This results in the shrinkage of the cells. However, in case of excess water, the cells can burst (Potts 1994).

**Cell bomb** In this method a very high pressure (up to 25,000 psi) is applied. After sometime, the pressure is released. Due to the pressure change, the dissolved gases present in the cells, get released in the form of bubbles which open up the cells (Goldberg 2008). It is a very effective method to homogenize the cells and tissues and to release the intact organelles. Cells are subjected to high pressure (in a pressure vessel) along with high levels of oxygen-free nitrogen which gets dissolved in the cells (~5500 kilopascals [kPa]), equivalent to 800 pounds per square inch (psi). Later, the pressure is suddenly released which forms nitrogen bubbles and ruptures the cell membrane resulting in the release of the cell contents. It is a very effective method for mammalian, plant cells, and fragile bacteria. There is no heat damage to proteins and other organelles during nitrogen cavitation. It uses an adiabatic expansion that cools the sample. Inert nitrogen gas can protect the labile cell components from oxidation. It helps to maintain the pH of the suspending medium.

(b) **Chemical Methods**

**Surfactants/detergents** Interface between hydrophobic and hydrophilic systems can be disrupted with surfactants/detergents (Rohini et al. 2020). The best examples are biological membranes and therefore they are the primary target of detergents. In *E. coli* when SDS is used as a detergent, it effectively targets the interface and separates the cells from the membrane. SDS can unfold cytosolic proteins and partition the membrane proteins, into small droplets which are called as micelles. The impact of the surfactant depends on the detergent and the concentration used. A detergent has a water-soluble hydrophilic head and a hydrophobic tail. This helps in the insertion and then dispersion of the membrane. The chemical composition of the hydrophilic and hydrophobic end decides the action on proteins and on the cell membranes.

Some surfactants can solubilize the membranes completely and can denature proteins. However, the mild surfactants can dissociate the loosely bound proteins. It is important to know if the hydrophilic group in the surfactant is ionic or nonionic. The ionic surfactants can solubilize the membranes easily and can denature the proteins. Usually, an anionic surfactant has a sulfate or carboxylic group, whereas a cationic surfactant has an ammonium group. The best example of an anionic detergent is SDS (sodium dodecyl sulfate), which has a sulfate hydrophilic head and 12-carbon tail (dodecyl or lauryl). Sodium deoxycholate derived from bile salts is utilized in lytic buffers. CTAB (cetyltrimethylammonium bromide) is a cationic detergent, which is widely used in the isolation of DNA from plants. Nonionizing detergents have a polar head without any charge. They do not denature easily. Triton X-100,

Brij-35, NP-40, and Nonidet P-40 come under this category and are utilized in immunoassay wash solutions at low concentrations and in lytic buffers at higher concentrations. Detergents are valued for their ability to form micelles at optimum concentration which is called as critical micelle concentration (CMC). Detergents form small particles which are known as micelles. In this, the hydrophilic heads orient outward and the hydrophobic tails inwards. The molecular weight varies from 1200 to 80,000 Daltons. Micelles dissolve and disrupt the cell membrane and disperse into membrane/micelle hybrid particles. Proteins embedded in cellular membranes are picked up by these micelles. Toluene, ether, benzene, methanol, surfactants, phenyl ethyl alcohol, and dimethyl sulphoxide (DMSO) are utilized to permeate the cell walls. For Gram-negative bacteria usually, EDTA is used. This disrupts the cell walls by chelating with the cations which makes holes in the cell walls.

(c) **Enzymatic methods** To obtain the periplasmic enzymes, like cell wall mannans or similar components, the plants are treated with cellulases, and filamentous fungi are treated with chitinase to obtain protoplasts (Zheng et al. 2011). To get single cells, or to obtain genomic DNA from the cells, the cells are treated with protease enzymes like trypsin, dispase, and collagenase. For protoplast generation, the cells are treated with buffered enzymes in the presence of an osmotic stabilizer like (1 M) sorbitol (Ezeronye and Okerentugba 2001). Digestive enzymes are helpful in decomposing the microbial cell wall. Different cell types and strains vary in the cell wall constituents and membranes, and thus the enzyme to be used for cell disruption depends on the type of the microbe cell. Lysozymes hydrolyze the bonds of β-1-4-glucoside in the peptidoglycan. The cell walls of yeast and fungi are different. However, zymolyase can degrade both yeasts and fungi as it has β-1,3 glucanase and β-1,3-glucan laminaripentao-hydrolase activities.

In cavitation technique, initially a cavity/bubble is formed by using ultrasonic vibration and then it is made to collapse by reducing the pressure. This helps in the disruption of cells which happens due to the pressure fluctuation. During the collapse, a lot of mechanical energy gets released which disrupts the cell wall. The major problem is the removal of cell debris after the procedure. The method also generates a large amount of heat which should be dissipated. To solve these problems, use of hydrodynamic cavities has been found good (Ioannis and Ioannis 2021). These are produced by pumping the cell suspension through a narrow channel which increases the velocity. Human neutrophils can be disrupted at a pressure of 420 $lb/in^2$. The amount of pressure required for cell disruption varies with the cell type. The subcellular organelles can be disrupted at a higher pressure (>1000 $lb/in^2$).

## 13.4 Approaches to Microbial Cell Disruption in Large Scale Production

Microbial cell disruption can be done by mechanical and nonmechanical techniques. Mechanical methods like bead milling and high-pressure homogenization (HPH) are used on the large scale. These help in fluid flow, particle–particle interaction, and pressure drop during the cell disruption. DNA being shear-sensitive cannot be recovered by these mechanical methods. Cryopulverization can be used for biological samples which contain water as they become brittle at cold temperatures. Tissue pulverizers are available and are helpful as they do not need manual effort, also the sample recovery is good and the cleaning is simple. Yields of proteins and nucleic acids obtained by this method are much higher. Pressure Cycling Technology (PCT) uses alternating cycles of hydrostatic pressure between ambient and ultra-high levels (up to 90,000 psi) to control the actions of molecules in biological samples. Microfluidizer is another method used for the disruption of cells. By using microchannels with fixed geometry, and an intensifier pump, high shear rates are generated that rupture the cells. It has an effective temperature control, and therefore, the yields of active proteins and enzymes are very high. When the viscosity of the cell suspension is high, then the downstreaming becomes difficult which includes procedures like filtration and accurate pipetting. While using nitrogen decompression, large quantities of nitrogen are dissolved initially in the cell under high pressure. When the gas pressure is suddenly released, the nitrogen comes out of the solution as expanding bubbles that stretch the membranes of each cell until they rupture and release the contents of the cell. In most of the disruptive methods, either friction or mechanical shearing is involved, and this generates heat. However, in the nitrogen decompression procedure, the sample gets cooled due to an adiabatic expansion. Gases like carbon dioxide, nitrous oxide, carbon monoxide, and compressed air can also be utilized, but nitrogen is preferred as it is nonreactive in nature, and it does not alter the pH of the suspending medium. The method is economical.

Cell disruption techniques vary in their effectiveness due to a considerable number of parameters, including cell morphology, cell wall characteristics, chemical compound properties, diffusivity in the solvent, and the intracellular locus of the molecule of interest among others. These techniques could be utilized individually or sequentially within a molecule recovery process in a way to ensure industrial viability. The success depends on the balance between cost, time, and yield. HVED is gaining momentum as a cell disruption method used to recover microbial molecules of interest and the breadth of its applicability is likely to increase (Harrison 2011).

## 13.5 Cell Disruption and Extraction of Lipids

Cell lysis is necessary for lipid extraction which can be done by homogenization, sonication, or microwave techniques. These techniques are energy intensive and therefore, newer approaches like supercritical fluid extraction (SFE), ultrasound-assisted extraction (UAE), and microwave-assisted extraction (MAE) are being used now.

In earlier times, soxhlet extraction or Folch and Bligh-Dyer extraction methods were used for lipid extraction. But due to environmental pollution, and high extraction time involved in these methods, they are not used much in recent times. Microalgae contains bioactive compounds that are commonly used in medicine, food, bioenergy, polysaccharides (carrageenan and fucoidans) and cannot be degraded with the help of enzymes. The microalgae usually contain polar lipids like phospholipids and glycolipids, and neutral lipids such as triacylglycerol and UFAs. Polar lipids are mainly found in cell membranes and organelles whereas glycerol and UFAs are used for energy storage. Fish, like whales and codfish, are good sources of lipids.

(i) **Traditional Methods of Extraction**

Soxhlet extraction This method was used for the quantification of total lipid amount in milk and for the quantification of food and pharmaceuticals. As extraction by this method requires a lot of time and consumes large amounts of reagent, generally it is not preferred (López-Bascón and Luque de Castro 2020).

Folch extraction In this method, the extraction is done by using a mixture of chloroform and methanol (2:1 v/v). Water is then added for phase separation. After the rotary evaporation, the lipids are extracted. This method is reliable and does not require high temperature or pressure (Ulmer et al. 2018).

(ii) Bligh-Dyer extraction This method involves two-phase solvent. Lipids come in the chloroform layer and can then be separated. The protein gets precipitated (Göran et al. 1998).

Extraction of lipids requires the use of toxic solvents like methanol, chloroform, and n-hexane. These solvents are harmful to human health and environment. Therefore, scientists had been working for greener reagents that could be utilized for extraction and be replaced in place of the toxic reagents. Single-cell oils (SCO) are edible oils that can be obtained from single-celled microorganisms such as yeast, fungi, and algae. A small percentage of microorganisms have the ability to synthesize and accumulate 20–87% of their total biomass as intracellular lipids, which are defined as oleaginous microorganisms. Microorganisms are considered to be oleaginous if they are able to supply acetyl-CoA into the cytosol and act as an important compound precursor for fatty acid synthetase (FAS), and as a source of NADPH, which is a reducing agent in fatty acid biosynthesis. During fermentation, lipid synthesis gets activated under nitrogen deprivation in oleaginous microorganisms. It converts substrates like carbon dioxide, sugars, and organic acids to SCO. The products include acylglycerides, free fatty acids (FFA), phospholipids, glycolipids, lipoproteins, sterols, and hydrocarbons. In yeasts, there are two methods for lipid synthesis. These are de-novo synthesis and ex-novo synthesis. In de-novo lipid biosynthesis, fatty acid precursors, such as acetyl-CoA and malonyl-CoA, are formed which get incorporated into the lipid storage biosynthesis. Carbon sources such as glucose, fructose, lactose, sucrose, whey, glucose-enriched wastes, and molasses are used as the substrates for de-novo lipid biosynthesis. While the cells are growing, the nitrogen gets consumed for cell propagation and the synthesis of proteins and nucleic acids. With the deprivation of nitrogen content in the medium, the metabolic pathways get inhibited which in turn lowers the growth rate. However, it starts the synthesis of fatty acids and triglycerides using the precursors like acetyl-CoA, malonyl-CoA, glycerol, and energy molecules such as ATP and NADPH. *Yarrowia lipolytica*, *Geotrichum*, *Rhodosporidium*, *Cryptococcus*, and *Trichosporon* synthesize lipids via ex-novo synthesis. Fatty acids can also be obtained by the breakdown of triacylglycerols or fatty esters catalyzed by lipase enzymes. In ex-novo synthesis, lipid accumulation is initiated independently of the nitrogen availability in the hydrophobic medium. Microbial oil extraction has been found to be difficult for large commercial exploitation. Production costs can be reduced if renewable feedstocks like low-cost lignocellulosic materials and other industrial residues can be used as substrates.

Hexane and a mixture of hydrocarbons are being used for the extraction of lipids from oleaginous microorganisms. Hexane has good solubilizing ability and narrow boiling points (between 63 and 69 °C). Hexane reacts with pollutants and forms ozone and photochemicals. Chloroform is a highly carcinogenic chemical and comes under hazardous air pollutant (HAP) in the United States of America.

(iii) **Extraction with Green Solvents**

Terpenes and esters have been used as green solvents for the extraction of lipids. Ethanol is the most economical green solvent (Young and Robert 2019). It has been found that cyclopentyl methyl ether, 2-methyl tetrahydrofuran, and ethyl acetate could replace hexane during lipid extraction from *Y. lipolytica* IFP29. Their fatty acid profiles; oleic acid (C18:1), linoleic acid (C18:2n6), and palmitic acid (C16:0), minor amounts of palmitoleic acid (C16:1) and stearic acids have similarity to hexane in terms of selectivity. It is not possible to use p-cymene, d-limonene, α-pinene, ethyl lactate, isopropanol, and ethanol in place of hexane, due to technical reasons and cost. N-ethylbutylamine, N-dipropylamine, and N, N-dimethylcyclohexylamine have been used for the extraction as all of these extract high amount of fats.

A newer method known as Automated pressurized liquid extraction method (APLE) has been found to be good for the extraction of lipids from the dried cells of oleaginous strains. In this method, the lyophilized yeast cells are mixed with diatomite in a mortar, which is then transferred into a stainless steel chamber through a funnel. The complete chamber is then placed inside the APLE system along with the solvents (chloroform and methanol in the ratio of 1:1). Under optimized conditions for extraction (temperature, pressure, and operation time), the APLE system is connected to the collection bottles in which the desired compound is collected.

Another method for the extraction of bioactive compounds is the supercritical fluid extraction (SFE) method (Molino et al. 2019). This method utilizes organic solvents for extraction. This is called as a liquid–liquid extraction method in which the various bioactive compounds can be extracted and later fractionated. Most of the times carbon dioxide gas ($CO_2$) is used as it is safe, nontoxic, inexpensive and can be separated easily after the process. The process variables (temperature, pressure, flow rate, and bed size) can be optimized. Adsorbent-based oil capturing agents can also be employed for in-situ lipid recovery during fermentation. The in-situ oil adsorptive fermentation system can be used repeatedly for extended periods and therefore becomes effective integration both for upstream and downstream processing that is industrially feasible. The fatty acid composition of the resin-adsorbed oil has been found to be different from the intracellular oil. The biofuel properties of the fatty acids, which are extracted from the adsorbed oil by the resins, produce smoother engine operation and a better performance at lower temperature along with reduced emissions which are similar to plant-based oils. Supercritical fluid extraction that utilizes $CO_2$ is an efficient, mild, and green alternative method for lipid extraction. Other newer methods like automated pressurized liquid extraction and simultaneous in-situ lipid recovery using adsorbent-based oil are methods where the recovery of lipids has been found very efficient. It can be done in less time without giving any harsh pretreatment.

When the cell disruption is done under hypo-osmotic or under iso-osmotic medium, morphological integrity is preserved. Use of sucrose has been found to be a good osmoticum as it prevents the swelling/ shrinkage in the subcellular organelles. Mannitol and sorbitol have also been found to be effective. However, they aggregate the subcellular organelles. Aqueous media is preferred for cell disruption as the non-aqueous media (ether/chloroform) can inactivate some enzymes and hamper the morphological integrity. In order to maintain the nuclear integrity, ions like $Mg^{2+}$ or $Ca^{2+}$ are added to the homogenization medium. However, these ions can decrease the respiratory control in mitochondria.

## 13.6 Cell Disruption, Extraction, and Purification of Proteins

### Protein Purification Methods

Protein purification involves four steps. These include cell lysis, binding of the protein to any particular matrix which is followed by washing and lastly elution of the protein. For maximum yields, the parameters like solubility, size, charge, and binding should be optimized. Proteins are the macromolecules and they help in maintaining the structure of the cell. In case the cellular protein does not function in a proper way, it may cause a number of diseases. Protein purification helps to understand not only the interaction with other proteins but also with the DNA and the RNA. For recombinant protein, the microorganism in which it is expressed plays a vital role. Purification becomes easier when purification tags (small protein) are fused on a recombinant protein. The fusion tags/immobilization of protein on the chip have been found to be effective which keeps the protein intact. Purification should be done in such a way that it does not damage the structure and function of the protein. *Escherichia coli* is a bacterial expression system that is widely used in industrial processes (Terpe 2006). The molecular weight (MW) of a protein is the sum of the masses of the individual AAs composing the protein and is expressed in daltons (Da) or kilodaltons (kDa). A Da is the same as an atomic mass unit which is approximately the mass of a nucleon and is equivalent to 1 g/mol. They can be assayed by spectroscopic method (using Bradford reagent or a chromogenic substrate) or by immunological technique (using an antibody that can recognize the protein of interest).

**Preparation of protein crude extract** This is prepared by grinding the selected material. For an intercellular protein, the cells should be broken down in a buffer along with an inhibitor.

**The fusion tags** Polyhistidine tag is commonly used. As it is small in size, it is not required to be removed during the down streaming. Interaction of the polyhistidine tag with the metal makes it possible to purify the other insoluble proteins. Another tag used is glutathione-S-transferase (GST) which has cytosolic protein, and it is present mostly in eukaryotic organisms. Another fusion tag is the HaloTag protein. This tag has the ability to increase the expression and solubility of recombinant proteins in *E. coli*. It has a size of 34 kDa, which is a monomeric protein tag obtained from *Rhodococcus rhodochrous* (Sun et al. 2015). This ligand interaction allows the protein complex to be boiled in SDS sample buffer, before analyzing it on SDS-PAGE. Polyhistidine-Tagged Proteins can be purified using magnetic resins without multiple steps of centrifugation.

### 13.6.1 Centrifugation

After disruption, centrifugation is an important step in protein /product isolation. A single centrifuge is provided with various sizes of rotors which have the capacity to centrifuge different sizes/volumes of tubes/bottles by the same instrument. The cells and different cell components can be separated by differential centrifugation. At the beginning, the cell debris and dead cells can be separated at 300 × g for 10 min at 4 °C. After the crude extract is formed, it is filtered or centrifuged (works on centrifugal force) which sediments the large particles as the more dense components of the mixture migrate away from the axis. However, the less dense components of the mixture migrate toward the axis of the centrifuge. The precipitate is known as pellet and the remaining solution is called as the supernatant which can be separated by decantation without disturbing the precipitate (details given in Chap. 12 of this book). When many rounds of centrifugation are made, it is known as differential centrifugation (Mikkelsen and Cortón 2004).

### 13.6.2 Dialysis

Dialysis is a method that removes contaminants and is helpful in the separation of macromolecules. There is a diffusion reaction that separates the macromolecules through a semi-permeable membrane. When a solvent is exchanged around a protein, the procedure is known as dialysis. It is basically based on the principles of size exclusion. Protein solution is placed inside a semi-permeable membrane which is permeable to water and ions, but not to the protein (dialyzing bag) which is suspended in a larger volume of buffered solution. The dialyzing bag is provided with pores. It is the pore size that determines the molecular-weight cutoff (MWCO) of the membrane. In the process, a sample and a buffer solution (called the dialysate) are taken. The amount of dialysate is much more as compared to the sample volume (around 200 to 500 times more). The sample is put inside the dialysis bag and the buffer solution is around it. It is a good method to remove the contaminants from the sample. The semi-permeable dialyzing membrane/bag is made up of a sheet of cellulose which has different sizes of pores. Molecules larger in size than that of the pores cannot pass through the membrane. Only the smaller molecules can pass through it freely. The size of the sample molecules is bigger than the membrane pores and therefore, they get retained in the bag. However, the small molecules diffuse out through the membrane till an equilibrium concentration is attained in the entire dialysate volume (Maria 2008).

**Diafiltration**

The method of diafiltration resembles the technique of dialysis as it also utilizes a semipermeable membrane which helps in separating the macromolecules from the compounds of lower molecular weight. However, in this method, the sample is pushed through the membrane with high pressure. The technique follows the principle of reverse osmosis.

There are diafiltration units which use the principle of centrifugation. These are called as concentrators. This helps in concentrating the samples. Both the solvent (water) and the solute (low molecular weight) are forced through the membrane filter. Macromolecules remain on the sample side of the membrane and get concentrated as the water gets forced out, across the membrane on the opposite side.

The development of chromatographic techniques has made the separation of biomolecules much easier. This technique helps in the identification, quantification, and also in the purification of biomolecules.

### 13.6.3 Column Chromatography

It is an excellent method to separate proteins and other components from mixtures based upon the size (gel filtration/size exclusion), charge (ion exchange), and affinity (a specific binding affinity). All the chromatographic techniques use a resin (solid phase) with special chemical properties which are held in a glass cylinder (column). A buffered solution (mobile phase) is poured through the column, and the fractions are collected in the tubes after they come out from the column. The migration of proteins varies based on the resin.

### 13.6.4 Gel Filtration/Size Exclusion

**Desalting and buffer exchange**

When the salts are removed from a sample it is known as desalting. When buffer salts are replaced with another set of buffer salts, then it is known as buffer exchange. Gel filtration chromatographic technique is commonly used for size exclusion chromatography (SEC). Initially, the chromatographic column is equilibrated with water. While using buffer exchange, equilibration of the column resin is done with the same buffer, which is to be used for final elution. Both, desalting and buffer exchange are separation processes that are based on the gel filtration. It is also known as molecular sieve chromatography. In this method, a solution containing macromolecules is passed through a column that is packed with a porous resin. When matched correctly, the macromolecules are too large to enter the pores of the resin and thus pass quickly through the column. In contrast, the buffer salts and other small molecules enter the pores of the resin, slowing their rate of migration through the resin bed. This reduction in flow rate causes the macromolecules to get separated from the slower, smaller molecules. The bigger molecules make a way (in the liquid phase; void volume) through the column, which is outside the beads. Therefore, they get eluted at the earliest. The void volume of buffer has no proteins, and it passes first through the column. Usually, the void volume consists of 20% of the total column volume size. The macromolecules are separated by collecting different fractions as they are eluted from the column. The macromolecules that come out from the column are in the equilibration buffer. As the original buffer is left in the resin, the technique is termed as buffer exchange. Among the commercially available desalting devises, products from the industry named as "Thermo Scientific Zeba" are very commonly used. These resins have high desalting and protein recovery properties. The resins used are porous in nature. The resins are available for processing the sample size between 2 μL and 4 mL. Longer time is taken by small molecules to go through the column as compared to the larger molecules which can easily travel around the resin.

### 13.6.5 Ion Exchange

The functional groups that are present in biomolecules ionize when they are in a solution. This can give a specific charge to the molecule. All proteins consist of many amino acids. The aminoacids have amino (-$NH_2$) as well as carboxylic acid (-COOH) functional groups. The surface charge of the molecule depends on the pH of the medium and the ionization reaction of the molecule. If the pH value is less than 7.0, in that case, more of the amino groups get protonated. Therefore, in such cases, the surface of the protein molecule gets a positive charge. However, at higher pH values, more of the carboxylic acid groups get deprotonated. This helps the anions to be on the surface of the protein molecules, and thus it becomes negatively charged. At the isoelectric point (pI), the protein is neutral in nature with no charge on the molecule. In ion exchange chromatography the separation of compounds is mainly based on the net charge. The media contains negatively or positively charged functional groups which are covalently bound to a support. Based on the charge, it is categorized as a cation or an anion exchanger. The charge on the biomolecule gets bound to the opposite charge sites of the stationary phase. This can be released from the column after optimizing the ionic strength or the pH of the mobile phase.

There are two types of exchangers which are known as anion or cation exchangers. The proteins directly interact with the resin. The column is equilibrated with a buffer solution to maintain a constant pH in the column. The protein mixture is loaded on to the column where the proteins interact with the resin depending upon their own charge. Buffer is poured on to the column until all the proteins which are not interacting with the resin get washed off.

A gradient of increasing salt concentration (disrupts ionic and hydrogen binding) in the buffer is applied on to the column. This allows the most weak interacting proteins to get released first, then followed gradually by the more stronger proteins. This can also be achieved by changing the buffer pH being applied to the column. This method comes under the class of liquid chromatography. For a better separation, the net surface charge is very specific for each molecule at a particular pH. It is a good method to separate polar or ionic species. Ion chromatography (IC) is a term that includes ion exchange chromatography (IEX) along with ion exclusion chromatography (IEC) and ion pair (IP) chromatography. In all these methods, elution is made using electrolytic solutions. In ion pair chromatography, the analytes are eluted from a neutral stationary phase by reagent ions which carry opposite charges. In IEC, the charged functional groups are covalently bound to the solid surface of the matrix. These functional groups are usually made of different materials which include; cellulose, silica, and styrene-divinylbenzene. Cation exchanger resins contain immobilized negatively charged functional groups ($RSO_3$–$RCO_2$– $RPO_4$– and $RO^-$) and anion exchangers contain immobilized positively charged functional groups (i.e., quaternary ammonium groups).

The strong ion exchangers include the ones in which the resins contain sulfonyl and quaternary ammonium groups. They can ionize at any pH. The weak ion exchangers contain functional groups like carboxyl and secondary or primary amines. They can ionize only at a certain range of pHs. When an ionized solute passes through an ion exchange column, the sample ions get adsorbed and displace the counterions on the surface. The adsorption process is reversible and adsorbed ions are eluted by a salt solution. Using a suitable mobile phase helps in regenerating the resins. The retention time of the analytes is decided on the concentration of the ion-pairing reagent in the mobile phase which helps in their separation.The charge on the analyte (positive or negative) decides the electrostatic bonding to the oppositely charged particles of the stationary phase present on the ion-exchange column. The separation of the enzyme depends on the packing material of the column, particle size and pore diameter of the support, column length, mobile phase, and temperature. If the column material is made up of small particle-size resins, then high amounts of enzyme molecules get diffused, but at a slower rate. Although the elution time is less, but, the resolution is much better. The surface area of a support in the column and the diameter of pores affects the penetration of enzyme molecules into the column matrix. This affects the mass transfer and the loading capacity. Pore diameters of ≈300 Å have been found to be good for most of the enzymes which have a molecular weight, up to 100,000. This also gives a good resolution. For higher molecular weight enzymes, resins with bigger pore diameter size result in better resolution. For resolution purposes, the length of the column does not play a major role. However, shorter columns are advantageous as they provide concentrated eluents at low pressure and at less column cost. Only disadvantage in using short column is that it needs to be reloaded at shorter intervals. The ion exchange resin for a particular enzyme separation depends on the isoelectric point of the enzyme to be separated. When the pH is above the isoelectric point, the enzyme is acidic in nature and then the anion exchange resins are used for separation. For the separation of basic enzymes, cation exchange resins are utilized. The sample and the number of interactions with the stationary phase play a major role in deciding the retention time of the sample. Usually, the elution of the sample is made by changing the ionic strength of the aqueous buffer used for elution.

The salt composition of the mobile phase, the pH, and the ionic strength are important parameters in the separation of enzymes. A combination of organic solvents with water is utilized as the mobile phase. The ratio of the two has to be optimized keeping the end product in mind. This needs to be done before taking the final batch for a specific enzyme/product separation. Usually in the combination, the amount of organic solvent should be less as the enzyme molecule can become unstable. The charge on the enzyme molecule is dependent on the pH of the solvent. However, the ionization of ammonium groups in any protein takes place at pH which is below the isoelectric point (pI). This ultimately gives a positive charge to the enzyme. If the pH is above the pI, then it gives a negative charge to the enzyme. The charge on the enzyme decides the retention capacity on the ion exchange column. Accordingly, for elution, the pH of the mobile phase is taken into account. These are very important factors and the pH of the mobile phase should be above the pI of the enzyme to be separated in the case of anion exchange columns. Whereas, in cation exchange columns, the pI of the enzyme that needs to be separated should be low. With higher valent ions, the retention is less as they are stronger displacers. Gradient elution at low temperatures helps in keeping up the stability of the enzyme.

The major steps in IEX elution include the equilibration in which the stationary phase has to be washed with the initial buffer to stabilize the baseline. The sample should be kept in the buffer of the same pH and is loaded on to the column. The ionic groups present on the column, then get bound with the oppositely charged sample molecules. The uncharged molecules can be removed from the column by washing it with the starting buffer. A salt gradient is used for the elution of the bound analytes from the sample. When the ionic strength is low at the beginning, then only the weak analytes (which do not bind tightly) get eluted out. Then, the salt concentration is increased slowly. This helps the strongly bound molecules with high surface charge density to get eluted out in a sequential manner. A pH gradient helps in eluting the bound analytes where they get released at their specific pI value from the column. However, this method is not good if a molecule precipitates at its pI value. At the end, the column is regenerated by using a high ionic strength buffer. This buffer is allowed to flow through the column until the baseline is achieved. Ultimately, the column can be conditioned with the starting buffer. For detection of the compounds, use of UV or fluorescence detectors is common. However, mass spectrometers, refractive index (RI), or multi-angle light scattering (MALS) detectors are also used.

**Anion Exchanger** This removes anions from protein mixture in order to have only positively charged moieties. During elution, initially the weakly negative proteins are knocked off which are followed by a stronger negatively charged, and finally with the most negatively charged proteins. The anion exchange columns have positively charged functional groups which are covalently bonded to the stationary phase particles and cation exchangers have negatively charged functional groups which are bonded to the stationary phase particles. As their names suggest, anion exchange chromatography is used for the separation of anions such as deprotonated or "acidic" protein molecules that have more carboxylic acid groups on their surface.

**Cation Exchanger** Cations are removed from the protein solution from the resins that are negatively charged. Before elution, all negatively charged and also uncharged proteins are passed out from the column. Then, the weak positive proteins get eluted, which are followed with the elution of the stronger positively charged cations, and finally the most positively charged proteins get eluted. Protonated basic proteins with more amino groups on their surface can be separated by this method.

**Resins used in Ion Exchange Columns**

These columns are majorly made by using polystyrene and cellulosic material. They differ in flow properties, access to ions, chemicals, and in their stability. Polystyrene resins are made when styrene and divinylbenzene are polymerized. In case, the concentration of divinylbenzene is high, then it produces more cross-linkages. Small molecular weight compounds/enzymes can be separated by these resins. If the cross-linkages are more, then the resin becomes rigid and does not swell much. This reduces the porosity capacity. Sulfonic acids are strong acids with good proton dissociation ability. Resins that contain sulfonic acid groups are cationic exchangers. However, an acidic exchanger is formed if carbohydrate groups get attached to the aromatic rings in place of a sulfonic acid group.

**Common Ion Exchanger Resins and Their Preparation**

Carboxymethyl Cellulose (CM-cellulose)—Cationic exchanger and DEAE cellulose—Anionic Exchanger

Preparation of resins involves the three steps which involve swelling of the resins which helps in bringing the functional groups to get exposed for ion exchange. For anion exchanger, the resins are initially treated with an acid (0.5 N HCl) and then with a base (0.5 N NaOH). In case of cation exchangers, the resins are first treated with a base (0.5 N NaOH) and then with an acid (0.5 N HCl). To remove the impurities, the resins are treated with EDTA. Very fine particles need to be removed as these can hamper the flow rate. Therefore, the resins are washed with a lot of water. Last step in the preparation of resins is the equilibration with the counter ions. The exchanger resins are washed with various reagents. This is based on the desired counterion to be introduced. These are given as follows:

1. For NaOH the counter ion to be introduced is $Na^+$
2. For HCl the counter ion to be introduced is $H^+$
3. For $NaNO_3$ the counter ion to be introduced is $NO_3$.

Buffers to be used

(a) Anionic exchange chromatography should be carried out with cationic buffers.
(b) Cationic exchange chromatography should be carried out with anionic buffers.
(c) The pK of the buffer should be as near as possible to the pH at which the system is buffered (Table 13.1).

Major applications of ion exchange chromatography

1. Analysis of amino acids by amino acid Auto analyzer.
2. Determination of the base composition of nucleic acids
3. Purification of water and softening of drinking water.
4. Protein purification.
5. Separation of vitamins, organic acids and bases.

### 13.6.6 Affinity Chromatography

In this separation method, the immobilized ligand and the substrate get bound. The method was discovered in 1951 (Campbell, Luescher, and Lerman) for the separation by using antigen on cellulose columns which have affinity for a specific antibody and need to be separated from a complex mixture. The column consists of an insoluble polymer/gel. It is generally known as matrix. On this, a specific

**Table 13.1** Common buffers used in anion and cation chromatography

| Anion exchanger | Buffer (pKa[a]) (mmol/L) |
|---|---|
| N-methyl piperazine | 4.75 (20) |
| Tris | 8.06 (20) |
| Cation Exchangers | Buffer (pKa) (mmol/L) |
| Maleic acid | 2.00 (20) |
| Phosphate | 7.20 (50) |

[a]pKa describes the acidity of the molecule

competitive inhibitor or ligand is attached covalently. Matrix should be inert both chemically as well as physically. It should not dissolve in the buffer used during the process and the molecule should get coupled to a ligand easily. The surface area must be large for the ease of attachment. The spacer arm on the matrix improves the binding qualities. As far as the selection of ligand is concerned, it has to be decided after studying the properties of the macromolecule to be eluted/ isolated as the ligand and the target molecules need to be bound. Hapten is a common antigen used for the elution of antibodies.

The resins in the column are equilibrated in the buffer prior to the loading of protein sample. The specific protein/enzyme binds with the ligand in the resin and the rest goes out of the column. The elution of the desired compound of interest is made by the addition of a salt/a metal chelator or a high concentration of the ligand. It can also be done by changing the pH. In recent times, this technique is integrated with HPLC for more specificity and sensitivity. The method has been found useful in the separation of nucleotides, nucleic acids, transfer RNA, and enzymes. The gel should be loose and porous, so that, the macromolecules can pass through them in a smooth way. The particles should be spherical and rigid. Sephadex (cross-linked) and sepharose have been found very effective resins. Many enzymes have been purified by combining molecules of carbohydrate moiety with cyanogen halides. Synthetic polyacrylamide gels without any charge have better porosity and physical properties. As they have polyethylene as base, it makes them very stable. The steric interference in the binding process can be reduced by introducing an inhibitor that consists of a long hydrocarbon chain (arm). This is attached to a soluble support. Affinity chromatography, which utilizes the reversible biological interaction has forces in different degrees. The bound target molecules can be eluted with a competing ligand in the mobile phase. Changing the pH/ionic strength/polarity also helps in the elution of target molecule.

Using small column is preferable. There are many columns available in the market. However, column by Bio-Rad Econo-Column (cat. no. 731–1550), with a bed volume from 0.2 to 10 ml, and the flow rate to 1–2 ml/min has been found effective for the separation/elution quality of the macromolecules. The column is cost-effective. Mixing the matrix beads on a magnetic stirrer is not advisable as it can fragment the beads. However, mixing on a rotating wheel is advisable. Reaction in batches and keeping a check on the mixing procedure is important. It should be seen that all of the liquid buffer interacts with the beads (Rodriguez et al. 2020). The most commonly used loading buffers are PBS, pH 7.4, or 0.1 M Tris-HCl, pH 8.0. Use of large columns are avoided as the matrix can bind non-specifically. This can increase the bed volume. Care needs to be taken while using beaded agarose (sepharose) gels, as they can neither be dried nor frozen and they shrink irreversibly. Also, these beads are not tolerant to most of the organic solvents. Solvents like dimethylformamide (50%, v/v) and ethylene glycol (50%, v/v) have been found to be good in keeping the structure of the beads. When the compounds are not soluble in water, the solvents prove to be useful for coupling. Low temperature of 4 °C for sepharose along with antibacterial preservatives have been found to be good. There is no adverse effect on agarose beads if they are kept in alkali (0.1 M NaOH) and in acid (1 M HCI) for a period of 2–3 hours at room temperature. The linked ligand remains intact under these conditions. Such treated beads can also be used repeatedly. However, exposure to high pH for prolonged periods may cause slight shrinkage to the gel. Coupling of the ligand with the beads is usually done in a hood which has ventilation. The washed Sepharose-4B beads are mixed with water (50:50) and then cyanogen bromide (50–300 mg/ ml of packed Sepharose) is added to the stirred suspension.

The pH is maintained at 11 using NaOH. The molarity of NaOH solution should be optimized based on the amounts of sepharose and cyanogen bromide used. Ice is added to keep the temperature around 20 °C. The washing process is done in a Buchner funnel, to which the ligand solution is added, and then the mixing should be completed within 90 seconds. The procedure must be conducted rapidly as the activated sepharose is not stable. This is then taken in a beaker in which mixing is done with the help of a magnetic bar at 4 °C. It is kept at 4 °C for a period of 16–20 hours so that all the reactive agarose groups come out. The gel is then washed with water many times till it is confirmed that ligand will not be removed from the gel. The amount of ligand coupled to the sepharose beads should be 20 to 30 times more than it is needed in the final product. This is usually 100–150 mg of cyanogen bromide/ml of packed sepharose at pH less than 10. However, the stability of the activated sepharose gets reduced if the pH is more than 10.5.

Affinity chromatography is very useful for separating of compounds from a mixture. The technique is very useful in removing the impurities, in enzyme assays, for the detection of substrates, to know the binding sites of the enzymes, in vitro antigen-antibody reactions, and in the detection of mutations in the nucleic acids. As the method is very specific, the target molecule can be obtained in a pure state within a single step. The matrix can be reused as it can be washed easily and then dried. All these properties make this method very useful and produce high yield of the purified product. The procedure can be improved by reducing the time duration of the procedure, using lesser quantities of the solvent, and using it at larger variation of pH.

### 13.6.7 Electrophoresis

This is a separation method of proteins from cells, subcellular fractions, column fractions, or immunoprecipitates. The proteins move with the help of a solvent to which an electric field is applied. The movement is based on the charge density. The mobility depends on the net charge on the molecule, the ionic strength, and the properties of the matrix (pore size and viscosity). Commonly two matrix materials are used, namely, polyacrylamide and agarose. In polyacrylamide gel electrophoresis, the proteins migrate through the pores. As the PEG becomes more viscous, the pore size gets reduced, due to the increase in the concentration. This affects the rate of the protein movement in the gel matrix which decreases as the PEG concentration increases. The movement of the protein also depends on the size, shape, and the net charge on the protein. Agarose gel matrix has a bigger pore size and therefore, the separation of nucleic acids and large protein molecules is done by using this matrix. Nondenaturing PAGE is also called as native-PAGE, which separates the proteins based on their mass/charge ratio. SDS is an ionic detergent that gets denatured and binds to the proteins. It makes the proteins negatively charged.

**Paper electrophoresis** This method is effective in separating the aminoacids/peptides from a mixture. The aminoacids with maximum charge move on the top.

**PAGE (Poly Acrylamide Gel Electrophoresis)** This method is commonly used to separate DNA and RNA fragments from a mixture. Initially, a gel is set (agar/ polyacrylamide) as per standard procedure. The gel is a crosslinked polymer of which the composition and porosity are chosen based on the specific weight and composition of the target to be analyzed. The gel is composed of different concentrations of acrylamide along with a cross-linker. This produces a network of polyacrylamide with different sizes. For the separation of large nucleic acids, pure agarose gel is utilized. Although the gel appears solid, it has a porous matrix. However, agarose is made up of long unbranched chains of uncharged carbohydrate with no cross-links. This results in the formation of a gel with large pores and makes the separation of macromolecules and macromolecular complexes very easy. For the separation of proteins or DNA the pH of the buffer and protein mixture is kept high (~9). With this, the proteins carry a net-negative charge.

The gel is placed in an electrophoretic chamber after the samples of DNA are dispensed into the wells (wells are made in the gel material). This is then connected to the electricity. The flow of electricity must have a negative charge at one end and a positive charge at the other end. Once the electric power is on, the DNA molecules get separated based on their sizes and mass. The molecules with higher mass move at a slower speed in the gel as compared to the molecules with the lower mass. Hence, the molecules with lower mass move upwards, and the molecules with higher mass stay downwards in the gel. These different-sized molecules form distinct bands in the gel.

Different types of electrophoretic systems are described as follows:

(a) **SDS PAGE** The molecules are separated by size in this procedure. SDS (sodium dodecyl sulfate) denatures the protein removing 2°, 3°, and 4° structures and the SDS coats the molecules giving them a uniform charge/mass ratio. Treatment with ß-mercaptoethanol breaks down the disulfide bonds present in the protein samples. This produces a linear chain (1° structure).
(b) **IEF (Iso Electric Focusing) electrophoresis** In this technique, different molecules get separated depending on their isoelectric point (pI). The overall charge on the molecule depends on the pH. As an IEF gel is poured it creates a pH gradient. A protein with a pH above its isoelectric point (pI) gets negatively charged and then migrates toward the anode (positive electrode). As it migrates through a gradient of decreasing pH, the overall charge of the protein becomes more till the protein reaches the pH that corresponds to its pI where it has no net

charge. At this point, the migration ceases and the protein forms a sharp stationary band/s in the pH gradient corresponding to its pI.

(c) **2D gel electrophoresis** In two-dimension (2D) PAGE, the proteins are separated two dimentionally. First, on the basis of their isoelectric point and then based on their mass.

**Antigen-antibody reaction** Each B cell produces unique antibodies, that recognizes a specific epitope on the antigen which starts an immune response. An antigen has an epitope with a distinct molecular surface. This gets bound to an antibody. They can be either conformational or sequential epitopes. This ability of antibodies to recognize and bind to an epitope on an antigen is the key in clinical analysis. Polyclonal antibodies (IgG) are those that are present in the blood serum of the host when it is collected. A monoclonal antibody is produced by one cell. When a tumor cell that can replicate endlessly is fused to B cells (from an animal), it gets stimulated with an antigen. This fusant cell is known as a hybridoma, which produces antibodies continuously.

**Western Blotting** When a biological sample is transferred from a gel on to a membrane and their detection is made on the surface of the membrane, then it is known as a blot. The technique is known as immunoblotting, as an antibody is used to detect the specific antigen. A target protein is identified from a complex protein mixture in a semi-quantitative manner. Initially, the macromolecules are separated using gel electrophoresis and then transferred to a membrane (usually nitrocellulose). However, the membrane should be blocked with a protein which is common and economical. This helps in preventing any nonspecific binding of antibodies to the membrane. This is labeled by a secondary antibody which recognizes the primary antibody. Biotin, fluorescent probes, and enzyme conjugates serve as labels which convert a substrate to a coloured product and therefore the membrane gets stained.

### 13.6.8 ELISA (Enzyme-Linked Immunosorbent Assay)

ELISA (enzyme-linked immunosorbent assay) is an immunoassay that detects the presence of antigens in biological samples. An antibody detects the target antigen. This happens using very specific antibody-antigen interactions. The immunologic reactions in this technique are detected and quantified by the catalytic properties of enzymes. It is a heterogeneous EIA technique which is used for clinical analyses. One component gets adsorbed either in a nonspecific manner or it gets covalently bound on the surface of a solid phase. This could be a microtiter well or a magnetic particle or a plastic bead. This helps to separate the bound and free labeled reactants.

Usually, the sample aliquot which has the antigen (Ag) is allowed to bind with a solid-phase antibody (Ab). This is then washed and then an enzyme-labeled antibody is added. This forms an antibody-antigen complex. This is called as a sandwich complex of solid-phase Ab-Ag-Ab enzyme. The antibody which does not bind is then washed away, and the substrate enzyme is added. This forms the product which depends upon the quantity of the antigen. Detection of antibodies to viruses and autoantigens in serum or whole blood are quite accurate by this method. The method of radioimmunoassay (RIA) was modified to develop the ELISA technique. Recent developments in ELISA methodology has led to the development of fluorogenic, quantitative PCR, and electrochemiluminescent reporters. For the assay, polystyrene plates (96-well plates) are being used which can strongly bind to the protein. The assay also requires a primary and/or a secondary detection antibody, analyte/antigen, coating antibody/antigen, buffer, wash, and substrate/chromogen. In the primary detection, the antibody binds only to the protein of interest. However, a second enzyme-conjugated antibody binds to a primary antibody which is not conjugated with the enzyme.

The following steps are required in an ELISA immunoassay:

- Coating (with either antigen or antibody).
- Blocking typically with the addition of bovine serum albumin (BSA).
- Detection.
- Final read.

Most common way of detection is a color-based method which is developed with the addition of a substrate. The common substrates are horseradish peroxidase (HRP) and alkaline phosphatase (AP). The substrate for HRP is hydrogen peroxide, which forms a blue color change, whereas AP (incubate at RT for 30 min) measures the yellow color of nitrophenol.

In order to remove the unbound material between each step the plate is washed with a buffer (PBS) and a nonionic detergent. In a usual ELISA protocol, a serial dilution of concentration is placed in the wells of the plate.

There are four major types of ELISA

- Direct ELISA (antigen-coated plate; screening antibody).
- Indirect ELISA (antigen-coated plate; screening antigen/antibody).
- Sandwich ELISA (antibody-coated plate; screening antigen).
- Competitive ELISA (screening antibody).

**Direct ELISA**

In this method, the antigen is coated on the ELISA plates which are incubated for one hour at 37 °C/4 °C overnight. The plates are then washed, which removes any unbound antibodies.

The second step prevents the binding of any non-specific antibodies to the plate and minimizes the false-positive results. Then the buffer is added and the plate is rewashed. After this a selected enzyme-antibody conjugate is added and this is incubated for one hour followed by a color reaction and analysis. Direct ELISA technique helps in removing the secondary antibody cross-reactivity. However, the sensitivity is low as compared to the other types of ELISA.

**Indirect ELISA**

In this method, two antibodies are required. First is the primary detection antibody. This antibody attaches itself with the sample protein. This is followed by a washing with saline buffer. Then, the enzyme-conjugated secondary antibody is added, which is then kept for incubation as per the optimized conditions. This is washed again and the substrate is added. The detection is done on a colorimeter/ spectrophotometer. The method is sensitive and cost-effective. However, cross-reactivity between the secondary detection antibodies needs to be avoided.

**Sandwich ELISA**

In this method, the wells of the plates are coated with the antibody. As the antigen is sandwiched between the two layers of antibodies the method is termed as sandwich. The plates are covered and incubated overnight at 4 °C after the addition of the capture antibody. Later, the plates are washed with PBS many times and are then buffered/blocked with BSA. Then the antigen is added which binds to the capture antibody. This is incubated for 90 minutes at 37 °C. The plates are rewashed, and the primary detection antibody is added to the plates and incubated for another 1–2 hours at room temperature. This is again washed with buffer and the secondary enzyme-conjugated antibody is added. This is then incubated for 1–2 hours. The plates are washed again before adding the substrate. Then, the change in color can be measured. Although the method is very sensitive and gives accurate results, it is very time-consuming and is expensive.

**Competitive ELISA**

This method analyses the presence of an antibody specific for antigens in the test serum. The method uses two specific antibodies. First is the enzyme-conjugated antibody and another one is present in the test serum.

When these two antibodies are combined into the same well, there is competition to bind with the antigen. In case there is a color change, it shows that the test is negative. The absence of color proves a positive test due to the presence of the antibodies in the test serum. The method lacks sensitivity but can be used for small sample volumes.

ELISA techniques are used in many diagnostic tests like, antibodies in blood, autoantibodies (anti-dsDNA, anti-dsg1,ANA), antibodies against infectious diseases, and Hepatitis A, B, C, HIV. They can also detect and estimate the levels of tumor markers, prostate-specific antigen (PSA) and Carcinoembryonic Antigen (CEA). The technique is also helpful in detecting the hormone levels.

There are factors that may interfere during testing. These include the quality and integrity of the assay plate, coating buffer, capture antibody, blocking buffer, target antigen, detection antibody, enzyme conjugate, and the substrate. ELISA technologies play a major role in clinical research for diagnostic and screening tests.

Mass Spectrometry This method helps in detecting very minute quantities of material (100 picograms or less). It determines the mass of a molecule by measuring the mass-to-charge ratio (m/z) of its ions which are generated by inducing either the loss or gain of a charge from a neutral species. Ions are electrostatically directed into a mass analyzer. There they get separated according to m/z and are detected.

## 13.7 Concentration

The concentration of proteins is done by precipitation using a solution of ammonium sulfate, polyethylene glycol, organic solvents, trichloroacetic acid, potassium chloride/sodium dodecyl sulfate, and three-phase partitioning. Solvents are usually removed through a semipermeable membrane or by using high pressure. In order to draw the solvent across the semipermeable membrane, it is surrounded by a hygroscopic reagent. Freeze-drying is a good method to concentrate proteins. The newer methods to remove the solvents include lyophilization, reverse extraction, solute precipitation along with dialysis (solvent exchange), ultrafiltration, and chromatographic techniques. Vivaspin filters have been found to be good for concentration in which not much protein is lost as compared to other filters. A concentrator provides communication between many low-speed asynchronous channels and some high-speed synchronous channels. Usually, low-speed can accommodate different speeds, codes, and protocols. The material used are mirrors for solar thermals. For the BIPV system it is either glass or transparent plastic. Rotary evaporator and freeze dryers can concentrate proteins and peptides easily after optimization of parameters like pressure, temperature and condenser loading. Using a larger flask provides a larger surface area which can increase the performance up to 50%. Increasing the rotation speed increases the surface for evaporation. This results in better efficiency of heat transfer from the bath. Foaming of proteins should be prevented as it results in the product loss. Use of a foam sensor or specialized glass assembly are helpful. The foam sensor is an optical sensor mounted on the condenser. During the concentration of proteins, the powder samples stick to the evaporating flask which causes the product loss. Therefore, it is better to use evaporation flask after drying. To obtain better yields of protein, it is important to take care during freeze drying of the sample also. It is advisable to keep the freezer and the freeze dryer in the same room so that the time of sample transfer and for maintaining the frozen sample at a low temperature is minimal. Performing a leak test on the freeze dryer is helpful to detect misconfigurations and defects. The drying rate of a product depends on the surface area and the volume of the product. The larger the surface area, the faster the product dries. The sublimation rates have been found much faster when shell freezing is used instead of bulk freezing. The lyophilization process gets affected by the container surface and the depth of the fill. Usually, a larger dried sample cake hinders the vapor from escaping from below to a large extent.

## 13.8 Product Recovery

**Product Purification**

Direct separation for downstream processing and purification is the best. Many downstream processes and purification operations are unit operations that are robust, well established, and can be used for scaling up operations. Therefore, available methods are often chosen in the development of purification methods. Since the selectivity of biocatalytic reaction is high, less side products and auxiliary reagents have to be removed. Recently, there has been a lot of development for the advanced isolation and purification. Newer technologies are being used based on the number of main products formed and the type of auxiliary compounds used. In the case of two very similar products, such as regioisomers, formed in an equimolar ratio, the SMB chromatography is a useful purification method which can be used at large scale operations.

**Metabolite Purification**

Biological cells represent a valuable source for metabolites, because the chemical synthesis in many cases is not possible or is not economically viable. As metabolites are small-molecular-weight compounds, many product purification processes in biotechnology are similar to that of chemistry and biochemistry. Due to the wide structural diversity of metabolites, a large series of purification procedures are needed. This can be a one step purification, utilizing one or more unit operations or the most challenging separations of homologues which may have similar molecular properties. The diversity of closely related metabolites created by biological cells in biocatalytic pathways can both be an opportunity for multiproduct bioprocesses or a challenge for the purification of a single product. Upstreaming process can be reduced. Recovery of the product depends on many factors such as when and where the product is formed, the purity of the product, use of the product, impurities within the broth, cost of the product and if it is intra-cellular or extra-cellular. It is the product recovery cost that decides the cost of the product in the market. Purification steps and intended use can increase the investment toward product recovery and purification. There are different techniques which are utilized for product recovery as filtration, centrifugation, precipitation, dialysis, solvent extraction, chromatographic, electrophoresis techniques, and also crystallization techniques. Once the filtrate is obtained, it is purified and then marketed. The choice of recovery process is based on the following criteria:

(i) The intracellular or extracellular location of the product.
(ii) The concentration of the product in the fermentation broth.
(iii) The physical and chemical properties of the desired product.
(iv) The intended use of the product.
(v) The minimal acceptable standard of purity.
(vi) The magnitude of biohazard of the product or broth.
(vii) The impurities in the fermentor broth.
(viii) The marketable price of the product.

The characteristics of the broth can be done using simple equipment and by using different techniques:

(a) Selection of a microorganism.
(b) Optimization of the fermentation conditions for maximum metabolite production.
(c) Exact time for harvesting.
(d) Control of pH during the process of harvesting.
(e) Temperature treatment after harvesting.
(f) Use of flocculating agent.
(g) Enzyme treatment.

In any fermentation process, the recovery of the product is extremely important. Also, it is important to check that the antifoam has been removed completely from the supernatant. The presence of foam can affect the centrifugation/ ultrafiltration processes. The ion-exchange resins can get affected which are utilized during the recovery process. Many times the ionic strength of the production medium is high. For this, the harvested supernatant can be diluted with demineralized water before it is processed. With the help of surfactants like long-chain fatty acids, amines, and quaternary ammonium compounds, some materials can be made surface active. These are known as colligends, and the surfactants are termed as collectors. Precipitation helps in the enrichment and concentration processes just in one step. The material volume gets reduced for processing and certain impurities can be removed from the broth.

The most common methods are as follows:

1. Use of acids and bases can help in changing the pH of a solution until the isoelectric point of the compound is reached and pH becomes equal to the pI. At this point, there is no overall charge on the molecule and its solubility goes low.
2. For the recovery and fractionation of proteins, salts of ammonium and sodium sulfate are utilized. Therefore, the water is removed from the protein surface and the hydrophobic parts come together which help the protein to precipitate. The most hydrophobic proteins precipitate first and then precipitate in a gradient manner. This makes the fractionation to take place. This technique is known as salting out.
3. Organic solvents such as methanol can be used to precipitate the dextrans out from the broth. Proteins can be precipitated by using chilled ethanol and acetone due to the changes in the dielectric properties of the solution.
4. Polyethylene glycol (PEG) which are nonionic polymers can be used to precipitate proteins.
5. Many compounds can be precipitated by using polyelectrolytes.
6. Triazine dyes can bind and precipitate proteins.
7. There are affinity precipitants that can bind and precipitate the compounds selectively.

8. For many thermostable products, heat treatment has been found to be effective in deactivating the cell proteases. The most common process used at all scales of operation to separate suspended particles from a liquid or gas is filtration. In this a porous medium is used which keeps the particles back, however, the liquid/ gas gets passed. There are many factors that help in making the process cost-effective. These are as follows:
   (a) The viscosity and density of the filtrate.
   (b) The size, shape, and packing characteristics of the solid particles.
   (c) Ratio of the solids/liquid.
   (d) Recovery of the solid or liquid fraction or both.
   (e) Scale of operation.
   (f) Batch or continuous operation.
   (g) Places required for sterilization/aseptic conditions.
   (h) If pressure or vacuum suction is needed for an adequate flow rate of the liquid.

While filtering bacteria/gelatinous suspensions usually filtration technique is used with the help of diatomaceous earth, in which a thin layer is applied to the filter, followed by mixing it with the harvested broth. The plate and frame filter works out in an industrial scale which requires the least floor space. However, to remove large quantities of solids from a broth it is not advisable as it increases the labour costs and also the time needed for dismantling, cleaning and reassembly. Scaling up of purified proteins is quite difficult as the proteins are utilized in various applications and are therefore, needed in pure and active form. It involves optimization of many parameters that can affect the quality and yield of the protein product.

- Choice of expression system: All the factors like protein level, solubility, stability, post-translational modifications, and compatibility depend on the expression system which could be a bacteria, yeast, insect, mammalian cell, or a plant cell.
- Selection of purification method: In this step, the proteins are separated from the other components like the cell debris, nucleic acids, lipids, salts. A method is selected out of the many purification methods discussed in this chapter. This can be selected for maximum production of the protein and the purity.
- Validation of protein product: The protein final product should be validated for the characteristics like the concentration, purity, identity, activity and the stability. This step is important as it helps in ensuring that the final protein product has the specifications and the standards for its intended use.

Troubleshooting during scaling up of protein: This is an important step during scaling up which can be helpful in overcoming and rectifying any step which would be going wrong otherwise. This could be done by further optimization and validation which would result in pure and higher yield of the product.

## FAQ's

1. What factors influence the selection of cell disruption method?
2. What is the most common method used for disruption of cells?
3. Which physical method is used for cell disruption?
4. What are the factors which affect the efficiency of cell disruption?
5. Explain which enzymes are used in cell disruption and why?
6. What is the function of EDTA in cell disruption?
7. For protein purification which are the best techniques to be practised?
8. What are the factors which induce cell damage?
9. Which chemicals are used for cell disruption?
10. How does microfluidizer work?

## References

Al-aqbi ZT, Albukhaty S, Zarzoor AM, Sulaiman GM, Khalil KAA, Belali T, Soliman MTA (2021) A novel microfluidic device for blood plasma filtration. Micromachines 12(3):336. https://doi.org/10.3390/mi12030336

de Bruin OM, Birnboim HC (2016) A method for assessing efficiency of bacterial cell disruption and DNA release. BMC Microbiol 16(1):197. https://doi.org/10.1186/s12866-016-0815-3

de Maria DLC (2008) Chapter 8 - Membrane based separation techniques: dialysis, gas diffusion and pervaporation. In: Spas DK, Ian DM (eds) Comprehensive analytical chemistry, vol 54. Elsevier, pp 203–234. ISSN 0166-526X.ISBN 9780444530943. https://doi.org/10.1016/S0166-526X(08)00608-9

Ezeronye O, Okerentugba P (2001) Optimum conditions for yeast protoplast release and regeneration in *Saccharomyces cerevisiae* and *Candida tropicalis* using gut enzymes of the giant African snail *Achatina achatina*. Lett Appl Microbiol 32:190–193. https://doi.org/10.1046/j.1472-765x.2001.00885.x

Goldberg S (2008) Mechanical/physical methods of cell disruption and tissue homogenization. Methods Mol Biol 424:3–22. https://doi.org/10.1007/978-1-60327-064-9_1

Göran E, Gudrun B, Anders K (1998) Differences between Bligh and Dyer and Soxhlet extractions of PCBs and lipids from fat and lean fish muscle: implications for data evaluation. Marine Pollution Bull 36(3):222–230. ISSN0025-326X. https://doi.org/10.1016/S0025-326X(97)00176-8

Grigorov EK, Boris M, Marin G, Vassil (2021) Review of microfluidic methods for cellular lysis. Micromachines 12:498. https://doi.org/10.3390/mi12050498

Halim RH, Razif D, Michael KW, Paul A (2012) Microalgal cell disruption for biofuel development. Appl Energy Elsevier 91(1):116–121

Harrison STL (2011) Cell Disruption. In: Moo-Young M (ed) Comprehensive biotechnology, 2nd edn. Academic Press, pp 619–640. ISBN 9780080885049. https://doi.org/10.1016/B978-0-08-088504-9.00127-6

Ioannis K, Ioannis A (2021) Chapter 6 - Cavitation in hydraulic turbines. In: Phoevos K, Manolis G (eds) Cavitation and bubble dynamics. Academic Press, pp 173–209. ISBN 9780128233887. https://doi.org/10.1016/B978-0-12-823388-7.00001-1

Kamineni S, Huang C (2019) The antibacterial effect of sonication and its potential medical application. SICOT J 5:19. https://doi.org/10.1051/sicotj/2019017

Lakatos P, Gross J, Thut G (2019) A new unifying account of the roles of neuronal entrainment. Curr Biol 29(18):R890–R905. https://doi.org/10.1016/j.cub.2019.07.075

López-Bascón MA, Luque de Castro MD (2020) Chapter 11 - Soxhlet extraction. In: Colin FP (ed) Hand books in separation science, liquid-phase extraction. Elsevier, pp 327–354. ISBN 9780128169117. https://doi.org/10.1016/B978-0-12-816911-7.00011-6

Middelberg A (1995) Process-scale disruption of microorganisms. Biotechnol Adv 13(3):491–551. https://doi.org/10.1016/0734-9750(95)02007-P

Mikkelsen SR, Cortón E (2004) Centrifugation methods. Bioanal Chem Hoboken, N J: U S John Wiley & Sons Inc. https://doi.org/10.1002/0471623628.ch13. ISBN 978-0-471-54447-0

Molino A, Larocca V, Di Sanzo G, Martino M, Casella P, Marino T, Karatza D, Musmarra D (2019) Extraction of bioactive compounds using supercritical carbon dioxide. Molecules 24(4):782. https://doi.org/10.3390/molecules24040782

Passos F, Hernández-Mariné M, García J, Ferrer I (2014) Long-term anaerobic digestion of microalgae grown in HRAP for wastewater treatment. Effect of microwave pretreatment. Water Res 49. https://doi.org/10.1016/j.watres.2013.10.013

Passos F, Gutiérrez R, Uggetti E, Garfí M, García J, Ferrer I (2017) Towards energy neutral microalgae-based wastewater treatment plants. Algal Res 28:235–243

Phothiset SC, Sanguansri (2014) Effects of freezing and thawing on texture, microstructure and cell wall composition changes in papaya tissues. J Sci Food Agri 94. https://doi.org/10.1002/jsfa.6226

Potts M (1994) Desiccation tolerance of prokaryotes. Microbiol Rev 58(4):755–805. https://doi.org/10.1128/mr.58.4.755-805.1994

Rodriguez EL, Poddar S, Iftekhar S, Suh K, Woolfork AG, Ovbude S, Pekarek A, Walters M, Lott S, Hage DS (2020) Affinity chromatography: a review of trends and developments over the past 50 years. J Chromatogr B Analyt Technol Biomed Life Sci 1157:122332. https://doi.org/10.1016/j.jchromb.2020.122332

Rohini PG, Shamal LC, Prashant SA (2020) Chapter 17-Polymers in cosmetics. In: Mariam AAA, Deepalekshmi P, Marcelo AC (eds) Polymer science and innovative applications. Elsevier, pp 545–565. ISBN 9780128168080. https://doi.org/10.1016/B978-0-12-816808-0.00017-2

Shehadul IM, Aryasomayajula A, Selvaganapathy PR (2017) A review on macroscale and microscale cell lysis methods. Micromachines(Basel) 8(3):83. https://doi.org/10.3390/mi8030083

Singh R, Kumar M, Mittal A, Mehta PK (2016) Microbial enzymes: industrial progress in 21$^{st}$ century. 3 Biotech 6(2):174. https://doi.org/10.1007/s13205-016-0485-8

Sun C, Li Y, Taylor SE, Mao X, Wilkinson MC, Fernig DG (2015) HaloTag is an effective expression and solubilisation fusion partner for a range of fibroblast growth factors. Peer J 3:e1060. https://doi.org/10.7717/peerj.1060

Terpe K (2006) Overview of bacterial expression systems for heterologous protein production: from molecular and biochemical fundamentals to commercial systems. Appl Microbiol Biotechnol 72(2):211–222. https://doi.org/10.1007/s00253-006-0465-8

Ulmer CZ, Jones CM, Yost RA, Garrett TJ, Bowden JA (2018) Optimization of Folch, Bligh-Dyer and Matyash sample-to-extraction solvent ratios for human plasma based lipidomics studies. Anal Chim Acta 1037:351–357. https://doi.org/10.1016/j.aca.2018.08.004

Young HC, Robert V (2019) Green solvents for the extraction of bioactive compounds from natural products using ionic liquids and deep eutectic solvents. Curr Opin Fd Sci 26:87–93. ISSN 2214-7993. https://doi.org/10.1016/j.cofs.2019.04.003

Zheng H, Yin J, Gao Z, Huang H, Ji X, Dou C (2011) Disruption of *Chlorella vulgaris* cells for the release of biodiesel producing lipids: a comparison of grinding, ultrasonication, bead milling, enzymatic lysis, and microwaves. Appl Biochem Biotechnol 164:1215–1224

MIX
Papier aus verantwortungsvollen Quellen
Paper from responsible sources
FSC® C105338

If you have any concerns about our products,
you can contact us on
**ProductSafety@springernature.com**

In case Publisher is established outside the EU,
the EU authorized representative is:
**Springer Nature Customer Service Center GmbH**
**Europaplatz 3, 69115 Heidelberg, Germany**

Printed by Libri Plureos GmbH
in Hamburg, Germany